"Farming for Our Fut for eliminating net agricult compelling book is packed with original research, well-sourced analyses, and innovative solutions. Lehner and Rosenberg are clear-eyed about the challenges presented by today's agricultural practices and policies, yet thoughtfully lay out the path forward. This book will be an invaluable resource for years to come."
—Emily M. Broad Leib, Clinical Professor of Law and Director, Food Law and Policy Clinic, Harvard Law School

"Resilient diversified farms are born from plenty of research, data, supportive policy, and an array of well-managed practices. So, too, a thriving world. In this book, Peter Lehner and Nathan Rosenberg show us why climate stability and resilience must be major goals in all agriculture law. With the right tools and an unwavering focus on programs and policies that reward ecosystem services, we can combat our current climate crisis, safeguarding not only our food, but our future."
—Sherri Dugger, Executive Director, Socially Responsible Agriculture Project

"Unless we transform farming and food production, many of the gains we make through transformation of our energy and transportation sectors will be wiped out by agriculture. *Farming for Our Future* provides a timely and clear-eyed take on the possibilities—and pitfalls—of a climate-smart food system."
—Scott Faber, Senior Vice President, Government Affairs, Environmental Working Group

"Agriculture has long occupied too small a role in U.S. climate policy discussions, despite its considerable direct and indirect contribution to greenhouse gas emissions and its tremendous capacity to absorb carbon dioxide from the atmosphere. This book should elevate agriculture to greater prominence in these discussions. Peter Lehner and Nathan Rosenberg have gone deeper and broader than anyone before in identifying the many legal levers that can be used to move agriculture toward carbon neutrality. Congress, the White House, USDA, other agencies, and the private sector should use this as their legal guide."
—Michael B. Gerrard, Professor and Director, Sabin Center for Climate Change Law, Columbia Law School

"Farms are a major contributor to climate change, are impacted by climate change in a big way, and, with practical farming system changes, can be a major contributor to climate change mitigation. Lehner and Rosenberg lay out the details in a highly readable and succinct manner and with a strong emphasis on the policies that need to be adopted to reach a net zero agriculture in the United States. Their prescriptions form a well-drawn blueprint for the White House and USDA to follow and for Congress to adopt in the 2023 federal farm bill. Adoption of the book's recommendations would put U.S. agriculture on a rapid path to decarbonization and resilience. Policymakers should pay heed!"

—Ferd Hoefner, Policy Consultant and Former Policy Director of the National Sustainable Agriculture Coalition

"Farming for our Future is a must read for anyone interested in agriculture and climate disruption. The authors provide a very readable and yet well documented review of the science, law, and policy on agricultural practices in the face of global change, including how agriculture drives greenhouse gas emissions and why we need to increase the resiliency of agriculture in the future. I have worked in this area as a scientist for decades, and yet still learned a great deal from their book, particularly on the law and policy aspects. Importantly, this is not a depressing book, although the topic can certainly be depressing. Rather, it is a detailed call for action that provides a blueprint for reimagining how society produces food and improves environmental quality and the quality of rural life as we move forward. I bet this book will dominate the discussion on agricultural policy for many years to come."

—Robert W. Howarth, Ph.D., David R. Atkinson Professor of Ecology & Environmental Biology, Cornell University

"The agriculture sector reflects both the peril and possibility of climate change: left unaddressed, greenhouse gas emissions related to food production will stymie efforts to avoid catastrophic warming, yet agriculture also offers great potential to aid in the transition to a negative-emissions economy. This book offers the first comprehensive scientific, legal, and policy treatment of agriculture and climate change in the United States. Accessible to anyone involved in agriculture or interested in agricultural policy regardless of current practices or political views, the book also will prove invaluable to the national and state policymakers who must chart the future of farming. In short, this is precisely the book we need at precisely the moment we need it."

—Douglas A. Kysar, Joseph M. Field '55 Professor and Co-Director, Law, Ethics & Animals Program, Yale Law School

"Every year in the United States, billions of federal dollars are directed to influence how we farm, but until recently almost none of that investment has addressed the climate impacts of the agricultural sector. While the spotlight on agriculture's potential to mitigate climate change has increased, a robust review of the strategic role of public policy to shape farm practices has lagged behind. This book makes a clear case for the policy market signals and incentives that shape our agricultural sector and leverage the private sector to greatly amplify the adoption of climate-smart practices in American production of food, fiber, and fuel."
—Nicole Lederer, Chair and Co-Founder, Environmental Entrepreneurs

"Climate change is wreaking havoc on our agriculture, triggering devastating wildfires, crippling droughts, freakish '500 year' storms, and infestations by novel pests and invasive species. This groundbreaking book documents how agriculture plays a unique role in the yin and yang of this threat: farmers suffer shocking blows from the climate crisis yet agribusiness contributes mightily to greenhouse gas emissions. Lehner and Rosenberg show how climate-friendly farming must play a crucial role in tackling climate change and propose innovative government and corporate policies—and improved eating habits—that can help us fight this ultimate peril for our species and planet."
—Erik D. Olson, Senior Strategic Director, Health & Food, Natural Resources Defense Council

"Every U.S. taxpayer is invested in subsidizing an agricultural system that, as Lehner and Rosenberg demonstrate, currently exists to benefit a powerful business sector and a small number of farmers who are quite a bit wealthier than most people in the United States yet have to do next to nothing to enjoy this unfettered public largesse. This is a result of complex policy history that few people have the time or expertise to untangle. But the stakes are high. The current system generates food and other outputs that are profitable for an array of industries but noxious to public health, contribute significantly to the climate crisis, pollute water and air, degrade soil and—worst of all—exploit vulnerable workers. Instead, different and eminently sensible economic and policy structures could make more healthful food plentiful and affordable; contribute toward eliminating and drawing down greenhouse gases; regenerate clean water, air, and healthy soil; and provide dignified livelihood to farmers, food chain workers, and rural economies. The authors have expertly parsed the complex policy history and clearly articulated attainable new goals and policy mechanisms to get us there. This is some of the clearest and most compelling writing on agricultural history, law, and policy that I've read, presented in concise and accessible form for any reader to digest. To this point,

every eater should read this to better understand why we must demand that policymakers reform a dated and ineffective agricultural system to one that meets the needs of all of society, today and in the future."

—Ricardo J. Salvador, Director and Senior Scientist,
Food & Environment Program, Union of Concerned Scientists

"This invaluable book provides a blueprint for building healthy soils and farms to cool our planet. It is a vital resource for policymakers and stewards of the land."

—Betsy Taylor, President,
Breakthrough Strategies & Solutions

"Agriculture is the most powerful way we use our planet, and farm policy can make the difference between farmers using good or destructive practices. The way we farmed led to the devastating dust bowl, and better farm policies caused farmers to change practices to help end it. Today, the way we farm is contributing significantly to global warming. The good news is that a growing number of farmers are using farming practices that sequester carbon and finding that they improve profitability and production as well as slow climate change. This book does an incredible job of explaining the practices and policies we need to help farmers build regenerative production systems that will protect our climate and our future."

—Seth Watkins, TEDx presenter, *Farming Evolved: Agriculture Through a Different Lens*, and fourth generation farmer, Pinhook Farm, Iowa

"In *Farming for Our Future: The Science, Law, and Policy of Climate-Neutral Agriculture*, Peter Lehner and Nathan Rosenberg tackle some of the most vexing problems plaguing our food system and their role in climate change. With its thorough examination of past and present U.S. agricultural policy, this book provides a detailed look at the climate-related challenges we face and offers credible recommendations for improvement. Our current agricultural system is, in many practical ways, isolated and disconnected from its impact on people and the environment. What this book does is show how this luxury is one near its end by highlighting the real-world consequences of decades of inaction. Moreover, using a narrative that highlights a wide variety of agricultural stakeholders, this book shows how our lopsided, monopolistic approach to agriculture needs to change and makes the case that the time is now."

—Craig Willingham, Deputy Director,
CUNY Urban Food Policy Institute

FARMING FOR OUR FUTURE

THE SCIENCE, LAW, AND POLICY OF CLIMATE-NEUTRAL AGRICULTURE

Peter H. Lehner & Nathan A. Rosenberg

ENVIRONMENTAL LAW INSTITUTE
Washington, D.C.

Copyright © 2021 Environmental Law Institute
1730 M Street NW, Washington, DC 20036

Published November 2021.

Printed in the United States of America
ISBN 978-1-58576-237-8

Cover design by Evan Odoms
Photo courtesy of New Forest Farm

Table of Contents

About the Authors vii
Acknowledgments ix
Foreword xi
Chapter I. Introduction 1
    Key Recommendations 8
Chapter II. The Stakeholders in Farm Policy 11
    A. Farmers and the Farm Economy 13
        1. Transformations in the Farm Economy 14
        2. Overstatements in USDA Census of Agriculture Data 15
        3. A More Accurate Assessment of Farm Income and Wealth 19
        4. The Legacy of Discriminatory Agricultural Policy 22
    B. New Constituencies 23
        1. Farms in the Rural Economy 24
        2. Non-White Farmers 26
        3. Rural Communities 28
        4. The Opportunity for Carbon Farming 31
    Key Recommendations 32
Chapter III. The Climate Crisis and Agriculture 35
    A. Climate Change's Impact on Agriculture 35
    B. Agriculture's Contribution to Climate Change 37
        1. Global Agricultural Greenhouse Gas Emissions 38
        2. U.S. Agricultural Greenhouse Gas Emissions 40
        3. State-Level Agricultural Greenhouse Gas Emissions 46

  C. Underestimates and Uncertainties 47
    1. Underestimates 49
    2. Uncertainties 54
  D. Agriculture's Dual Opportunity 58
  Key Recommendations 60

Chapter IV. Climate-Friendly Agricultural Systems and Practices 63
  A. Agricultural Systems and Practices for Reducing
    Greenhouse Gas Emissions 66
    1. Cropland 66
    2. Grazing Land 88
    3. Animal Feeding Operations 94
    4. On-Farm Fuel Combustion and Electricity 102
  B. Agriculture's Maximum Contribution to Curbing
    Climate Change 103
  Key Recommendations 107

Chapter V. Transforming Farm Policy Toward Climate-Neutral
Agriculture 111
  A. Research, Extension, and Technical Assistance
    Programs 113
    1. Federal Research Programs 113
    2. Public-Private Research Programs 120
    3. State Research and Demonstration Programs 121
    4. Data Collection and Analysis 123
    5. Extension Service 129
    6. Technical Assistance 132
    7. Improving Coordination Among Research,
      Extension, and Technical Assistance Programs 134
  B. Public Subsidy and Conservation Programs 135
    1. Crop Insurance 137

Table of Contents

       2. Commodity Programs 141
       3. The Commodity Credit Corporation 143
       4. Conservation Payments 145
       5. Conservation Easements 154
       6. Conservation Compliance Requirements 156
       7. Lending Programs 160
       8. Trade Policy 162
       9. Transforming the Farm Safety Net Through Legislative Action 163
   C. Grazing Practices on Government Land 166
   D. Perennial Agriculture 168
       1. Land Tenure 172
       2. Research, Development, and Extension 172
       3. Input, Distribution, and Marketing Infrastructure 174
       4. Farm Finance and Support 174
       5. Coordinating Efforts 176
   E. Toward Climate-Neutral Agriculture 176
   Key Recommendations 177

Chapter VI. Public Policy Pathways Beyond USDA for Advancing Climate-Neutral Agriculture 181
   A. Regulatory Options 181
   B. Tax Policy 193
   C. Small Business Administration Lending Programs 197
   D. Biogas Subsidies 198
   E. Greenhouse Gas Pricing 201
   Key Recommendations 203

Chapter VII. Private- and Nonprofit-Sector Opportunities for Advancing Climate-Neutral Agriculture 205
   A. Research 205

- B. Financing Options 206
- C. Easements and Other Conservation Tools 207
- D. Carbon Measurement Tools.. 209
- E. Carbon Markets 210
- Key Recommendations 217

Chapter VIII. Off-Farm Food System Emission Reduction Opportunities 219
- A. Upstream: Greenhouse Gas Emissions From Farm Inputs 219
  1. Emissions From Fertilizer Production..................... 220
  2. Fuel Economy Standards for Agricultural Equipment and Reduction of On-Farm Energy Use 222
- B. Downstream: Emissions From Food Processing, Packaging, Marketing, and Waste.................................. 223
  1. Processing, Packaging, Distribution, and Marketing Emissions 224
  2. Landfill Waste Emissions 224
- Key Recommendations 228

Chapter IX. Changing Consumption Patterns............................ 229
- A. Dietary Guidelines 229
- B. Federal Procurement and Food Assistance 232
- C. Private-Sector Strategies 235
  1. Certification Systems and Supply Chain Commitments 235
  2. Plant-Forward Alternatives 237
- Key Recommendations 239

Chapter X. Conclusion 241
Index 243

About the Authors

Peter H. Lehner is Managing Attorney of the Sustainable Food & Farming (SFF) Program at Earthjustice, the country's largest nonprofit public interest environmental law organization. The SFF Program deploys strategies to promote a more just, environmentally sound, and climate-resilient agricultural system and to reduce the health, environmental, and climate harms from production of our food. From 2007–2015, Lehner was the Executive Director of the Natural Resources Defense Council (NRDC) and the NRDC Action Fund. There, he grew the organization and particularly the climate change and clean energy programs, opened several new offices and programs, and expanded the food system work. From 1999–2006, he served as Chief of the Environmental Protection Bureau of the New York State Attorney General's office, supervising all environmental litigation by and against the state. Earlier, Lehner started and led the Environmental Prosecution Unit at the New York City Law Department. Lehner holds an A.B. in Philosophy and Mathematics from Harvard College and is an honors graduate of Columbia University Law School, where he now teaches a class on agriculture and environmental law. Lehner also helps manage two family farms. He is on several NGO boards and has been honored with numerous awards by EPA and many environmental groups.

Nathan A. Rosenberg is a visiting scholar at the Food Law and Policy Clinic at Harvard Law School and an attorney consulting for Earthjustice. Nathan has taught agricultural law and policy at the University of Iowa College of Law, the University of Arkansas School of Law, and the New York University Steinhardt School of Culture, Education, and Human Development. He has also worked as a legal fellow for the Natural Resources Defense Council and as director of the Delta Directions Consortium.

Acknowledgments

Several years ago, Professors John Dernbach and Michael Gerrard asked us to write the agriculture chapter for a book they were editing, *Legal Pathways to Deep Decarbonization in the United States* (ELI Press 2019). Because prior Deep Decarbonization Pathways Project reports did not address the U.S. agriculture sector in depth, our chapter outlined both the technologies and practices that could be implemented to reduce the climate change impact of agriculture, and the legal options to accelerate their adoption. After the *Legal Pathways* book was released, the publisher, the Environmental Law Institute, noted that agriculture's impact on climate change was a rich but largely under-addressed field, and asked us to update and expand the chapter into a stand-alone book. This book is the result of that effort.

We would like to extend our deepest gratitude to John Dernbach and Michael Gerrard for inviting us to contribute to *Legal Pathways* and for their many ideas and insightful edits and comments. We also thank the participants of the Legal Pathways to Deep Decarbonization workshop at Columbia Law School for their suggestions. And through the writing of both the chapter and the book (as well as an intermediate article), our sharp-eyed editor at ELI, Rachel Jean-Baptiste, has been attentive, helpful, and full of good ideas.

We are very grateful to all of our Earthjustice colleagues who have provided tremendous input to this book, the prior article, and the many documents that we have relied on. They have carefully reviewed our scientific and legal statements and helped shape our policy proposals. We particularly thank Alexis Andiman, Carrie Apfel, Claire Huang, Ranjani Prabhakar, Mustafa Saifuddin, and Tyler Smith, as well as Sorangel Liriano, Greg Loarie, Mekela Panditharatne, Surbhi Sarang, Dawa Sherpa, and Patrice Simms. We are also very grateful to Earthjustice for providing us time and support so we could share what we've learned about this critical topic.

We wish to thank Ben Anderson, Andrew Bowman, Michael Castellano, Adam Chambers, Alyssa Charney, Craig Cox, Marcia DeLonge, Graham Downey, Thomas Driscoll, Mark Easter, William Eubanks, Scott Faber, Jonathan Gelbard, Elizabeth Hanson, Claire Horan, Karen Hudson, Wendy Jacobs, Barbara Jones, Allen Olson, Keith Paustian, Margot Pollans, Kiley Reid, Charles Rosenberg, Susan Rosenberg, Susan Schneider, Edward Strohbehn, Bryce Stucki, Margaret Torn, and Jim Williams, for voluntarily taking the time to review our drafts of this book and earlier articles and documents that formed its foundation. We are grateful for their many helpful comments.

Many thanks to Becca Bartholomew, Sherri Dugger, and Sarah Rodman-Alvarez whose earlier research helped shape these recommendations. Their interviews for us with over 100 sustainable farmers, advocates, and community leaders contained a wealth of expertise and experience that deeply enriched the scientific and legal analyses. We are grateful also for the generosity of those interviewed for their time, trust, and openness.

We are indebted to the participants and organizers of the Policy Pathways to Perennial Agriculture seminar, hosted by the Radcliffe Institute for Advanced Study and the Harvard Law School Food Law and Policy Clinic, for helping us better understand the challenges and opportunities presented by perennial systems. We are especially grateful to Emily Broad Leib, Fred Lutzi, Sarah Lovell, Emma Scott, and Eric Toensmeier for their feedback and suggestions. We also thank Lingxi Chenyang, Andrew Currie, and Hannah Darin for generously sharing their research on agroforestry and farm policy, which greatly informed our understanding of perennial agriculture.

We are also deeply indebted to the many advocates and allies in the environmental and sustainable agriculture movements whose insights have made our own work possible. There are far too many to name but we would be remiss if we failed to mention Carlos Borgonovo, Stephanie Cappa, Alyssa Charney, Callie Eideberg, Madeleine Foote, Aviva Glaser, Elizabeth Henderson, Ferd Hoefner, Greg Horner, Mark Izeman, Allison Johnson, Eric Kamrath, Michael Lavender, Claire O'Connor, Erik Olson, Mary Pfaffko, Diego Robelo, Virginia Ruiz, and Seth Watkins.

And finally, and most important, we thank the thousands of researchers, farmers, and farmworkers who have worked for years to develop, implement, and refine the practices we discuss. They have set the stage for a nationwide effort to transform agriculture into a climate solution.

Foreword

In this book, we examine the agricultural strategies, practices, and technologies that can make agriculture in the United States climate-neutral or better, and thus help to curb climate change. Agriculture both contributes to climate change and is severely affected by the more frequent extreme weather, increased pests and heat, and other challenges produced by it. Not only does our food security depend on an agricultural system that can continue and thrive in changed conditions, but our overall security depends on limiting climate change. And while agricultural greenhouse gas emissions may seem modest compared with those from the electricity, transportation, or industrial sectors, agriculture's contribution to climate change is substantial (far more than generally realized) and we will not be able to achieve our mitigation goals unless agricultural emissions sharply decline. Fortunately, agriculture can be a major part of the climate solution, and in the process improve rural communities and the lives of those who work on farms and ranches.

Many visionary farmers, researchers, and advocates have already started down this path. Here, we summarize the many agricultural practices that have been demonstrated to reduce greenhouse gas emissions and increase carbon sequestration in soil, including cover cropping, more varied crop rotations, agroforestry and silvopasture (adding trees into cropping or grazing systems), perennial crops, prescribed rotational grazing, dry manure management, and others. Of course, there are nuances in impact depending on region, climate, and how farmers implement these practices—and there are often conflicting scientific studies and over-simplified advocacy claims—yet it is clear that agriculture itself can become climate-neutral without offsets. Numerous scientific studies as well as the actual experience of many farmers (including one of the authors) have shown that these practices also reduce soil erosion and water pollution—a significant challenge in many parts of the country—and increase soil fertility and productivity, and thus often profitability. They also help farms and ranches build climate resiliency and better withstand the increased stresses that climate change is already bringing, such as floods, droughts, heat waves, pests, and more.

Despite these proven benefits, only a small portion of farmers in the United States implement these practices on only a small fraction of U.S. agricultural land. It is clear that policy change is needed in order to accelerate their adoption. The heart of this book examines policy pathways to accelerate adoption of climate-friendly practices by amending existing federal and state legal regimes and enacting new ones. We recommend improving public agricultural research, development, and technical assistance efforts, especially for climate-friendly practices; reforming federal subsidy and conservation programs; and revising agricultural lending programs, trade policy, grazing practices on government land, programs for perennial agriculture, regulatory strategies, tax policy, biogas subsidies, and greenhouse gas pricing. We also describe how the private and philanthropic sectors can stimulate carbon farming. Moreover, because agriculture is the foundation of our national food system, we also look to upstream and downstream strategies to move toward climate-neutrality, including reducing emissions that stem from production of farm inputs and on-farm energy and from food processing, distribution, consumption, and waste. We also discuss the potential of encouraging consumption of climate-friendly foods through national dietary guidelines, procurement at all levels of government, and private-sector initiatives such as certification schemes and healthier menu options.

Most people think about food often; our food choices both shape and reflect our values and our communities. More than 20 million people work in agriculture and the food system, some in capacities they love and many in jobs with low pay, few protections, and difficult working conditions. We intend this book to help decisionmakers, farmers, consumers, and others take advantage of this opportunity to transform such a critical and central part of people's lives into a solution for climate change. And while there are many aspects of farm policy that this short book does not and cannot address, we hope that the demonstration of the enormous beneficial potential impact of the policy changes we do recommend will inspire and lead to additional policy exploration. After all, we are what we eat and we reap what we sow.

Chapter I.
Introduction

The U.S. agricultural system, one of the most productive in the world, has been profoundly shaped by government policies, especially over the past 100 years. While very successful at producing inexpensive commodities, agriculture in the United States also contributes to significant environmental and health harms, employs millions of people in often unsafe and poorly paid positions, and promotes chronic disease as much as good health. Agriculture also contributes substantially to climate change, the focus of this book. To serve this country's true needs, agriculture must change significantly, and we need new policies to catalyze and accelerate that change.

Agriculture uses our most fertile land to produce what we need most—often known as the four Fs: food, feed, fuel, and fiber. While we continue to need those, we must add a fifth F: our future. This represents both the future of agriculture, which depends on stable weather patterns, and the future of humanity. We must see climate stabilization—through the zero-emission production of plants and animals that can sequester carbon in soil and biomass—as an appropriate aim of agriculture and agricultural policy.

The U.S. agricultural system, shaped by geography, markets, culture, and hundreds of years of federal policy, generates an enormous amount of food. Agriculture in the United States produces about 430 billion pounds of food each year, amounting to more than 3,600 calories per person per day—far more than the approximately 2,300 calories per person per day recommended by the U.S. Dietary Guidelines. The United States exports about 20% of the food it produces, mostly grains and meats, and imports about 15% of the food it consumes, mostly produce and seafood. In addition to being abundant, our food is cheap—Americans pay about one-third less for their food than they did in 1980.[1]

---

1. Jean C. Buzby et al., Economic Research Service, USDA, The Estimated Amount, Value, and Calories of Postharvest Food Losses at the Retail and Consumer Levels in the United States 11 (2014) (EIB-121); Economic Research Service, USDA, *Calorie Availability and Importance of Food in Household Spending Are Inversely Related*, https://www.ers.usda.gov/data-products/chart-gallery/gallery/chart-detail/?chartId=77755 (last visited Jan. 23, 2021); Elizabeth Mendes, *Americans Spend $151 a Week on Food; the High-Income, $180*, Gallup, Aug. 2, 2012, https://news.gallup.com/poll/156416/americans-spend-151-week-food-high-income-180.aspx; Economic Research Service, USDA, *Exports Expand the Market for U.S. Agricultural Products*, https://www.ers.usda.gov/

Despite this success, however, our current food system presents some of the most pressing challenges of the 21st century. The wealth created by agriculture is overwhelmingly distributed to large corporations—largely food processors, meat and dairy companies, and agrochemical companies—and wealthy farm owners. A small group of landowners controls the majority of farmland in the United States, exacerbating growing inequality. The workers who perform the majority of on-farm labor are excluded from basic labor protections and often work in inhumane conditions for low wages. Meanwhile, land consolidation, poverty wages, and monoculture production systems deprive communities in farm country of their tax base and their natural resources.

The wealth produced by agriculture also comes at great cost to the environment and to public health. Agriculture occupies more than half of the contiguous United States, thus excluding other uses, including non-agricultural grasslands and forests that, of course, have different ecological and climate impacts. Agriculture is one of the largest sources of water quality impairment, contributing to the Gulf of Mexico dead zone; eutrophication and algae blooms in Chesapeake Bay, the Great Lakes, and thousands of other water bodies; and contamination of drinking water supplies. Industrial agriculture's use of pesticides exposes many millions to toxic chemicals through their jobs, communities, and food. Diet-related disease now costs Americans more than one trillion dollars a year—more than smoking—while food insecurity plagues millions of families.

These impacts have left agriculture as the only major sector of the U.S. economy where estimated externalities significantly exceed the sector's earnings.[2] The coming years promise even more challenges. Agriculture is threatened by more extreme and erratic weather and other hazards brought about by climate change; agricultural production increasingly relies on marginalized foreign labor; and the power of agribusiness continues to grow.

Agricultural policy has traditionally focused on the narrow interests of commercial farmers and ranchers. Then-Senator Barack Obama's advisors took note of this in a 2008 internal campaign memorandum on the U.S.

---

data-products/chart-gallery/gallery/chart-detail/?chartId=58396 (last visited Jan. 23, 2021); U.S. Food & Drug Administration, *FDA Strategy for the Safety of Imported Food*, https://www.fda.gov/food/importing-food-products-united-states/fda-strategy-safety-imported-food#:~:text=To%20help%20meet%20these%20consumer,of%20its%20overall%20food%20supply (last visited Jan. 23, 2021).

2. KPMG International, Expect the Unexpected: Building Business Value in a Changing World 9 (2012). *See also* The Rockefeller Foundation, True Cost of Food: Measuring What Matters to Transform the U.S. Food System, July 2021 (finding that, while Americans spend $1.1 trillion on food each year, the true cost of food, including the impact on human health, climate change, biodiversity, and livelihoods, is $3.2 trillion per year), https://www.rockefellerfoundation.org/wp-content/uploads/2021/07/True-Cost-of-Food-Full-Report-Final.pdf.

Figure 1. Agriculture's Environmental Impact

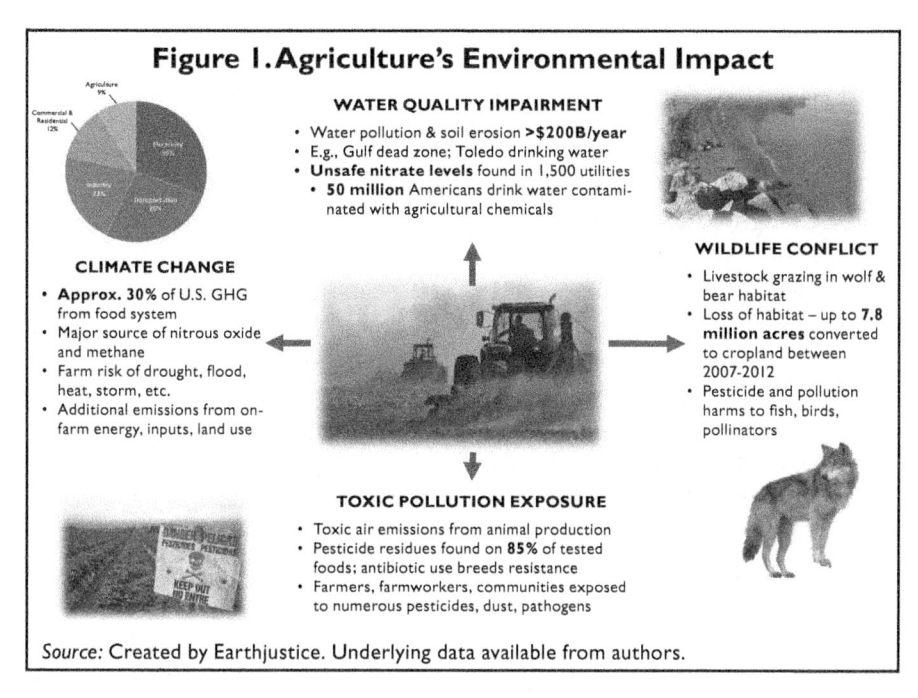

WATER QUALITY IMPAIRMENT
- Water pollution & soil erosion >**$200B/year**
- E.g., Gulf dead zone; Toledo drinking water
- **Unsafe nitrate levels** found in 1,500 utilities
- **50 million** Americans drink water contaminated with agricultural chemicals

CLIMATE CHANGE
- **Approx. 30%** of U.S. GHG from food system
- Major source of nitrous oxide and methane
- Farm risk of drought, flood, heat, storm, etc.
- Additional emissions from on-farm energy, inputs, land use

WILDLIFE CONFLICT
- Livestock grazing in wolf & bear habitat
- Loss of habitat – up to **7.8 million acres** converted to cropland between 2007-2012
- Pesticide and pollution harms to fish, birds, pollinators

TOXIC POLLUTION EXPOSURE
- Toxic air emissions from animal production
- Pesticide residues found in **85%** of tested foods; antibiotic use breeds resistance
- Farmers, farmworkers, communities exposed to numerous pesticides, dust, pathogens

Source: Created by Earthjustice. Underlying data available from authors.

Department of Agriculture (USDA), explaining that the cabinet-level department "primarily serves a single (albeit important) constituency: farmers and ranchers (and, to a lesser extent, their customers)."[3] Recent lawsuits, congressional hearings, and media reports have also demonstrated that the farmers and ranchers served by USDA are overwhelmingly white, male, and wealthy. We found through a Freedom of Information Act request that of the billions of dollars distributed to farmers through the Trump Administration's trade war bail-out payments, 91% went to male farmers and 99.4% went to non-Hispanic white ones. If agriculture is to achieve climate neutrality, agricultural policy must serve a much broader constituency. Such changes will not only reduce emissions, but they will also help improve public health and reduce inequality.

While most climate change-focused policy debate, research, and advocacy has concentrated on the energy and transportation sectors, there is growing recognition that agriculture also can—and indeed must—be an integral part of the climate change solution. This book seeks to help accelerate that transformation.

---

3. E-mail from Lisa Brown to Adam Hitchcock, John Podesta, William Daley, Christopher Edley Jr., Valerie Jarrett, and Federico Peña, AR—Executive Summaries—Energy & Natural Resources Cluster (Oct. 21, 2008) (published by WikiLeaks), https://wikileaks.org/podesta-emails/emailid/26312 (last visited Oct. 23, 2020).

Many other books have focused on the technologies and practices that can help food producers reduce greenhouse gas emissions, increase carbon stored in soil and biomass, and become more resilient to the ravages of climate change. However, few focus on the changes in federal, state, and private policies that will be necessary to ensure that producers adopt these practices widely enough and fast enough to meet the challenge presented by the climate crisis. That is the goal of this book.

Unlike almost all other sectors of the economy, agriculture is both a source and a sink for greenhouse gases. In other areas, the goal is to reduce emissions through lowered demand (such as energy-efficient buildings and appliances) or clean supply (such as wind or solar power). While reducing demand (especially for animal products that have a particularly high climate impact) is a critical strategy for agriculture, there are also opportunities both to reduce greenhouse gas emissions and to store carbon in soil and plants. For that reason, throughout this book we focus on net emissions—that is, the quantity of greenhouse gases released into the atmosphere less the quantity sequestered in soil and plants. Policymakers can take full advantage of agriculture's potential to decarbonize the economy only by making use of the sector's dual capacity to minimize emissions and maximize carbon storage.

We use two terms throughout this book to describe agricultural methods that reduce net agricultural emissions. The first, "climate friendly," refers to practices or strategies that reduce greenhouse gas emissions or increase soil carbon sequestration when compared to conventional methods. While superior to standard practices, climate-friendly practices are not necessarily optimal, either in terms of their climate benefits or their overall benefit to society. By contrast, "carbon farming" describes a suite of climate-friendly practices and strategies designed to result in optimal environmental, societal, and climate outcomes.[4] For example, since anaerobic digesters—expensive systems that compost animal waste and burn the gas it produces for energy—reduce greenhouse gas emissions from industrial livestock facilities known as concentrated animal feeding operations (CAFOs), they may be climate-friendly. But because they help sustain a system of agricultural production with significant emissions and other negative externalities—and do nothing

---

4. "Carbon farming" includes grazing and animal husbandry. As Eric Toensmeier notes in *The Carbon Farming Solution*, there are "several, sometimes conflicting, definitions of carbon farming." However, it is generally described as a system of agricultural economics and practices organized around carbon sequestration. ERIC TOENSMEIER, THE CARBON FARMING SOLUTION 6 (Brianne Goodspeed & Laura Jorstad eds., 2016). "Regenerative agriculture" is another term for largely overlapping agricultural practices. *See generally* RODALE INSTITUTE, REGENERATIVE ORGANIC AGRICULTURE AND CLIMATE CHANGE (2014).

to improve ecoystem or community health—they do not meet our definition of "carbon farming."

There is a further difference between climate-friendly and carbon-farming approaches: the former seeks to reduce emissions while optimizing productivity, whereas the latter adds increased carbon sequestration as a goal. As discussed below, the United States now uses hundreds of millions of acres of land to grow crops that are used inefficiently to produce corn ethanol, sweeteners, or highly processed products or animal feed. While climate-friendly practices could reduce emissions generated by these processes, a federal farm policy that supports carbon farming would make carbon sequestration one of this land's primary uses and lead to wider changes in how and where we produce food, feed, fuel, fiber, and our future. These changes, as this book demonstrates, will also bring about significant improvements in water and air quality and life in rural commuities.

Throughout the book, we call for "climate-neutral" agriculture. This is not because we think that agriculture cannot feasibly become a net sink. On the contrary, as our emphasis on carbon farming indicates, we think that agricultural production should be organized to ensure that farms sequester more than they emit. Nonetheless, there is a risk that treating agriculture as a carbon sink may make it easier to forego much-needed emissions reductions in other sectors. In addition, standard approaches significantly undercount agriculture's climate change impact and there are numerous uncertainties regarding the science, practice, and permanence of carbon sequestration in agriculture. We must be exceedingly careful not to overestimate agriculture's longterm potential to reduce net emissions—even while we work to maximize its sequestration capacity.

One final note regarding terminology: we use "farmer" throughout the book to refer to owners or managers of all agricultural operations, including ranches.[5] This differs from the increasingly common practice of separating farmers and ranchers (or graziers) into distinct categories and using a third term, such as "producer" or "operator" when referring to both. Our primary motivation is to avoid clunky and unnecessary formulations, but this decision also represents a philosophical conviction. While distinguishing ranchers and graziers from other farmers may be appropriate at times, doing so as a rule elides the role that plant production and biodiversity play in animal agriculture. As we discuss, practices such as establishing perennial pastures, planting rows of trees on grazing land, or integrating livestock into cropland can transform an environmentally destructive operation into an

---

5. Likewise, we use "farm" to refer to all agricultural operations, including ranches.

environmentally beneficial one. In short, ranchers don't just manage grazing livestock, they also manage plant life.[6]

Agricultural production is one of the least regulated industries in the country, enjoying a near-comprehensive exemption from environmental regulations, which scholars refer to as the "anti-law" of farms and the environment.[7] This allows large-scale industrial operations to adopt highly polluting, inexpensive practices that drive smaller or more sustainable farms out of business or force them to forego more responsible practices in order to be able to compete. Policymakers will need to regulate the worst practices in order to allow farms to escape this vicious cycle. Although relatively few farms use carbon-farming practices—which often are not supported by agrochemical companies or other agribusiness institutions—these practices almost always improve soil health. The most promising practices increase biological diversification and incorporate perennial plants, dramatically improving nutrient management, water-holding capacity, pollination, and weed and pest control, in addition to soil health. They thus can increase yields, enhance resilience to climate change, and, in the right policy environment, increase farm profitability. If farmers were to adopt these practices, they would improve their businesses, help their communities, and protect our environment.

This book contains the only comprehensive overview of the science, existing law, and policy proposals on agricultural emissions in the United States. Although most of the book is on agriculture, we also include a discussion of emissions in other parts of the food system to help readers develop a broader understanding of potential emissions reductions.

Chapter II looks at the stakeholders in farm policy. It argues that farm policy must not only meet the needs of farmers, but should also focus on the interests of non-white farmers, farmworkers, and rural residents.

Chapter III discusses how agriculture can contribute to massive decreases in emissions, sometimes called "deep decarbonization." It outlines the threats climate change poses to agriculture and the need to build resilience in order to build food security for our country. It further details the sector's current levels of emissions from production, explains how the U.S. Environmental Protection Agency's standard analysis significantly understates agriculture's climate change impact, and identifies uncertainties inherent in measuring

---

6. In recognition of this, many ranchers and graziers refer to themselves as "grass farmers." *See, e.g.,* VERMONT GRASS FARMERS ASSOCIATION, *About Us,* https://www.vtgrassfarmers.org/about-us/ (last visited Jan. 23, 2021). *See also, e.g.,* Seth Watkins, *Farming Evolved: Agriculture Through a Different Lens,* TEDxDesMoines (July 30, 2021), https://www.youtube.com/watch?v=WJ0wP9FJU1s.
7. J.B. Ruhl, *Farms, Their Environmental Harms, and Environmental Law,* 27 ECOLOGY L.Q. 263, 265-328 (2000).

the impacts of wide-scale biological systems. Finally, the chapter outlines agriculture's capacity to sequester substantial amounts of carbon.

Chapter IV examines the on-field strategies, practices, and technologies available to increase soil carbon sequestration and reduce agricultural emissions while maintaining productivity. (Although they are important parts of the food system, we do not discuss fisheries and aquaculture in this book because they present very different greenhouse gas and legal issues.)

Chapters V and VI detail public law pathways—both amending existing federal and state legal regimes and enacting new ones—for reducing net agricultural emissions. Chapter V begins by identifying pathways for improving public agricultural research, development, and extension efforts, and then considers opportunities to reform federal subsidy, conservation, credit, and trade programs. It also provides recommendations to improve grazing practices on government land and expand perennial agriculture. Chapter VI then evaluates public policy pathways outside of USDA, focusing on regulatory strategies as well as tax policy, Small Business Administration lending programs, biogas subsidies, and greenhouse gas pricing.

Chapter VII describes non-public law approaches, focusing on how the private sector, including philanthropic organizations, can stimulate carbon farming. The topics covered include agricultural research, financing for carbon farming, measuring carbon content, conservation tools, and carbon markets.

Chapter VIII looks at overall food system emissions. It provides an overview of strategies for reducing upstream emissions—those that stem from the manufacture of farm inputs—and downstream emissions—those that result from food processing, distribution, consumption, and waste.

Chapter IX examines the potential to encourage the consumption of climate-friendly foods through national dietary guidelines and through procurement at all levels of government, as well as through private-sector initiatives, such as certification schemes and healthier menu options. Chapter X concludes.

For many years climate-change policy almost entirely ignored agriculture, while agricultural policy has only served to intensify climate change. Fortunately, the United States is now energetically exploring how agriculture can play a major role in curbing climate change. This is exciting, yet challenging since the sector is decades behind. This book hopes to help policymakers, advocates, farmers, and others to shape effective and equitable policies toward climate-neutral agriculture.

Key Recommendations

- The agriculture sector in the United States is highly productive, turning out about 3,600 calories per person per day, far more than the approximately 2,300 per day needed. In addition, U.S. agriculture produces over 4,000,000 metric tons of cotton and 16 billion gallons of biofuel each year.

- Agriculture occupies over half of the land in the contiguous United States, contributes to air and water pollution and climate change, and has undergone significant changes in the last decades. Now, most food, feed, fuel, and fiber is produced on highly concentrated, heavily mechanized, specialized, and chemical dependent operations.

- Unlike almost all other sectors of the economy, agriculture is both a source and a sink for greenhouse gases.

- While a fair amount is already known about agricultural practices that can help farmers reduce greenhouse gas emissions and increase carbon sequestration, significant policy changes are needed to accelerate the widespread adoption of these practices, and focused research could dramatically improve the productivity and environmental outcomes of these practices.

- Policymakers can take full advantage of agriculture's potential to decarbonize the economy only by making use of the sector's dual capacity to minimize emissions and maximize carbon storage.

- Agricultural production should be organized to ensure that farms sequester more carbon than they emit while at the same time maintaining productivity.

- There is a risk that treating agriculture as a carbon sink may be used to justify forgoing much-needed emissions reductions in other sectors.

- Studies show a wide range of results regarding the ability of agricultural practices to increase soil carbon or reduce emissions. We must not overestimate agriculture's long-term potential to reduce net emissions, and should not suggest not-yet-achievable precision in calculations.

Chapter I. Introduction

- Agricultural production is subject to few environmental regulations; policymakers will need to consider regulating the most polluting practices at the largest operations to enable more sustainable farmers to compete.

Chapter II.
The Stakeholders in Farm Policy

We cannot implement effective policies to reduce agricultural emissions without an accurate understanding of the primary constituencies. While analysts often make broad statements about "farmers," "ranchers," and "rural communities," careful analysis of who actually produces our food, where and how they live, and how they are doing is much more rare. For example, while mainstream news reports suggested that 2019 was yet another crisis year for farmers, "when farm families wondered how they were going to keep the farm afloat,"[1] farmers overall, in fact, saw their 11th highest per farm profits since 1929. Many of these commentators do not include in their assessment the profound impact on producer income of federal counter-cyclical subsidies, favorable tax treatment, and exports supported or even mandated by trade agreements. These programs can even make the costs of climate change harder to discern, as relief programs often ensure that the costs of increased weather variability or extreme weather events do not fall on farmers. And, over and above the established farm bill conservation, commodity, and crop insurance programs, the president can,

---

Note: This chapter is adapted from an unpublished manuscript by Nathan A. Rosenberg, Bryce W. Stucki, and Peter H. Lehner.

1. Robert Leonard, *Trump Has Sucker-Punched Farmers. America Will Suffer*, N.Y. TIMES, Jan. 14, 2019, https://www.nytimes.com/2019/01/14/opinion/trump-shutdown-government-farmers-aid.html. *See also* Annie Gowen, *Left Behind: Farmers Fight to Save Their Land in Rural Minnesota as Trade War Intensifies*, WASH. POST, Aug. 3, 2019, https://www.washingtonpost.com/graphics/2019/national/farm-bankruptcies-rise-as-trumps-trade-war-grinds-on/; John Muyskens et al., *Midwestern Farmers' Struggles With Extreme Weather Are Visible From Space*, WASH. POST, July 2, 2019, https://www.washingtonpost.com/business/2019/07/02/midwestern-farmers-struggles-with-extreme-weather-are-visible-space/; Laura Reiley, *Weather Woes Cause American Corn Farmers to Throw in the Towel*, WASH. POST, June 18, 2019, https://www.washingtonpost.com/business/2019/06/18/weather-woes-cause-american-corn-farmers-throw-towel/; Amber Phillips, *Trump's Trade War Could Cost Him With a Key Constituency: Farmers*, WASH. POST, Aug. 28, 2019, https://www.washingtonpost.com/politics/2019/08/28/will-trumps-trade-war-cost-him-with-key-constituency-farmers/; Daniel W. Drezner, *Donald Trump Has Emasculated the American Farmer*, WASH. POST, Aug. 13, 2019, https://www.washingtonpost.com/outlook/2019/08/13/donald-trump-has-emasculated-american-farmer/; Tory Newmyer, *The Finance 202: Trump's Trade War Keeps Punishing Farmers. But Farmers Remain Optimistic*, WASH. POST, Aug. 7, 2019, https://www.washingtonpost.com/news/powerpost/paloma/the-finance-202/2019/08/07/the-finance-202-trump-s-trade-war-keeps-punishing-farmers-but-farmers-remain-optimistic/5d4a0020602ff17879a188d5/; Annie Gowen, *"I'm Gonna Lose Everything": A Farm Family Struggles to Recover After Rising Debt Pushes a Husband to Suicide*, WASH. POST, Nov. 9, 2019, https://www.washingtonpost.com/nation/2019/11/09/im-gonna-lose-everything/.

and often does, provide significant additional assistance through the Commodity Credit Corporation—in 2019 the Trump Administration doubled government direct payments to farmers up to almost $24 billion.[2] Without this full picture, policymaking is hobbled.

Unfortunately, very few journalists have the time or expertise to assess the actual contours of farm economics or to examine in detail who these producers are and what they are doing. The number of reporters in rural America has rapidly declined and fewer and fewer reporters live in farm regions or have a background in agriculture.[3] Without experience, it is hard to disentangle the interplay of weather, markets, and government programs that affect farmers.

And who does the work on farms? Again, writers will often make broad references to farmers and sometimes to farmworkers, but rarely do they distinguish among different farm actors or rigorously analyze the lives and work of people in the different groups. In fact, lawmakers have excluded the agricultural industry from many labor and environmental regulations,[4] which makes the distinctions between employers and employees, producers and their neighbors, starker than in almost any other context in the United States.

Policymakers created the foundations of modern farm policy at a time when a substantial portion of the American population lived on farms and the average farm family was more likely to be poor than the average non-farm family. But that is not the case today, and modern agriculture policy should be revised to reflect our current reality. As it exists now, U.S. farm policy largely benefits a small number of almost entirely white producers who are substantially wealthier than the average American, and who are required

---

2. U.S. Department of Agriculture (USDA) Economic Research Service, *Net Cash Income*, https://data.ers.usda.gov/reports.aspx?ID=17831 (last updated Sept. 2, 2020).
3. *See, e.g.*, Elizabeth Grieco, *For Many Rural Residents in U.S., Local News Media Mostly Don't Cover the Area Where They Live*, Pew Res. Center: Factank, Apr. 12, 2019, https://www.pewresearch.org/fact-tank/2019/04/12/for-many-rural-residents-in-u-s-local-news-media-mostly-dont-cover-the-area-where-they-live/; April Simpson, *As Local News Outlets Shutter, Rural America Suffers Most*, Stateline, Oct. 21, 2019, https://www.pewtrusts.org/en/research-and-analysis/blogs/stateline/2019/10/21/as-local-news-outlets-shutter-rural-america-suffers-most; Jim Boren, *Agriculture Is Huge Story in California, but Newsrooms Around the State Aren't Paying Attention to the Details*, Fresno St. Inst. for Media & Pub. Tr., Dec. 9, 2019, https://mediaandpublictrust.com/2019/12/09/agriculture-is-huge-story-in-california-but-newsrooms-around-the-state-arent-paying-attention-to-the-details/; Chris Clayton, *The Agriculture Beat Is a Crucial Lens on a Changing Climate*, Colum. Journalism Rev., Oct. 24, 2017, https://www.cjr.org/special_report/climate-change-agriculture-food.php.
4. Labor law scholars have dubbed the phenomenon "agricultural exceptionalism" and environmental law scholars refer to the "anti-law" of farming and the environment due to the sector's almost total exclusion from environmental regulations. *E.g.*, Juan Perea, *The Echoes of Slavery: Recognizing the Racist Origins of Agricultural and Domestic Worker Exclusion From the National Labor Relations Act*, 72 Ohio St. L.J. 95-138 (2011) (discussing agricultural exceptionalism); J.B. Ruhl, *Farms, Their Environmental Harms, and Environmental Law*, 27 Ecology L.Q. 263, 295-305 (2000) (arguing that the environmental law of agriculture constitutes an "anti-law").

to do little to protect the health and environmental concerns of their workers and neighbors or to address climate change.

Sound, sustainable, and fair agriculture policy should be built on an accurate understanding of the affected constituencies, rather than on assumed and outdated images and narratives. Here, we try to provide this foundation, looking not only to the "farmer" and agribusiness, but also to other larger rural constituencies. These include farm laborers, who do most of the work on farms and outnumber farmers by a wide margin, and rural residents, who, according to numerous polls, support environmental reforms by substantial margins. These other constituencies, of course, should also include those who consume our food and all those affected by climate change.

To achieve climate-neutral agriculture in the United States, as well as to make it more just and sustainable, we must engage all these groups. They are the ones who will live with—and see through—these policies.

A. Farmers and the Farm Economy

The answer to the question "who farms?" for most people is simple: farmers. But it is hired farmworkers who do most of the work on American farms.[5] Farmers are better understood as business owners or managers who hire or employ their own labor to turn a profit. We discuss the characteristics of farmworkers in a later section, but here we clarify who farmers are, including their economic positions. This discussion will inform the recommended policies set out later in the book.

We first provide a brief overview of the farm economy since the New Deal. We then analyze U.S. Department of Agriculture (USDA) data to disaggregate different groups of "farmers" and "ranchers," thus providing a clearer picture of the farm community and economy. We show that more than half of those who USDA includes as "farmers" are actually retirees, hobbyists, and taxpayers with "paper farms" (so-classified for tax purposes), whose economic output distorts general, commonly reported statistics on actual farm businesses. The last subsection explains the origins, extent, and significance of the modern agricultural "safety net" that supports farmers and informs their politics. After our discussion of farmers, we turn to the rural constituencies that current farm policy largely ignores—workers, non-white farmers, and rural people in general—who are already pressing for many of the environmental reforms proposed in this book.

---

5. Philip L. Martin, Giannini Foundation, Immigration and Farm Labor: Challenges and Opportunities 2 (2017).

1. Transformations in the Farm Economy

Writers who discuss farm policy tend to rely on images and conceptions from the 1930s. At that time, almost a fifth of the population farmed, farms produced a fraction of what they do today, and farm household incomes were less than half that of non-farm households.[6] The Great Depression caused widespread foreclosures and tax sales, which the government stopped with massive New Deal spending programs.[7] These programs inaugurated a new regime in the farm economy.[8] From 1929 to 1940, government payments increased from 3% of net farm income to 29%.[9] Most of these funds went to large farms, a trend that has continued and intensified to the present day.[10] Between 1930 and 1992, the number of white farmers fell by 65% and Black farmers by 98%,[11] while the average farm size more than doubled from 199 to 464 acres.

As a result of these broad historical trends, the average farm household now has a higher annual income and more non-farm wealth than the average household. There are very few farm households below the poverty level, and the majority of the remaining ones are composed of elderly people who sell little to no agricultural products.[12] The farm operations that produce the vast majority of our food now more closely resemble small factories. The operator is often dependent on the modern financial system for both loans and current and future sales, and manages their operation from an office while hired labor works the land or livestock. These businesses rely heavily on a seasonal work force, assisted by heavy machinery, to produce a huge amount of standardized product for minimal cost. Our discussion below distinguishes between farm businesses and nonbusinesses, and examines the implications for farm policy and analysis.

---

6. Craig Gundersen & Susan E. Offut, *Farm Poverty and Safety Nets*, 87 AM. J. AGRIC. ECON. 885, 885 (2005).
7. *See* Nathan A. Rosenberg & Bryce Wilson Stucki, *The Butz Stops Here: Why the Food Movement Needs to Rethink Agricultural History*, 13 J. FOOD L. & POL'Y 12, 13-14 (2017).
8. *Id.* at 20-22.
9. *Id.* at 14.
10. *Id.*
11. *Id.* at 20-22.
12. Seventy-five percent of limited-resource farmers sold less than $10,000 of agricultural products in gross sales. Agricultural Resource Management Survey Special Tabulation Request From USDA ERS to Nathan Rosenberg (June 25, 2019) (on file with authors).

2. Overstatements in USDA Census of Agriculture Data

USDA reports many commonly cited statistics about farms and farm income that are distorted by the way the department's Census of Agriculture counts farms. For example, the *Wall Street Journal* reported in 2019 that "more than half of U.S. farms lost money farming in recent years."[13] But statistics like median farm income are skewed by the huge number of retirement, "lifestyle," paper farms created for tax purposes, and other "farms" that raise very little or no agricultural products that USDA nonetheless counts as farms.

Many of these problems relate to the Census of Agriculture's methodology. Even as fewer farms came to control more and more acreage, the agricultural census began to register a sharp increase in farms after USDA took over the survey in 1997. At that time, the department introduced a series of methodological changes designed to increase the survey's counts.[14] By 2017, these changes had brought the total count of farms back to roughly the same level as 1987, even as other sources of data on farms continued to show a decline. The number of households filing Schedule F forms with the Internal Revenue Service (IRS), which are used to report farm income and expenses, declined by 34% between 1978 and 2017, and the number of farm households identified by the Current Population Survey (CPS), a federal survey conducted by the Census Bureau that is the source of national employment statistics, declined by 35%. Meanwhile, the agricultural census showed a decline of only 17% during this period. All three data sources showed similar trends until 1997. Today, the agriculture census shows twice as many "farms" as the CPS.[15]

USDA changed the agricultural census, at least in part, to address its historic undercounts of small-scale and non-white farmers—especially Black and indigenous farmers. But the department overcorrected and now counts a large number of non-farms as farms. One major source of this overcorrection comes from USDA's definition of "farm," which has not changed since 1974: "A farm is defined in the census as any place from which $1,000 or more of agricultural products were produced and sold, or normally would

---

13. Jesse Newman & Jacob Bunge, *"This One Here Is Gonna Kick My Butt"—Farm Belt Bankruptcies Are Soaring*, WALL ST. J., Feb. 6, 2019, https://www.wsj.com/articles/this-one-here-is-gonna-kick-my-buttfarm-belt-bankruptcies-are-soaring-11549468759.
14. Nathan A. Rosenberg, *Farmers Who Don't Farm: The Curious Rise of the Zero-Sales Farmer*, 7 J. AGRIC. FOOD SYS. & COMMUNITY DEV. 2-4 (2017), *available at* https://papers.ssrn.com/sol3/papers.cfm?abstract_id=3104703.
15. Calculated by the authors from CPS March supplement data. Nat'l Bureau of Econ. Res., *NBER CPS Supplements*, *available at* https://data.nber.org/data/current-population-survey-data.html (last visited Jan. 23, 2021).

have been sold, during the census year."[16] Had USDA adjusted the $1,000 income threshold for inflation, that alone would have excluded almost half of all "farms" in the 2017 Census of Agriculture.[17]

Not only did USDA not adjust the income threshold, it also broadened its interpretation of "normally would have been sold." The department devised a point system to estimate how much income a plot of land could produce if it were used to raise or grow agricultural products—even if the operator had never used the land as agricultural land and the owners had no intention of using it that way.[18] Rural homes with berry bushes (at least one-fifth of an acre), vegetables (one-fifth an acre), horses (10 acres of pasture), cattle (one acre), or other potential "agricultural products" all count as farms under the official definition.[19]

Since 1997, the agricultural census has included a greater and greater share of "point farms": properties that met the definition of "farm" because USDA estimated that they *could* have sold, but did not sell, $1,000 of agricultural products.[20] By 2017, almost 30% of farms in the agricultural census were point farms[21] and more than 20% of census "farms" did not sell any agricultural products whatsoever.[22] In fact, well over half of all farms reported in the agricultural census are, by the department's own definitions,[23] not farm businesses, but retirement or "lifestyle" farms; this latter category was so-named "because many of the operators on these farms view their farms largely as an avocation or a place to live where they can enjoy a rural lifestyle."[24] As one

---

16. NATIONAL AGRICULTURAL STATISTICS SERVICE, USDA, 2017 CENSUS OF AGRICULTURE, U.S. NATIONAL LEVEL DATA VII (2019) [hereinafter 2017 CENSUS OF AGRICULTURE].
17. A total of 47.9% of all farms in the 2017 Census of Agriculture sold less than $5,000 in agricultural products.
18. *See* Freedom of Information Act (FOIA) Response No. 2019-REE-02265-F From USDA to Authors (Sept. 30, 2019) (on file with authors) (showing the point values used by USDA); Maggie Koerth, *Big Farms Are Getting Bigger and Most Small Farms Aren't Really Farms at All*, FIVETHIRTYEIGHT, Nov. 17, 2016, https://fivethirtyeight.com/features/big-farms-are-getting-bigger-and-most-small-farms-arent-really-farms-at-all/.
19. *See* FOIA Response No. 2019-REE-02265-F, *supra* note 18.
20. *See* Rosenberg, *supra* note 14 (describing methodological changes to the census that lead to higher numbers of zero-sales and other point farms).
21. 2017 CENSUS OF AGRICULTURE, *supra* note 16, at 9 tbl.2.
22. Calculated by the authors using Special Tabulation Request From USDA to Nathan Rosenberg (Nov. 20, 2019) (on file with authors) and 2017 CENSUS OF AGRICULTURE, *supra* note 16. In comparison, less than 6% of farms had zero-sales in 1992, the last agricultural census administered by the Census Bureau. Rosenberg, *supra* note 14, at 3.
23. *See* NATIONAL AGRICULTURAL STATISTICS SERVICE, USDA, 2012 CENSUS OF AGRICULTURE, FARM TYPOLOGY 1 tbl.1 (2015) [hereinafter 2012 CENSUS OF AGRICULTURE].
24. ROBERT A. HOPPE & JAMES M. MACDONALD, USDA, UPDATING THE ERS FARM TYPOLOGY 11 (2013) (EIB-110).

Chapter II. The Stakeholders in Farm Policy

journalist put it, most small farmers in the census "aren't the farms of the poor; they're the yards of the upper-middle-class."[25]

Another important, but harder to quantify source of overstatement of the number of farms comes from people who define their property as a farm for tax purposes. Farm operations receive numerous tax benefits, notably lower property taxes, which encourage property owners to classify their land as "agricultural." All 50 states offer "use-value assessment" for agricultural land, which allows owners to assess their property at rates well below market value, often by 90% or more.[26] Many states have extraordinarily broad definitions of "agricultural land" that make it easy for non-farms to qualify, and state and local governments often do not enforce the few restrictions that do exist or check if former agricultural operations are active.[27] Rural property owners can count their land as "farmland" with nominal, and sometimes less than nominal, gestures at agricultural production. In Florida, landowners can take advantage of the state's greenbelt law, designed to protect grassy, forested, and farming land, through a variety of well-known and inexpensive strategies, some as simple as renting cows.[28] New Jersey requires that landowners have five acres and sell $500 of goods a year.[29] In South Carolina, property owners only need five acres of trees to qualify for the agricultural land use benefit.[30] While there is no comprehensive study on how many landowners create paper farms for tax purposes, federal tax data suggest the number is considerable. Almost 75% of the 1.8 million taxpayers filing IRS Schedule F forms in 2017 reported net losses from their agricultural business, allowing them to collectively deduct $30 billion from their taxes.[31] More than 150,000 taxpayers submitted a Schedule F form despite not receiving

---

25. Koerth, *supra* note 18.
26. JOHN E. ANDERSON, AGRICULTURAL USE-VALUE PROPERTY TAX ASSESSMENT: ESTIMATION AND POLICY ISSUES 1 (2011).
27. For example, a 2005 *Miami Herald* investigation into Florida's property tax expenditures for farmland found that local property appraisers awarded tax breaks on land that had been purchased for more than three times its agricultural value, rezoned for development, and not kept up to farming standards. Beth Reinhard & Samuel P. Nitze, *Law Fails to Save Florida Farmland*, MIAMI HERALD, Sept. 8, 2014, https://www.miamiherald.com/latest-news/article1928900.html.
28. *See, e.g.*, Jordan Weissman, *America's Dumbest Tax Loophole: The Florida Rent-a-Cow Scam*, ATLANTIC, Apr. 17, 2012, https://www.theatlantic.com/business/archive/2012/04/americas-dumbest-tax-loophole-the-florida-rent-a-cow-scam/255874/.
29. Richard Rubin, *Goat Herd Helps Trump Lower Tax Bite*, WALL ST. J., Apr. 20, 2016, https://www.wsj.com/articles/goat-herd-helps-trump-lower-tax-bite-1461191607.
30. S.C. CODE ANN. §12-43-232 (2020).
31. More than 73% of taxpayers filing Schedule F reported net losses. IRS, *Table 1.3. All Returns: Sources of Income, Adjustments, Deductions, Credits, and Tax Items, by Marital Status, Tax Year 2017 (Filing Year 2018)*, https://www.irs.gov/pub/irs-soi/17in13ms.xls (last updated Sept. 3, 2020).

any gross income *at all* from agricultural products, allowing them to collectively deduct $6.8 billion from their returns.[32]

As a result of these and other factors, Census of Agriculture data overstate the number of actual farms. At the same time, the data largely ignore many other aspects of the farm economy such as farmworkers, consumers, and residents affected by neighboring farms. This not only distorts economic data such as average income, but it also distorts politics and policymaking. A critical first step in more effective policymaking is the development of a more accurate assessment of actual farm businesses. The Economic Research Service has produced reports on these entities, but the department should make the distinction between business and nonbusiness farms central to its census reports.

We can develop a far more accurate understanding of farm businesses if we use multiple data sources—including the CPS, IRS tax data, and detailed census statistics—rather than relying only on summary agricultural census numbers. In 2017, the USDA census reported more than 2 million "farms" but around 950,000 "farm businesses"[33] and between 1 and 1.4 million operators who said their primary occupation was farming.[34] Almost 1.8 million households filed Schedule F forms in 2017, but only about 1 million farms reported gross sales over $50,000. The CPS reported around 900,000 farm households and around 1 million farmers—or people who said their longest job in the previous year was as a farmer—in 2017.[35] These three sources all show that there are about half as many farms and farmers as generally reported by USDA and in most press accounts.[36]

---

32. *Id.*
33. USDA Economic Research Service, *Tailored Reports: Farm Structure and Finance*, https://data.ers.usda.gov/reports.aspx?ID=17882 (last updated Dec. 10, 2019).
34. *Id.* (reporting approximately one million operators who said their primary occupation was farming); 2017 Census of Agriculture, *supra* note 16, at 62 tbl.52 (reporting more than 1.4 million operators who said their primary occupation was farming).
35. Calculated by the authors using 2018 CPS March Supplement data. Nat'l Bureau of Econ. Res., *supra* note 15. To produce the estimates, the authors used the standard methodology. U.S. Census Bureau, Current Population Survey 2018 ASEC Supplement, *available at* https://www2.census.gov/programs-surveys/cps/techdocs/cpsmar18.pdf. We calculated "farm households" by calculating the number of households with someone who said their longest job in the previous year was as a farmer in the household. We calculated farmers as the count of people who said their longest job in the previous year was as a farmer.
36. 2017 Census of Agriculture, *supra* note 16, at 7 tbl.1, 62 tbl.52. Farm groups sometimes argue most low-sales farms like those excluded from USDA's farm business count would expand their operations if they had the financial wherewithal. However, the evidence strongly suggests that this is not the case. USDA data show that "farms" with low and very low sales are almost exclusively owned by households with comfortable incomes and above average wealth (even excluding farm assets).

3. A More Accurate Assessment of Farm Income and Wealth

People who work on and write about farm policy are heavily influenced by the widespread belief, noted above, that farmers face an almost continuous financial crisis. In fact, recent years have, by and large, been lucrative for farmers. None of this is to say that there are not farmers who struggle, including many small-scale and sustainable farmers. But even when their incomes vary, farmers' substantial wealth helps get them through. This more accurate understanding suggests that many farmers have some or all of the resources needed to shift agricultural practices, so that regulatory, educational, and outreach programs could be effective tools for accelerating climate-friendly farming, even if not linked to changes in subsidies.

First, it is critical to look at recent farm income data in context. From 2011 to 2013, farmers saw some of their highest total profits since 1929,[37] so comparisons of current incomes against this peak can be misleading.[38] Viewed more broadly, farmers' total profits are projected to be at their 24th and 23rd highest ever in 2019 and 2020, far above average.

Second, it is important to look to income figures based on the number of actual farm businesses. Standard USDA annual ranking figures of total incomes understate the incomes of individual farmers, since almost all of the most profitable farm income years were in the 1940s, when there were two to three times as many farms sharing these profits. Using USDA's figures for the number of farms, as in Figure 1 on the next page, 2019 and 2020 are projected to be the 11th and 9th most profitable years ever. Measured by net income per farm, five of the best ten years happened in the past decade. And in absolute terms, these figures translated to a net median income of $195,000 for commercial farms in 2017. Using the more accurate figure of about one million farms, the net income per farm—of active farms—in recent years would be about double that shown using USDA's official tally, and the rankings of the years' profitability would go even higher.

Third, farm policy must reflect an accurate understanding of farm wealth. The median farm household, which includes retirement and lifestyle farms, had a total net worth of $1 million in 2019 according to USDA, about eight times median household wealth, and of that total their non-farm net wealth is about $370,000, which alone is three times median household

---

37. The series began in 1929. *See* USDA Economic Research Service, *supra* note 2.
38. Ana Swanson & Jim Tankersley, *As Trump Appeals to Farmers, Some of His Policies Don't*, N.Y. Times, Jan. 7, 2018, https://www.nytimes.com/2018/01/07/us/politics/trump-farmers-agriculture-trade-taxes.html.

Figure 1. Net Farm Income

Source: Calculated by authors using data from U.S. Department of Agriculture (USDA) Economic Research Service, *Net Cash Income*, https://data.ers.usda.gov/reports.aspx?ID=17831 (last updated Sept. 2, 2020).

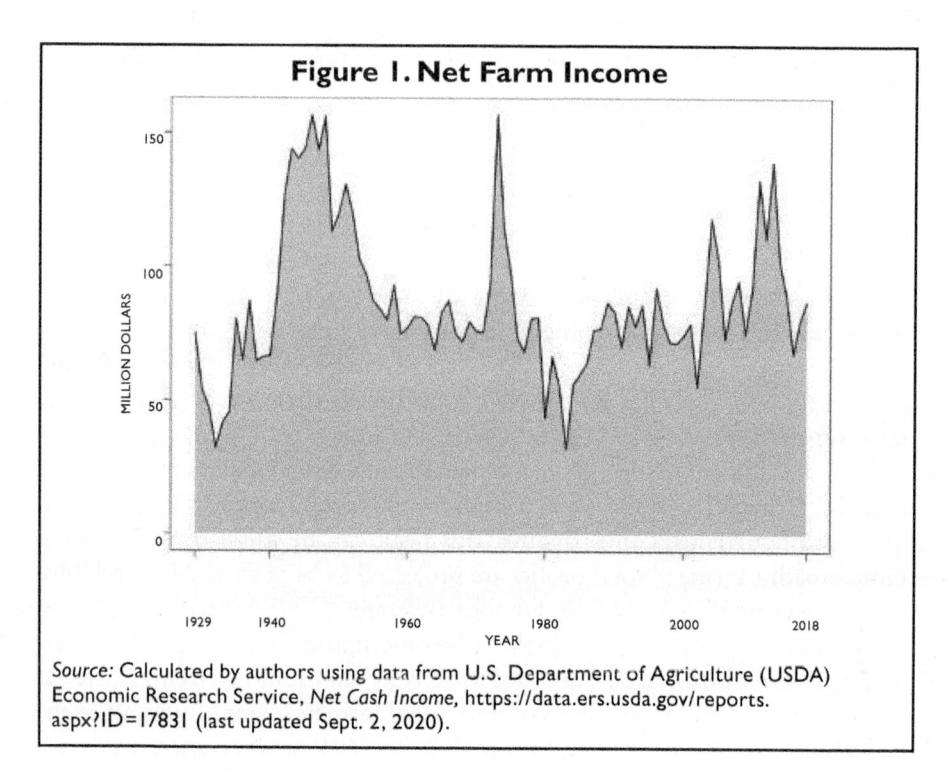

wealth.[39] Moreover, "intermediate" farm businesses (gross sales less than $350,000) had a median net worth around $1 million and commercial farms had a median net worth around $2 million.[40] Farm owners also benefit from land appreciation, which has been positive in every year since 1990 and which has had a greater rate of return than the S&P 500 since the post-war period.[41] As a result of these trends, farm wealth significantly exceeds that of non-farm households in every decile (as seen in Figure 2).

---

39. USDA Economic Research Service, *Tailored Reports: Operator Household Balance Sheet*, https://my.data.ers.usda.gov/arms/tailored-reports (last updated Dec. 18, 2020).
40. *Id.* Note that "low-sales farms" are sometimes called "intermediate farms" in USDA's collapsed typology.
41. Calculated by authors using National Council of Real Estate Investment Fiduciaries (NCREIF), *NCREIF Farmland Property Index*, https://www.ncreif.org/data-products/farmland/ (last visited Jan. 23, 2021); David A. Lins et al., *Institutional Portfolios: Diversification Through Farmland Investment*, 20 Real Estate Econ. 549-71 (1992); Roger G. Ibbotson & Laurence B. Siegal, *Real Estate Returns: A Comparison With Other Investments*, 12 Real Estate Econ. 219-42 (1984). A study of tax returns suggests farmers also seriously understate their incomes to save even more money. Naomi E. Feldman & Joel Slemrod, *Estimating Tax Noncompliance With Evidence From Unaudited Tax Returns*, 117 Econ. J. 327, 347 tbl.7 (2007) (estimating substantially lower rates of tax compliance among farm businesses than other types of businesses).

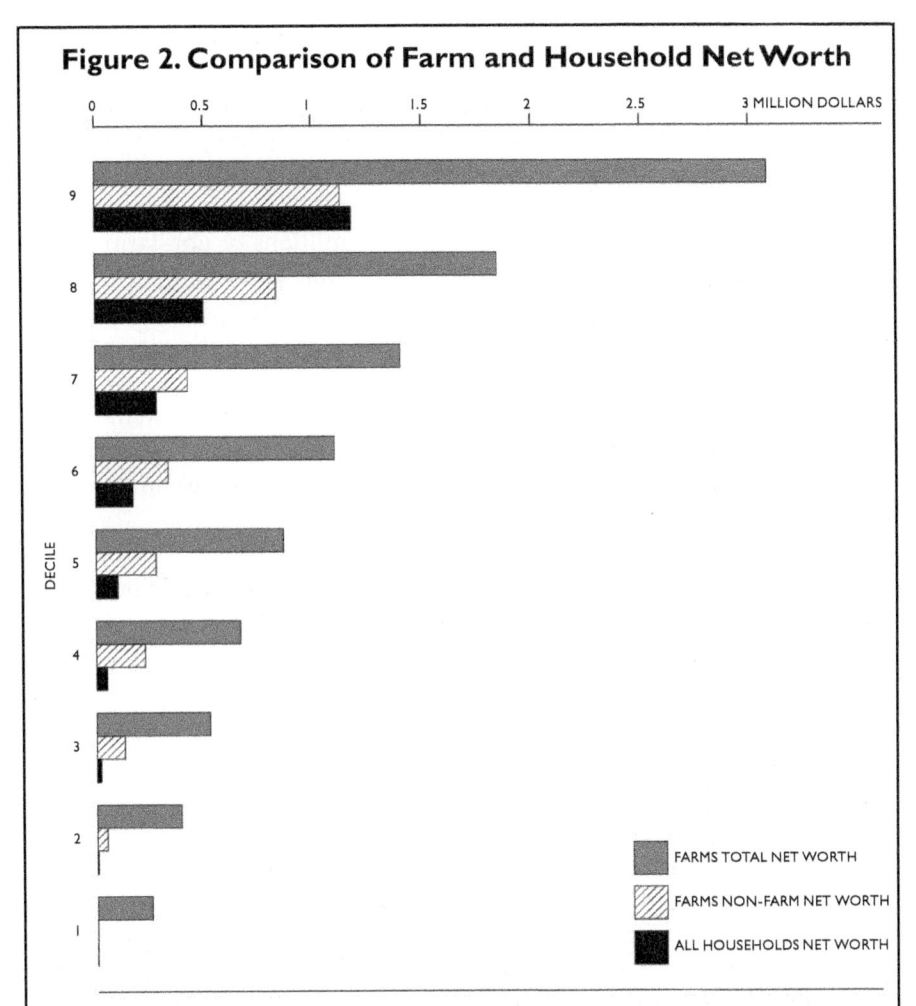

Figure 2. Comparison of Farm and Household Net Worth

Sources: Agricultural Resource Management Survey, Special Tabulation Request From USDA ERS to Nathan Rosenberg (Dec. 11, 2019) (on file with authors). Federal Reserve Board, *2016 Survey of Consumer Finances*, https://sda.berkeley.edu/sdaweb/analysis/?dataset=scfcomb2019 (last visited Jan. 23, 2021).

4. The Legacy of Discriminatory Agricultural Policy

Almost all farmers are old (66% 55 or older; 83% 45 or older), white (95%) men (76%) who live in rural areas.[42] But while the popular press and general political conversation tends to conflate these farmers with the entire rural community, in fact there are many other farm constituencies. Farmworkers, non-white and female farmers, and the many millions of non-farmworkers who live near farms are all critical parts of rural communities. Whereas farm owners enjoy federal subsidy payments and tax exemptions, these other groups largely do not. And while farm owners benefit from the exemption for farms from most environmental, antitrust, child labor, overtime, workplace safety, minimum wage, bankruptcy, motor carrier, and animal welfare laws, these other rural constituencies are often harmed by these same exemptions. As a result of these differences, these other constituencies have proven to be more open to changes to agricultural policy.

This stark contrast in attitudes among different rural groups is in part the result of a long and consistent history of farm policy that has favored white landowning farmers over others. The first Civil War Congress in 1861 created USDA, passed the Morrill Act—which provided funding for a nationwide system of colleges for "the Benefit of Agriculture and the Mechanic Arts"—and enacted the Homestead Act. The federal government eventually granted 246 million acres to 1.5 million families through the Homestead Act and its successors.[43] Effectively closed off to most Blacks and other minorities, homesteading gave European immigrants and other white families an opportunity to acquire considerable property and assets.[44]

In the first decades of the 1900s, Congress expanded the land-grant university system and federal funding for agricultural research, extension, and education. The New Deal increased assistance for large-scale, capital-intensive farms through an array of ambitious new subsidies and federal credit, crop insurance, and technical assistance programs.

The New Deal coalition that passed these new farm laws was heavily reliant on Jim Crow legislators. These legislators killed programs for and research on small farmers and sharecroppers, and ensured federal funds remained under "local control," by which they meant white landowners. These same

---

42. Statistics are for primary producers. 2017 CENSUS OF AGRICULTURE, *supra* note 16, at 62 tbl.52.
43. Trina Shanks, *The Homestead Act: A Major Asset-Building Policy in American History*, *in* INCLUSION IN THE AMERICAN DREAM: ASSETS, POVERTY, AND PUBLIC POLICY 29 (Michael Sherraden ed., Oxford University Press 2005).
44. *Id.* at 36; Keri Leigh Merritt, *Land and the Roots of African-American Poverty*, AEON, Mar. 11, 2016, https://aeon.co/ideas/land-and-the-roots-of-african-american-poverty.

legislators also excluded domestic workers and farmworkers, the two most common occupations for Black people, from key statutory benefits and labor protections.[45] (Congress extended minimum wage requirements to farmworkers—with some important exceptions—in 1966, but federal law still denies farmworkers the right to unionize or earn overtime pay.) This political alliance of big money agriculture and white supremacy enacted policies that pushed hundreds of thousands of Black tenants and sharecroppers off the farm and into cities, in what one historian called one of the "largest government-impelled population movements in all our history."[46]

B. New Constituencies[47]

While farmers have a cabinet-level agency devoted to their interests, there are also millions of other people affected by farm policy who generally have little to no say in it and receive few benefits. Indeed, all too often current farm policy acts against the interests of farmworkers, non-white farmers, and rural people. As a result, these farm policy stakeholders are open to changes to agricultural policy, and many have already been advocating for reforms along the lines of those urged here.

Part of the outsized influence that the small group of farmers has on policy is the belief that they dominate rural economies. A close look shows that is not the case, and that instead most farm income goes elsewhere. The concern for food security also underlies part of the influence on policy of farm owners, but again, the evidence makes clear that employed farmworkers do most of the work on American farms and ensure our food supply. Despite this, they are often denied basic rights by federal policy. Black, indigenous, and Hispanic farm groups have also largely been denied the benefits of farm programs, resulting in most being driven out of farming, and yet they have a history of interest in more sustainable approaches. Finally, farm policy largely ignores the actual expressed interests of rural communities, who, contrary to assumptions, consistently list clean air and water as among their top priorities. All of these constituencies, in addition to farmers and food consumers, must be active and empowered stakeholders to design and implement effective, just, and climate-friendly farm policy reform.

---

45. *E.g.*, Perea, *supra* note 4. Thus, when Congress passed the National Labor Relations Act of 1935, which extended organizing and collective bargaining rights to workers, and the Fair Labor Standards Act of 1938, which extended minimum wage and overtime requirements, farmworkers were excluded.
46. Donald H. Grubbs, *Lessons of the New Deal*, *in* THE PEOPLE'S LAND 19, 20 (Peter Barnes ed., Rodale Press 1975).
47. We refer to these groups as "new constituencies" not because they are new to agriculture but rather because farm policy does not currently serve their interests.

1. Farms in the Rural Economy

Many commentators argue that farmers are central to the rural economy, or conflate the farm economy with the rural economy in general.[48] And this leads many of them to wonder why, with "more farming wealth than ever, farming communities are poorer."[49] The typical answer is that monopolies and financial interests have siphoned off farm wealth. However, as discussed above, farmer income and wealth are far above the median. In fact, farmers play less of a role in the rural economy than is generally assumed.

Today, a small number of capital-intensive farms manage hundreds of millions of acres and produce a tremendous amount of commodity calories with a relatively small number of workers. From 1991 to 2015, farms with at least $1 million in sales (adjusted for inflation) increased their share of total production from 31% to 51%.[50] This increasing dominance of large, industrialized operations has been a major long-term factor in rural depopulation,[51] and researchers have associated the arrival of large-scale industrialized farms with increases in local income inequality and community conflict, as well as pollution.[52] A study on midwestern counties in the late 2000s found almost no relationship between farm revenues and the non-farm economy.[53] A USDA analysis also found that operators on larger farms are more likely to bypass local towns to acquire machinery, farm inputs, and credit.[54]

Not only do most farms appear to have a limited relationship with their surrounding community, their own role in the rural economy is very small. In 2018, farmers made up about 2% of the rural population,[55] agricultural jobs—including both jobs on farms and those providing goods and services

---

48. *See, e.g.*, Tim Marema & Bill Bishop, *White House Adviser Erroneously Calls Ag the "Primary Driver" of Rural Economy*, Daily Yonder, Apr. 25, 2017 (explaining that agriculture is not the "primary driver" of the rural economy, contrary to the statement of a White House adviser), https://dailyyonder.com/chief-white-house-adviser-erroneously-calls-ag-primary-driver-rural-economy/2017/04/25/.
49. Nick Shaxson, *Rural America Doesn't Have to Starve to Death*, Nation, Feb. 18, 2020, https://www.thenation.com/article/economy/big-agribusiness-finance-farming/.
50. James M. MacDonald, USDA, Three Decades of Consolidation in U.S. Agriculture 37 (2018) (EIB-189).
51. Kenneth M. Johnson & Daniel T. Lichter, *Rural Depopulation: Growth and Decline Processes Over the Past Century*, 84 Rural Soc. 3, 14-15 (2019).
52. Linda Lobao & Curtis W. Stofferahn, *The Community Effects of Industrialized Farming: Social Science Research and Challenges to Corporate Farming Laws*, 25 Agric. Hum. Values 219-40 (2008).
53. Jeremy G. Weber et al., *Crop Prices, Agricultural Revenues, and the Rural Economy*, 37 Applied Econ. Persp. 459-76 (2014).
54. J. Michael Harris et al., USDA, Agricultural Income and Finance Outlook 69 (2009) (AIS-88).
55. Calculated by the authors using Economic Research Service, USDA, Rural America at a Glance: 2019 Edition 1 (2019) (EIB-212) (reporting approximately 45 million people in non-metro areas), and our estimate, *supra* Chapter II.A.2, of approximately one million farmers.

Chapter II. The Stakeholders in Farm Policy

to farms—accounted for less than 6% of all jobs in rural counties, and farm jobs produced only about 3% of personal earnings.[56] The number of counties defined as "farming dependent" by USDA fell about 10% during the 2000s, and the number of farm jobs fell by 14%.[57]

In contrast to the relatively small role that commercial farms and farmers play in rural communities, farmworkers play a larger one. Yet mainstream news reports tend to ignore farmworkers and their challenges.[58] For example, in 2018, news outlets ran hundreds of articles about a suicide crisis among farmers, relying on a study by the Centers for Disease Control and Prevention (CDC) that found that "farming, fishing, and forestry" workers had the highest suicide rate of any occupational group.[59] Journalists and the CDC assumed that this category included farmers, even though in fact it is composed predominantly of farmworkers. As a result, Congress set up a program to address the "farmer suicide crisis," complete with grant applications that do not mention or otherwise include farmworkers, denying them the opportunity for relief.

There are about 2.5 million farmworkers—about twice as many as active farmers.[60] Farmworkers do two-thirds of the work on commercial farms, where almost all production happens, but receive only a quarter of the wages. Crop workers reported a median annual income between only $17,500 and $20,000 in 2015-2016—a third with family incomes below the poverty line.[61] USDA data suggest that the farmworker's share of the food

---

56. U.S. Bureau of Economic Analysis, *Regional Data*, https://apps.bea.gov/iTable/iTable.cfm?reqid=70&step=1&isuri=1 (last visited Oct. 26, 2020).
57. Timothy Parker, *Updated ERS County Economic Types Show a Changing Rural Landscape*, AMBER WAVES, Dec. 7, 2015, https://www.ers.usda.gov/amber-waves/2015/december/updated-ers-county-economic-types-show-a-changing-rural-landscape.
58. In 2019, for example, the *Washington Post* ran at least seven articles on the financial fortunes of farmers, but only three on farmworkers, and each of those articles discussed farmworkers from their employers' perspective. Annie Gowen & David Nakamura, *Rallying Farmers, Trump Pushes Border Wall but Opens Door to More Immigrants in Agriculture Jobs*, WASH. POST, Jan. 14, 2019, https://www.washingtonpost.com/politics/rallying-farmers-trump-pushes-border-wall-but-opens-door-to-more-immigrants-in-agriculture-jobs/2019/01/14/0c8062a6-1835-11e9-8813-cb9dec761e73_story.html; Danielle Paquette, *Farmworker vs. Robot*, WASH. POST, Feb. 17, 2019, https://www.washingtonpost.com/news/national/wp/2019/02/17/feature/inside-the-race-to-replace-farmworkers-with-robots/; Kevin Sieff & Annie Gowen, *With Fewer Undocumented Workers to Hire, U.S. Farmers Are Fueling a Surge in the Number of Legal Guest Workers*, WASH. POST, Feb. 21, 2019, https://www.washingtonpost.com/world/the_americas/with-fewer-undocumented-workers-to-hire-us-farmers-are-fueling-a-surge-in-the-number-of-legal-guest-workers/2019/02/21/2b066876-1e5f-11e9-a759-2b8541bbbe20_story.html.
59. Bryce Wilson Stucki & Nathan Rosenberg, *How a Simple CDC Error Inflated the Farmer Suicide Crisis Story—And Led to a Rash of Inaccurate Reporting*, COUNTER, June 21, 2018, https://thecounter.org/farmer-suicide-crisis-cdc-study/.
60. MARTIN, *supra* note 5.
61. JBS INTERNATIONAL, INC., FINDINGS FROM THE NATIONAL AGRICULTURAL WORKERS SURVEY (NAWS): A DEMOGRAPHIC AND EMPLOYMENT PROFILE OF UNITED STATES FARMWORKERS 2015-2016, at 36

dollar is about a tenth of the farmer's. Since there are more farmworkers than farmers, this money gets split up more for the former group, so that their incomes wind up an even smaller fraction than those of farmers.

Farmworkers also see little future under current farm policies. Even though the average farmworker is 41 and has been in farm work for more than 10 years, and though half of crop workers say they want to remain in agriculture until they retire, land is far too expensive and the threshold acreage for commercial agriculture far too high for them to have a realistic chance of becoming operators in the United States. Farmworkers face other barriers as well. About 83% of farmworkers are Hispanic, the average crop worker has only an eighth grade education, and almost two-thirds of crop workers say they cannot even "somewhat" speak English.

These workers are critical stakeholders and their greater involvement in farm policy would likely support more climate-friendly approaches. Farmworker groups have led campaigns for clean water, pesticide protections, and other environmental reforms since the 1960s and they are now at serious risk from climate change, with heat stress as the deadliest threat for farmworkers.[62] Farmworkers are also vulnerable to wildfires due to the physically demanding outdoor nature of their work,[63] and the fact that they often cannot afford to stop working.[64] In 2018, wildfires burned 1.8 million acres in California, dispersing unsafe levels of smoke for hundreds of miles,[65] and 2020 was vastly worse. Their interests must be reflected in effective climate-friendly farm policies, and their support will be critical to achieving those policies.

2. Non-White Farmers

The modern farm system also excludes almost anyone who is not a landowning white farmer. European settlers took land from the Native Americans and eventually forced them onto reservations. Even after that, the federal government mandated the sale of reservation land to non-Native Americans when tribal lands were deemed to be "surplus" or when the landowner was deemed not "competent" to hold property. Native Americans lost roughly

---

(2018).
62. S.E. Smith, *Heat Is Now the Deadliest Threat to Farmworkers. Only Two States Protect Them From It*, TALK POVERTY, June 20, 2019, https://talkpoverty.org/2019/06/20/farmworkers-heat-illness-deaths/.
63. Heather E. Riden et al., *Wildfire Smoke Exposure: Awareness and Safety Responses in the Agricultural Workplace*, 25 J. AGROMEDICINE 1 (2020).
64. Danielle Paquette, *During California Wildfires, Farmworkers Say They Felt Pressure to Keep Working or Lose Their Jobs*, WASH. POST, Nov. 20, 2018, https://www.washingtonpost.com/business/economy/during-california-wildfires-farm-workers-felt-pressured-to-keep-working-or-lose-their-jobs/2018/11/20/757f92a0-ec06-11e8-baac-2a674e91502b_story.html.
65. Riden et al., *supra* note 63.

80 million acres in this way between 1871 and 1928.[66] By 1910, freed Black people and their descendants had acquired at least 16 million acres of land, almost all of it in the South.[67] Through a variety of means, white families, often with federal assistance, deprived them of almost all of their acreage,[68] so that by 2012 there were only about 300 Black commercial farmers, or about 0.2% of the total.[69] Black farmers' lost wealth and income since 1910 would be worth hundreds of billions of dollars today according to recent estimates.[70] Federal and state governments passed a series of laws in the late 19th and early 20th centuries that barred Asians from owning land. Many Japanese American farmers who had been interned in camps during World War II returned to their farms to find white farmers had stolen them.[71]

Because of this and other widespread discrimination in landownership and agriculture, almost all farmers are white. USDA also still discriminates against Black and other non-white farmers on a systemic basis. For example, more than 99% of the 2019 tariff bail out, the single largest farm subsidy that year, went to white farmers.[72] These payments further entrench the positions of those who received them.

Many non-white farmers have responded to this system by working toward reform. Black farmer groups have led numerous campaigns against discrimination at USDA, often connecting their problems with the problems of department employees who have protested harassment, abuse, and mismanagement within USDA itself. Black farmers have also led campaigns in regions throughout the South to prevent the construction of concentrated animal feeding operations (CAFOs) and to support safer modes of production. Numerous leaders in the sustainable agriculture movement are non-white, and a wide variety of non-white farmer groups use sustainable and

---

66. Jessica A. Shoemaker, *An Introduction to American Indian Land Tenure: Mapping the Legal Landscape*, 5 J.L. Prop. & Soc'y 1, 22-23 (2020).
67. Jess Gilbert & Gwen Sharp, *The Loss and Persistence of Black-Owned Farms and Farmland: A Review of the Research Literature and Its Implications*, 18 J. Rural Soc. Sci. 1, 2 (2002).
68. *See, e.g.*, Pete Daniel, Dispossession: Discrimination Against African American Farmers in the Age of Civil Rights (2013); Nathan Rosenberg & Bryce Stucki, *How USDA Distorted Data to Conceal Decades of Discrimination Against Black Farmers*, The Counter, June 26, 2019, https://thecounter.org/usda-black-farmers-discrimination-tom-vilsack-reparations-civil-rights/.
69. Calculated by the authors using 2012 Census of Agriculture, *supra* note 23, at 9 tbl.1.
70. Vann R. Newkirk II, *The Great Land Robbery*, The Atlantic, Sept. 29, 2019, https://www.theatlantic.com/magazine/archive/2019/09/this-land-was-our-land/594742/.
71. Megan Horst & Amy Marion, *Racial, Ethnic, and Gender Inequities in Farmland Ownership and Farming in the U.S.*, 36 Agric. & Hum. Values 3 (2019).
72. Nathan Rosenberg & Bryce Wilson Stucki, *USDA Gave Almost 100 Percent of Trump's Trade War Bailout to White Farmers*, Counter, July 29, 2019, https://thecounter.org/usda-trump-trade-war-bailout-white-farmers-race/.

traditional practices to produce food and connect with their culture.[73] These leaders are already creating a vision of a more sustainable agriculture, and policymakers should ensure also that farm policy reflects their interests.

3. Rural Communities

While the interests of rural communities are usually assumed to be identical to those of larger commercial farmers, most rural people are not attached to the farm economy. In rural areas, construction provides about as many jobs as farming, with about twice the total earnings. Health care and social assistance, retail, and manufacturing all provide about twice as many jobs as farming, and government provides almost three times as many.[74] In total, these industries account for about 50% of the rural jobs and about 60% of rural income.[75]

Moreover, most rural residents are supportive of heightened environmental protections, even those opposed by farmers. Since virtually all of the climate-friendly practices recommended in this book will also reduce air and water pollution, rural residents' demonstrated concern for clean air and water makes it likely that they will be supportive of policies designed to encourage these practices. A national survey conducted in 2019 found that rural voters were more likely to be concerned about environmental and conservation issues that concerned farmland than urban and suburban voters.[76] Rural voters said clean water was their highest environmental priority among the seven listed options and only 26% of the respondents opposed government regulations to ensure clean water.[77] A majority of rural respondents (52%) also agreed with the statement that environmental protection should be prioritized, "even at the risk of curbing economic growth," in comparison to a small minority (28%) who agreed that economic growth should be prioritized, "even if the environment suffers to some extent."[78] A follow-up survey of rural voters in the upper Midwest found that rural voters in the region were much more likely to prioritize "ensuring clean water" and "ensuring clean air" than "conserving farmland/range lands."[79] Sixty-

---

73. *See, e.g.*, Nadra Nittle, *The People's Agroecology Process Brings a Global Lens to U.S. Food Justice Work*, Civ. Eats, Sept. 10, 2020, https://civileats.com/2020/09/10/the-peoples-agroecology-process-brings-a-global-lens-to-u-s-food-justice-work/.
74. U.S. Bureau of Economic Analysis, *supra* note 56.
75. *Id.*
76. Robert Bonnie et al., Duke Nicholas Institute, Understanding Rural Attitudes Toward the Environment and Conservation in America 15 fig.4 (2020).
77. *Id.* at 15 fig.4 & 19 fig.9.
78. *Id.* at 21 fig.11.
79. *Id.* at 19 fig.9.

eight percent of voters said that "ensuring clean air" was very important to them personally.[80]

Recent state-level surveys have also found that rural voters' main agricultural priority is the regulation of pollution from farms. A 2019 survey in Pennsylvania found that voters rated "safe drinking water," "protection and conservation of the environment," and "development of alternative energy sources" highly among environmental and agriculture issues.[81] A 2015 survey in Iowa found that the highest-ranked priority for agricultural policy was "protecting drinking water quality," followed by "protecting water quality for aquatic life."[82] A majority (55%) agreed or strongly agreed that "Iowa agriculture has some negative impacts on the environment," while only 25% disagreed or strongly disagreed with the statement that "farmland use should be regulated to ensure that it does not negatively impact the general public."[83] In addition, 79% of surveyed residents said they were concerned or very concerned by "water pollution from livestock production."[84]

Rural people who live near industrial livestock facilities have strong reasons to oppose them. CAFOs depress property values,[85] and various studies link living near a CAFO with respiratory problems, MRSA, hypertension, and other health problems. Local residents and activist groups often oppose CAFOs because of their foul odors, pollution, and public health risks. These demonstrations rarely make the national news, but they are significant events in the places where they happen. In 2018, more than 150 people turned out to oppose a CAFO in Mercer County, Ohio, home to about 40,000 people; a demonstration of the same relative size in New York City would have had 70,000 people. In 2020, a group called KnowCAFOs in Polk County, Wisconsin, fought off an ordinance that would have allowed new CAFOs in the county.[86]

So concerned are industry leaders about nuisance cases against animal production facilities that they have successfully urged many state legislators

---

80. *Id.*
81. Daniel J. Mallinson et al., Center for Rural Pennsylvania, Attitudinal Survey of Pennsylvanians, 2019, at 105 (2020).
82. J. Gordon Arbuckle et al., Iowan's Perspectives on Targeted Approaches for Multiple-Benefit Agriculture 2 fig.1 (Iowa State University, Sociology Technical Report No. 1038, 2015).
83. *Id.* at 4 tbl.2.
84. *Id.* at 6 tbl.3.
85. Kelley Donham et al., *Community Health and Socioeconomic Issues Surrounding Concentrated Animal Feeding Operations*, 115 Env't Health Persp. 317, 319 (2007).
86. *Polk County: Thank You! CAFO Ordinance Stalls Under Mountain of Negative Comments*, KnowCAFOs, July 9, 2020, https://knowcafos.org/2020/07/09/polk-county-thank-you-cafo-ordinance-stalls-under-mountain-of-negative-comments/.

to enact laws limiting such cases.[87] When the cases are allowed to proceed, the facts make clear that many rural residents would strongly support policies to limit pollution. For example, people living near a swine CAFO in eastern North Carolina filed a nuisance suit in 2014 against a subsidiary of Smithfield alleging that odors, pests, and truck traffic from a CAFO unreasonably interfered with their use and enjoyment of their properties.[88] In 2018, a jury found in favor of these neighbors, awarding them millions of dollars of compensatory and punitive damages, and in 2020, the U.S. Court of Appeals for the Fourth Circuit largely affirmed the awards.[89] A judge, concurring in the judgment, wrote:

> It is well-established—almost to the point of judicial notice—that environmental harms are visited disproportionately upon the dispossessed—here on minority populations and poor communities. But whether a home borders a golf course or a dirt road, it is a castle for those who reside in it. It is where children play and grow, friends sit and visit, and a life is built. Many plaintiffs in this suit have tended their hearths for generations—one family for almost 100 years. They are exactly whom the venerable tort of nuisance ought to protect. Murphy-Brown's interference with their quiet enjoyment of their properties was unreasonable. It was willful, and it was wanton. The record fully supports the jury's finding that punitive damages were warranted.[90]

Local rural groups have led similar fights against pesticides. A 2019 literature review found that people who live closer to agricultural land have higher levels of pesticide exposure.[91] Both acute and chronic exposure to pesticides is associated with cancer, depression, Parkinson's disease, diabetes, respiratory diseases, and other chronic ailments.[92] In 2012, a coalition of environmental, farmworker, and local California groups filed suit in response to the approval of methyl iodide, one of the most toxic pesticides used in

---

87. *See, e.g.,* Emily Moon, *Missouri Outlaws Rural Residents' Last Line of Protection Against CAFOs*, Pac. Standard, May 17, 2019, https://psmag.com/news/missouri-outlaws-rural-residents-last-line-of-protection-against-cafos; Morgan Niezing et al., *Government Eases Up on CAFOs as Residents Fight Their Expansion*, Missourian, Aug. 9, 2018, https://www.columbiamissourian.com/news/local/government-eases-up-on-cafos-as-residents-fight-their-expansion/article_d758dcf2-9c3d-11e8-82ff-271a9483e031.html; Chris Braun, *"Keep Makin' Bacon": Indiana's Right to Farm Act Statute Upheld as Constitutional*, American College of Environmental Lawyers, Jan. 11, 2021, http://www.acoel.org/post/2021/01/11/Keep-Makin%E2%80%99-Bacon%E2%80%9D-Indiana%E2%80%99s-Right-to-Farm-Act-Statute-Upheld-As-Constitutional.aspx.
88. Barry Yeoman, *Jury Awarded Hog Farm Neighbors $3.25 Million. Will Three-Wuarters of That Be Erased?*, Charlotte Observer, Jan. 31, 2020, https://www.charlotteobserver.com/news/business/article239694633.html.
89. McKiver v. Murphy-Brown, LLC, No. 19-1019 (4th Cir. 2020).
90. *Id.*
91. Clementine Dereumeaux et al., *Pesticide Exposures for Residents Living Close to Agricultural Lands: A Review*, 134 Env't Int'l (2019), https://doi.org/10.1016/j.envint.2019.105210.
92. Aaron Blair et al., *Pesticides and Human Health*, 72 Occupational & Env't Med. 81 (2015).

agriculture. In response, the manufacturer pulled the pesticide from the U.S. market and it was eventually banned.

There are broad swaths of the rural population, far more numerous than farmers, who see the need for stronger environmental protections and would support policies to accelerate adoption of climate-friendly practices. Policymakers must include them fully as stakeholders.

4. The Opportunity for Carbon Farming

The modern agricultural system produces a vast amount of food cheaply, but with significant environmental and social costs. It externalizes the costs of water, air, and climate pollution, depends on resource extraction, and relies on an immigrant work force with few rights. And it is built on a system that must be changed, having been so shaped by people opposed to the interests of farmworkers, non-white farmers, and the rural poor.

The system of industrialized monocultures supported by federal policy depopulated rural towns, while polluting the air and water throughout the countryside. Farmers were once a large and diverse group, both in their backgrounds and in the operations they ran; now, in large part due to policy choices, the farmers that dominate policymaking are a small and largely homogeneous group of conservative, wealthy, and white families.

Just as federal policy has largely shaped today's system, federal policy can change it. These changes could benefit and would be supported by many groups who should have a role in policymaking. While our recommendations to accelerate regenerative farming will also benefit farmers, we need to enlist everyone in the debate. Policies that expand carbon farming can create new constituencies with Black and other non-white farmers, agricultural workers, and rural residents. These constituencies, together with farmers already on the forefront of change, will, in turn, ensure these policies are successful, that they will endure, and that they are spread across the entire agricultural system. Together these constituencies will create a more just and sustainable farm economy.

Key Recommendations

- U.S. farm policy now largely benefits a small number of almost entirely white producers who are substantially wealthier than the average American, and who are required to do little to protect the health of their workers or neighbors or to address environmental impacts or climate change. Current agricultural policy should be revised to ensure that it benefits not just the large and wealthy, but rather the public broadly. Moreover, policies must be put into place to protect the health of workers and communities surrounding industrial agricultural facilities, and to limit environmental and climate harms.
- The average farm household now has a higher annual income and more non-farm wealth than the average household.
- More than one-half of those the USDA includes as "farmers" are retirees, hobbyists, or taxpayers with paper farms.
- There are about 2.5 million farmworkers, about twice as many as active farmers. They do two-thirds of the manual labor on farms and have a median income under $20,000.
- Sound, sustainable, and fair agriculture policy should be built on an accurate understanding of the affected constituencies rather than on assumed and outdated images and narratives centered on the "family farmer." In addition to "farmers" and agribusiness, policies must also consider farmworkers and rural residents, consumers, and those affected by climate change.
- Climate-friendly farm policies must consider the interests of and have the support of farmworkers, non-white farmers, and rural people as well as farmers to be successful.
- Farmworkers and rural communities demonstrate strong support for policies to reduce agricultural pollution and advance climate-friendly practices.
- Many farmers have some or all of the resources needed to shift agricultural practices. As such, regulatory, education, and outreach programs can be effective tools for accelerating climate-friendly farming, even if not linked to changes in subsidies.

- Policies that expand carbon farming can create new constituencies with Black and other non-white farmers, agricultural workers, and rural residents.

Chapter III.
The Climate Crisis and Agriculture

This chapter begins by describing how the climate crisis threatens to disrupt agricultural production at immense cost to society. We then outline agricultural emissions at the global, national, and state levels, demonstrating the need for quick and ambitious action to change agricultural practices. We explain why official figures significantly underestimate agricultural emissions and why, compounding the problem, agricultural emissions are difficult to estimate with precision. We conclude by explaining the need to transform agriculture from one of the world's largest emitters of greenhouse gases into a net sink.

A. Climate Change's Impact on Agriculture

Weather—the patterns of which make up the climate—profoundly affects our food system. The growing of crops requires certain amounts of water, heat, and sun; temperature and other conditions influence the growth and health of animals. Yet, climate change is dramatically altering the weather patterns in the United States. Figure 1 (page 36) shows a few categories of harm out of many possible examples. Floods, droughts, and heat waves are more frequent and more extreme; wildfires are increasing due in part to climate change. The range of many pests is expanding as warmer weather moves north.

These changing weather patterns and increased extreme weather events are exacting a heavy toll on American agriculture. The 2016 droughts in California led to more than $600 million in losses. Hurricane Maria flattened farm fields throughout Puerto Rico in 2017, causing almost $800 million in losses. The 2019 flooding in the Midwest left 5 to 10 million bushels of corn and soy to rot and 19 million acres unable to be planted. Heat stress causes kidney disease and other harms to farmworkers and can weaken animals and slow their growth. As climate change gets more severe, so will these impacts.

While crop insurance, generously funded disaster assistance, and other programs largely shield producers themselves from the economic impacts of climate change, the societal costs are immense. For example, a 2021 study using detailed information from county-level crop insurance claims found

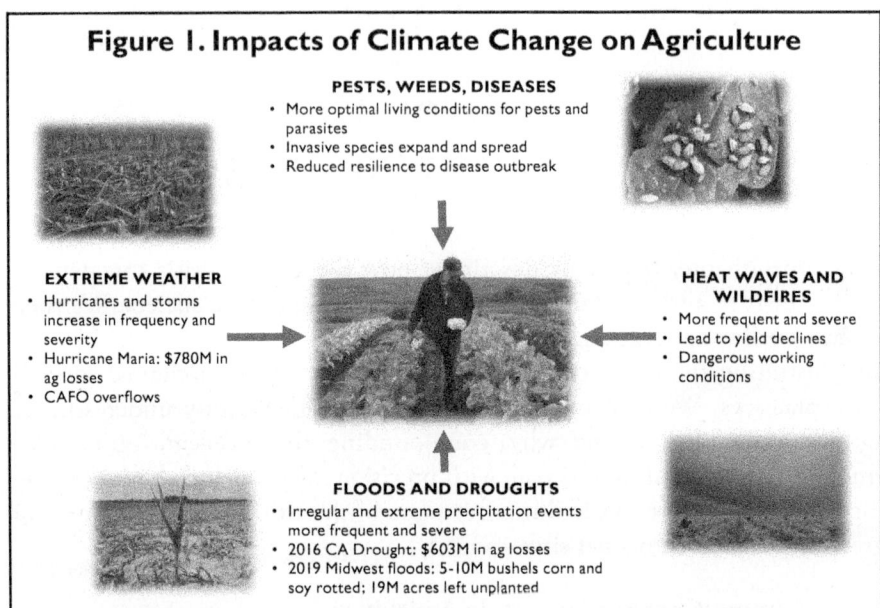

Figure 1. Impacts of Climate Change on Agriculture

PESTS, WEEDS, DISEASES
- More optimal living conditions for pests and parasites
- Invasive species expand and spread
- Reduced resilience to disease outbreak

EXTREME WEATHER
- Hurricanes and storms increase in frequency and severity
- Hurricane Maria: $780M in ag losses
- CAFO overflows

HEAT WAVES AND WILDFIRES
- More frequent and severe
- Lead to yield declines
- Dangerous working conditions

FLOODS AND DROUGHTS
- Irregular and extreme precipitation events more frequent and severe
- 2016 CA Drought: $603M in ag losses
- 2019 Midwest floods: 5-10M bushels corn and soy rotted; 19M acres left unplanted

Source: EPA, Climate Change Impacts on Agriculture and Food Supply, https://19january2017snapshot.epa.gov/climate-impacts/climate-impacts-agriculture-and-food-supply_.html (last visited Jan. 23, 2021); GLOBAL CHANGE RESEARCH PROGRAM, IMPACTS, RISKS, AND ADAPTATION IN THE UNITED STATES: FOURTH NATIONAL CLIMATE ASSESSMENT, VOLUME II 391-437 50-174 (2018); JOSUÉ MEDELLÍN-AZUARA ET AL., CENTER FOR WATERSHED SCIENCES, ECONOMIC ANALYSIS OF THE 2016 CALIFORNIA DROUGHT ON AGRICULTURE (2016); Curtis A. Deutsch et al., *Increase In Crop Losses to Insect Pests In a Warming Climate*, 361 NATURE 916-19 (2018); Matthew R. Smith et al., *The Impact of Rising Carbon Dioxide Levels on Crop Nutrients and Human Health*, FEED THE FUTURE (2018) (GCAN Policy Note 10); Deepak Ray, *Climate Change Is Affecting Crop Yields and Reducing Global Food Supplies*, THE CONVERSATION, July 9, 2019, http://theconversation.com/climate-change-is-affecting-crop-yields-and-reducing-global-food-supplies-118897; Tom Polansek, *U.S. Disaster Aid Won't Cover Crops Drowned by Midwest Floods*, REUTERS, Apr. 2, 2019, https://www.reuters.com/article/us-usa-weather-iowa/u-s-disaster-aid-wont-cover-crops-drowned-by-midwest-floods-idUSKCN1RE0BU; John Schwartz, *A Wet Year Causes Farm Woes Far Beyond the Floodplains*, N.Y. TIMES, Nov. 21, 2019, https://www.nytimes.com/2019/11/21/climate/farms-climate-change-crops.html?smid=nytcore-ios-share.

that increased temperatures "contributed $27 billion—or 19%—of the national-level crop insurance losses over the 1991–2017 period" and concludes with "very high confidence that anthropogenic climate forcing has increased U.S. crop insurance losses."[1] In addition, food prices are becoming more volatile as climate change interrupts both food production and our

---

1. Noah Diffenbaugh et al., *Historical Warming Has Increased U.S. Crop Insurance Losses*, 16 ENV'T RES. LETTERS 084025 (2021), https://iopscience.iop.org/article/10.1088/1748-9326/ac1223. The study also found "that observed warming contributed almost half of total losses in the most costly single year (2012)."

ability to transport food around the world.[2] Elevated levels of atmospheric carbon dioxide (CO_2) impact plant physiology and the relative availability of nutrients. As a result of these shifts, protein concentrations in staple crops are expected to fall by 6%-14%,[3] while also reducing micronutrient levels in vegetables and other crops.[4] These changes will disproportionately affect the food insecure, who must contend with a food system where healthy foods are already more difficult to find and more expensive to purchase.[5] We must not only eliminate agriculture's net emissions, but also make agricultural production more resilient.

American farms must employ practices that will better enable them to withstand the more frequent extreme weather that climate change will bring. They will also face changes in climate conditions—the temperature, length of growing season, and rainfall patterns among other factors—that in large part determine whether a crop is suited for a specific region. Fortunately, many of the same practices that can reduce the contribution of agriculture to climate change will also make agriculture more resilient to climate change. The practices described below, and the policies that can accelerate their adoption, will benefit those who implement them. Trees and perennial crops with larger roots can better withstand floods, droughts, and heat waves; cover crops or untilled lands contain more organic matter that is less susceptible to erosion; crops that are planted in a rotation provide less purchase for pests; enhanced location-specific management improves forage; and adding trees through silvopasture protects livestock and provides additional sources of income.

B. Agriculture's Contribution to Climate Change

"Agriculture" refers to the cultivation of crops and the raising of animals for the "4Fs": food, feed, fuel, and fiber. It accounts for 52% of the country's total landmass, including 62% of the landmass of the contiguous 48 states, making it the single largest type of land use in the United States (including

---

2. *See* Margot Pollans, *Food Systems*, in CLIMATE CHANGE, PUBLIC HEALTH, AND THE LAW 272 (Justin Gundlach & Michael Burger eds., Cambridge University Press 2018).
3. Danielle E. Medek et al., *Estimated Effects of Future Atmospheric CO_2 Concentrations on Protein Intake and the Risk of Protein Deficiency by Country and Region*, 125 ENV'T HEALTH PERSP. 087002-1 (2017).
4. Jinlong Dong et al., *Effects of Elevated CO_2 on Nutritional Quality of Vegetables: A Review*, 9 FRONTIERS PLANT SCI. 1 (2018) (finding that elevated atmospheric carbon dioxide (CO_2) levels decreased magnesium, iron, and zinc levels); Samuel S. Myers, *Rising CO_2 Threatens Human Nutrition*, 510 NATURE 139 (2014) (reporting that most grains and legumes have lower levels of iron and zinc under elevated levels of atmospheric CO_2).
5. *See* Nathan Rosenberg & Nevin Cohen, *Let Them Eat Kale: The Misplaced Narrative of Food Access*, 45 FORDHAM URB. L.J. 1091 (2018) (discussing the root causes of food insecurity and diet-related health disparities).

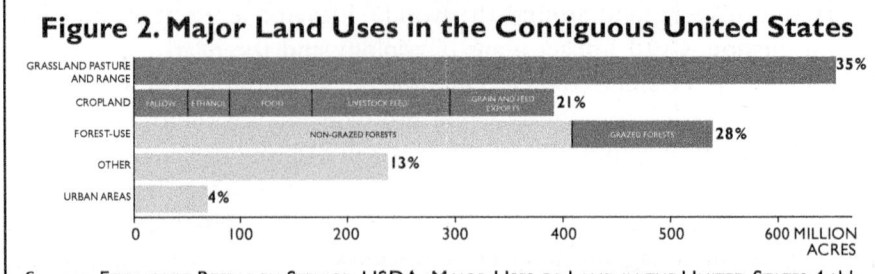

Figure 2. Major Land Uses in the Contiguous United States

Category	Percentage
GRASSLAND PASTURE AND RANGE	35%
CROPLAND (FALLOW, ETHANOL, FOOD, LIVESTOCK FEED, GRAIN AND FIELD EXPORTS)	21%
FOREST-USE (NON-GRAZED FORESTS, GRAZED FORESTS)	28%
OTHER	13%
URBAN AREAS	4%

Scale: 0 to 600 Million Acres

Sources: ECONOMIC RESEARCH SERVICE, USDA, MAJOR USES OF LAND IN THE UNITED STATES 4 tbl. 1 (2017) (EIB-178). Dave Merrill et al., *Here's How America Uses Its Land*, BLOOMBERG, July 31, 2018, at https://www.bloomberg.com/graphics/2018-us-land-use/. Percentages may not sum to 100 due to independent rounding. Darker shading indicates agricultural uses.

forested grazing lands).[6] Of the country's total 2.3 billion acres, approximately 392 million acres are now cropland, 655 million acres are grassland pasture and range, and 130 million acres are grazed forestland.[7] (See Figure 2 above.) Since agriculture uses so much land, modest reductions in emissions per acre can have an enormous cumulative effect when adopted across large numbers of farms. Moreover, as discussed below, the lost carbon sequestration capacity of land already converted to agriculture must be considered. Such changes can also help farmers adapt to the changing climate.

A central argument in this book is that carbon sequestration should be an essential function of agriculture—the fifth "F" for the future[8]—supported by federal agricultural programs and policies. Freedom from the worst effects of climate change is at least as critical as any other function of modern agriculture, including crop, animal, timber, and biofuels production. By reducing greenhouse gas emissions while also increasing soil carbon stores, agricultural operations can make a substantial contribution to decarbonization in the United States. The following subsections analyze global, national, and state-level agricultural emissions.

1. Global Agricultural Greenhouse Gas Emissions

Agriculture's contribution to climate change is expected to grow rapidly as other sectors decarbonize, the global population grows, and industrial animal production becomes more pervasive. Unless there are significant changes in the food system, agricultural emissions alone will make it impossible to

---

6. DANIEL P. BIGELOW & ALLISON BORCHERS, U.S. DEPARTMENT OF AGRICULTURE, MAJOR USES OF LAND IN THE UNITED STATES, 2012, at 4 tbl.1 (2017) (EIB-178).
7. *Id.*
8. *See* Peter Lehner, *Feed More With Less*, 34 ENV'T F. 42, 49 (2017).

Chapter III. The Climate Crisis and Agriculture

achieve the Paris Agreement goal of limiting warming to no more than 2 degrees Celsius above pre-industrial levels, let alone the safer target of 1.5 degrees Celsius. And if meat consumption continues to grow, climate change will be dramatically accelerated.

The Global Calculator, developed by research institutions across several countries, shows that even with the most aggressive mitigation in energy production and use, transportation, industry, and housing, the world will greatly exceed the two-degree Celsius target—if we do not reduce food system emissions.[9] Global food systems contribute about one quarter to one-third of total greenhouse gas emissions,[10] and numerous extensive scientific studies confirm that shifts in agricultural practices are critical for achieving international climate targets.[11] The vast majority of agricultural emissions now derive from animal agriculture, so significantly changing how animal products are produced and consumed will be critical. Stopping the conversion of native grasslands or forests to croplands is another important factor. The model sees a particularly big impact in reducing long-term greenhouse gas emissions through multi-cropping and agroforestry, where trees and shrubs are integrated on land with crops and/or livestock production. Other

---

9. *See* Global Calculator, *Home Page*, tool.globalcalculator.org (last visited Oct. 28, 2020). *See also* Michael A. Clark et al., *Global Food System Emissions Could Preclude Achieving the 1.5° and 2°C Climate Change Targets*, 370 SCIENCE 705-08 (2020) (noting that the global food system is responsible for about 30% of global greenhouse gas emissions and that "current trends in global food systems would prevent the achievement of the 1.5°C target and, that . . . [m]eeting the 1.5°C target requires rapid and ambitious changes to food systems as well as to all nonfood sectors").
10. Francesco Tubiello et al., *Greenhouse Gas Emissions from Food Systems: Building the Evidence Base*, 16 ENVTL. RES. LETTERS 065007 (2021), https://iopscience.iop.org/article/10.1088/1748-9326/ac018e/pdf (food system emissions contribute a third of total global greenhouse gases and three quarters of these emissions were generated either within the farm gate or in pre- and post-production activities); Monica Crippa et al., *Food Systems Are Responsible for a Third of Global Anthropogenic GHG Emissions*, 2 NATURE FOOD 198-209 (2021) (food system emissions amounted to 34% of total greenhouse gas emissions and agriculture and land use changes contributing 71% of that or 24% of total emissions); Cynthia Rosenzweig et al., *Climate Change Responses Benefit From a Global Food System Approach*, 1 NATURE FOOD 94-97 (2020) (finding food system greenhouse gas emissions to contribute 21-37 percent of total emissions); Joseph Poore et al., *Reducing Food's Environmental Impacts*, 360 SCIENCE 987-92 (2018) (based on over 1,500 studies, finding that "today's food supply chain creates—13.7 billion metric tons of carbon dioxide equivalents (CO_2eq), 26% of anthropogenic GHG emissions").
11. *See id. See also* Sinead Leahy et al., *Challenges and Prospects for Agricultural Greenhouse Gas Mitigation Pathways Consistent With the Paris Agreement*, 4 FRONTIERS SUSTAINABLE FOOD SYSTEMS 69 (2020); Eva Wollenberg et al., *Reducing Emissions From Agriculture to Meet the 2C Target*, 22 GLOBAL CHANGE BIOLOGY 12 (2016); Stefan Frank et al., *Agricultural Non-CO_2 Emission Reduction Potential in the Context of the 1.5C Target*, 9 NATURE CLIMATE CHANGE 66 (2019); BRENT KIM ET AL., JOHNS HOPKINS CENTER FOR A LIVABLE FUTURE, THE IMPORTANCE OF REDUCING ANIMAL PRODUCT CONSUMPTION AND WASTED FOOD IN MITIGATING CATASTROPHIC CLIMATE CHANGE (2015); Walter Willett et al., *Food in the Anthropocene: the EAT–Lancet Commission on Healthy Diets From Sustainable Food Systems*, 393 THE LANCET 447-92 (2019); TOMAS NAUCLÉ & PER-ANDERS ENKVIST, MCKINSEY & COMPANY, PATHWAYS TO A LOW-CARBON ECONOMY: VERSION 2 OF THE GLOBAL GREENHOUSE GAS ABATEMENT COST CURVE (2009).

Figure 3. Industrial Agriculture Contributes to Climate Change

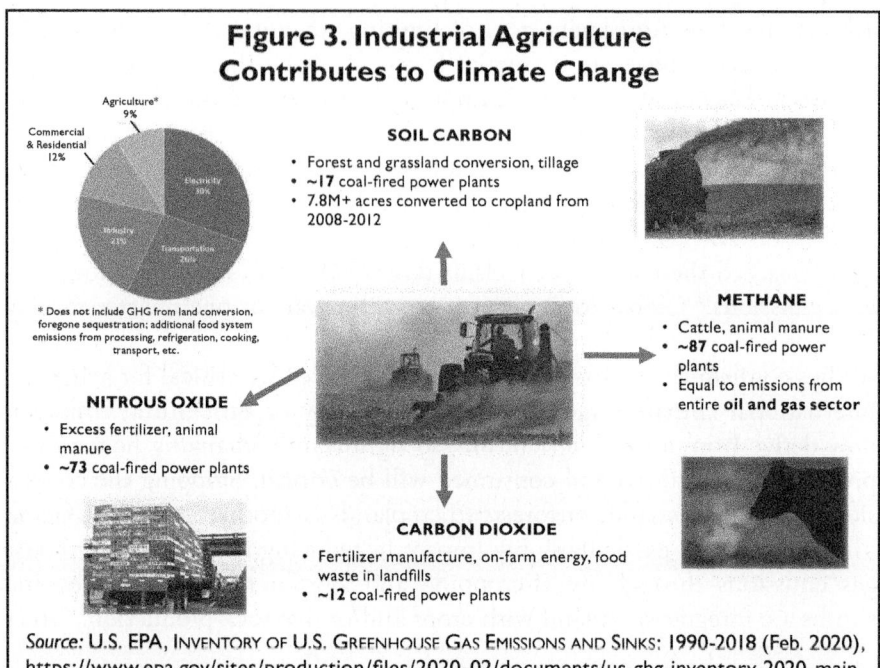

SOIL CARBON
- Forest and grassland conversion, tillage
- ~17 coal-fired power plants
- 7.8M+ acres converted to cropland from 2008-2012

METHANE
- Cattle, animal manure
- ~87 coal-fired power plants
- Equal to emissions from entire **oil and gas sector**

NITROUS OXIDE
- Excess fertilizer, animal manure
- ~73 coal-fired power plants

CARBON DIOXIDE
- Fertilizer manufacture, on-farm energy, food waste in landfills
- ~12 coal-fired power plants

Pie chart: Agriculture* 9%, Commercial & Residential 12%, Electricity 38%, Industry 22%, Transportation 28%

*Does not include GHG from land conversion, foregone sequestration; additional food system emissions from processing, refrigeration, cooking, transport, etc.

Source: U.S. EPA, INVENTORY OF U.S. GREENHOUSE GAS EMISSIONS AND SINKS: 1990-2018 (Feb. 2020), https://www.epa.gov/sites/production/files/2020-02/documents/us-ghg-inventory-2020-main-text.pdf (tables 2-1 & 2-2, trends in U.S. greenhouse gas emissions; 2-4, energy; 2-5, combustion; 2-10, industrial; 5-1, agriculture; 6-1, land use and land use change; and 7-1, 7-3 & 7-6, waste & landfills); coal-fired power plants equivalencies from U.S. EPA, *Greenhouse Gas Equivalencies Calculator*, https://www.epa.gov/energy/greenhouse-gas-equivalencies-calculator. Percentages are of total gross U.S. anthropogenic GHG emissions.

countries are already investing significant sums into agroforestry research and production, yet the United States has lagged, despite robust research demonstrating its significant potential to sequester carbon while producing ample food. Moreover, food production needs will increase as the global population continues to grow. Thus, perhaps even more than agriculture's current contribution, its long-term determinative factor in climate stability demands careful policy attention.

2. U.S. Agricultural Greenhouse Gas Emissions

The U.S. Environmental Protection Agency (EPA) estimates that 2019 emissions from agricultural activities—growing crops and raising livestock and poultry—totaled about 629 million metric tons of carbon dioxide equivalent (MMT CO_2 eq.), accounting for more than 10% of total U.S. greenhouse gas emissions.[12]

These agricultural emissions are at a minimum roughly equivalent to that produced by 136 million automobiles in a typical year.[13] However, unlike the greenhouse gas emissions of most other sectors of the economy, which consist of CO_2 released from the burning of fossil fuels, agriculture greenhouse gas emissions consist largely of nitrous oxide (N_2O) from soils and manure and methane (CH_4) from livestock and manure (as shown in Figure 3 on page 40). Agriculture also produces CO_2 from fossil fuel combustion (both on farms and off-site for on-farm electricity) and CO_2 from land conversion, neither of which are included in the EPA sector total. Agriculture is responsible for approximately 80% of U.S. N_2O emissions and 40% of U.S. CH_4 emissions—the same as the entire oil and gas sector's production emissions.[14]

The largest source of U.S. agricultural greenhouse gas emissions according to EPA is agricultural soil management. Activities by microorganisms in soil naturally result in emissions of N_2O, while agricultural practices, management, and land use can stimulate and accelerate these emissions by increasing the availability of nitrogen. Agricultural greenhouse gas emissions are dominated by emissions resulting from application of fertilizer as well as emissions associated with the breakdown of soil organic matter.[15] Soil management generates approximately half of all U.S. agricultural emissions and 93% of all U.S. N_2O emissions from agriculture.[16] Seventy-three percent of N_2O emissions from agricultural soil management come from cropland and 27% come from grazed grasslands.[17]

The next largest source of agricultural emissions is enteric fermentation, which results from the digestive process of ruminants (largely cows and sheep in the United States) (see Figure 4, page 42). Enteric fermentation creates CH_4, which animals subsequently release into the atmosphere through

---

12. U.S. EPA, INVENTORY OF U.S. GREENHOUSE GAS EMISSIONS AND SINKS: 1990-2019, at 5-1 (2021), https://www.epa.gov/sites/production/files/2020-02/documents/us-ghg-inventory-2020-main-text.pdf. The 2019 emissions are more than 1% higher than those from just the prior year, 2018.
13. *Compare id., with* U.S. EPA, GREENHOUSE GAS EMISSIONS FROM A TYPICAL PASSENGER VEHICLE (2018) (a typical passenger vehicle emits 4.6 metric tons of CO_2 annually).
14. *See* U.S. EPA, *Overview of Greenhouse Gases*, https://www.epa.gov/ghgemissions/overview-greenhouse-gases (last updated Sept. 8, 2020).
15. U.S. EPA, INVENTORY OF U.S. GREENHOUSE GAS EMISSIONS AND SINKS: 1990-2018, Box 5-3 at 5-36 (2020) (EPA 430-R-20-002), https://www.epa.gov/sites/production/files/2020-04/documents/us-ghg-inventory-2020-main-text.pdf). In the methodology used by EPA to calculate nitrous oxide emissions, microorganisms in soil produce nitrous oxide at rates dependent on the availability of nitrogen, land use and management, weather, soil types, and other environmental conditions.
16. *See* U.S. EPA, *supra* note 15, at 5-2 tbl.5-1.
17. *See id.* at 5-25 tbl.5-15.

Figure 4. Major Sources of Agricultural Greenhouse Gas Emissions in the United States

AGRICULTURE = 40% OF US TOTAL METHANE (CH₄) EMISSIONS

AGRICULTURE = 82% OF US TOTAL NITROUS OXIDE (N₂O) EMISSIONS

CH₄ — Enteric Fermentation | Rice | Manure (CH₄) | Manure (N₂O) | Soil — N₂O | CO₂

0 — 100 — 200 — 300 — 400 — 500 — 600 MMT CO2eq

Equivalent to: Methane emissions from natural gas systems

Equivalent to: Emissions from 77.1 coal-fired power plants

Source: U.S. EPA, INVENTORY OF U.S. GREENHOUSE GAS EMISSIONS AND SINKS: 1990-2018, at 5-2 tbl.5-1 (Feb. 2020), https://www.epa.gov/sites/production/files/2020-02/documents/us-ghg-inventory-2020-main-text.pdf.

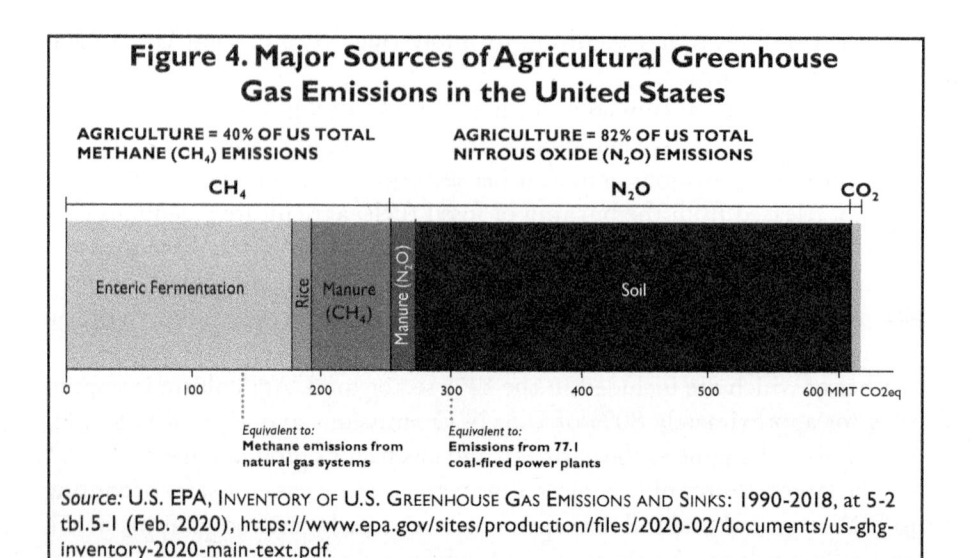

belching and exhalation.[18] Enteric fermentation is responsible for 32% of all agricultural emissions and 27% of CH_4 emissions in the United States.[19] (As discussed below, the relative impact of bovine exhalation would be much greater using more appropriate approaches for calculating CH_4 emissions.)

Manure management activities are the third major category of U.S. agricultural emissions, releasing N_2O and CH_4 in quantities that total 13% of total U.S. agricultural emissions.[20] The largest animal facilities—those with over 1,000 cattle on feed, 1,000 dairy cows, 2,000 finishing hogs, 100,000 turkeys sold, 300,000 broilers sold, or over 50,000 laying hens —generate the substantial majority of these emissions. Greenhouse gas emissions from enteric fermentation and manure are largely dependent on the number of animals raised in these facilities, which are heavily concentrated in a small proportion of the largest operations: over 50% of dairy cows in the United States are in the 4% of operations that stock 1,000 or more dairy cows.[21]

---

18. Andy Thorpe, *Enteric Fermentation and Ruminant Eructation: The Role (and Control?) of Methane in the Climate Change Debate*, 93 CLIMATE CHANGE 407, 411 (2009).
19. *See* U.S. EPA, *supra* note 15, at ES-7 tbl.ES-2, 5-2 tbl.5-1.
20. *See id.* at 5-2 tbl.5-1.
21. *See* NATIONAL AGRICULTURAL STATISTICS SERVICE, USDA, 2017 CENSUS OF AGRICULTURE, U.S. NATIONAL LEVEL DATA tbl.12 (2019). NASS uses different class sizes than does EPA in its definition of a "concentrated animal feeding operation" under the Clean Water Act. Pursuant to EPA rules, a large concentrated animal feeding operation (CAFO) is one with more than 1,000 animal units. An animal unit is defined as an animal equivalent of 1,000 pounds live weight and equates to 1,000 head of beef cattle, 700 dairy cows, 2,500 swine weighing more than 55 pounds, 125,000 broiler chickens, or 82,000 laying hens or pullets. *See* 40 C.F.R. §122.23(b)(4).

More than 90% of hogs in the United States are in the 12% of facilities that stock 2,000 or more hogs,[22] and more than three-quarters of all cattle on feed in the United States are in the 5% of facilities that stock 1,000 or more cattle.[23] As shown in Figures 5 and 6 (see page 44), the climate footprint of animal agriculture is directly correlated to this concentration of inventory, with a few large facilities responsible for the majority of greenhouse gas emissions.

Methane emissions released from soils flooded for rice cultivation and the field burning of crop residues make up more than 2% of total U.S. greenhouse gas emissions from agriculture.[24] In addition, CO_2 emissions from urea fertilization and liming—included by EPA in its estimate of agricultural emissions for the first time in 2015[25]—account for just under 2% of agricultural emissions.[26] We compare emissions of various agricultural activities in Figure 4.

The vast majority of agricultural emissions are from animal production, particularly beef and dairy. In the United States, meat and dairy production—including emissions related to production of their feed (which is about half of U.S. crop production), grazing, enteric fermentation, and manure—accounts for almost 80% of agriculture's greenhouse gas emissions.[27] If the global cattle population were a country, it would constitute the second largest greenhouse gas emitter after China.[28] Both the grazing stage for cows and the feedlot stage for beef, dairy, swine, and poultry production, as practiced now, produce substantial greenhouse gas emissions.[29] Because cows produce only one calf at a time, which nurse for months and then must graze until their bodies can take a grain diet in a feedlot, there are about five cows grazing for each of the close to 30 million cows in feedlots.[30] This requires vast amounts of land—almost 800 million acres or about 40% of the contiguous United States is devoted to grazing.[31]

---

22. *See* National Agricultural Statistics Service, *supra* note 21, at tbl.25.
23. *See id.* at tbl.12.
24. *See id.*
25. *See id.*
26. *See id.*
27. These sources were responsible for 421.8 MMT CO_2 eq. or 78% of agricultural emissions in 2017. *Compare* U.S. EPA, *supra* note 15, at 5-2 tbl.5-1 (showing annual emissions from agriculture by source), *with infra* note 35 (calculating emissions from agricultural soils devoted to feed crop production or grazing).
28. Justin Ahmed et al., McKinsey & Co., Agriculture and Climate Change: Reducing Emissions Through Improved Farming Practices 5 (2020).
29. *See* C. Alan Rotz et al., *Environmental Footprints of Beef Cattle Production in the United States*, 169 Agric. Systems 1 (2019).
30. Matthew N. Hayek & Rachael D. Garrett, *Nationwide Shift to Grass-Fed Beef Requires Larger Cattle Population*, 13 Env't Res. Letters 0845005 (2018).
31. *See* Bigelow & Borchers, *supra* note 6.

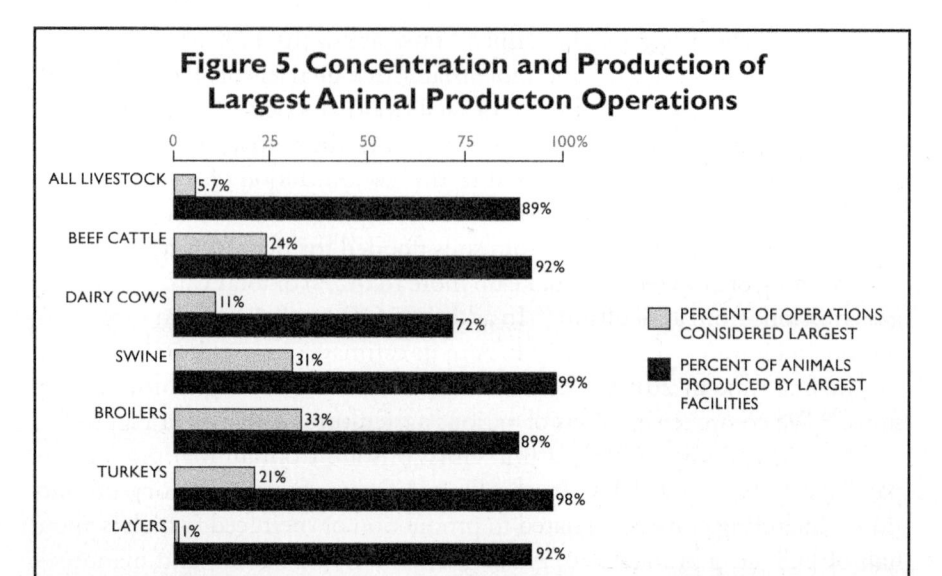

Figure 5. Concentration and Production of Largest Animal Production Operations

Category	Percent of Operations Considered Largest	Percent of Animals Produced by Largest Facilities
ALL LIVESTOCK	5.7%	89%
BEEF CATTLE	24%	92%
DAIRY COWS	11%	72%
SWINE	31%	99%
BROILERS	33%	89%
TURKEYS	21%	98%
LAYERS	1%	92%

Source: USDA, NATIONAL AGRICULTURE STATISTICS SERVICE, 2017 CENSUS OF AGRICULTURE, UNITED STATES SUMMARY AND STATE DATA, tbls.12-30 (April 2019); CLAIRE HUANG, EARTHJUSTICE, AIR AND GHG EMISSIONS CAFOS: QUANTIFYING A NATIONAL ESTIMATE 25 (Aug. 19, 2020). For purposes here, the largest facilities are those with over 1,000 cattle on feed, 1,000 dairy cows, 2,000 finishing hogs, 100,000 turkeys sold, 300,000 broilers sold, or over 50,000 laying hens.

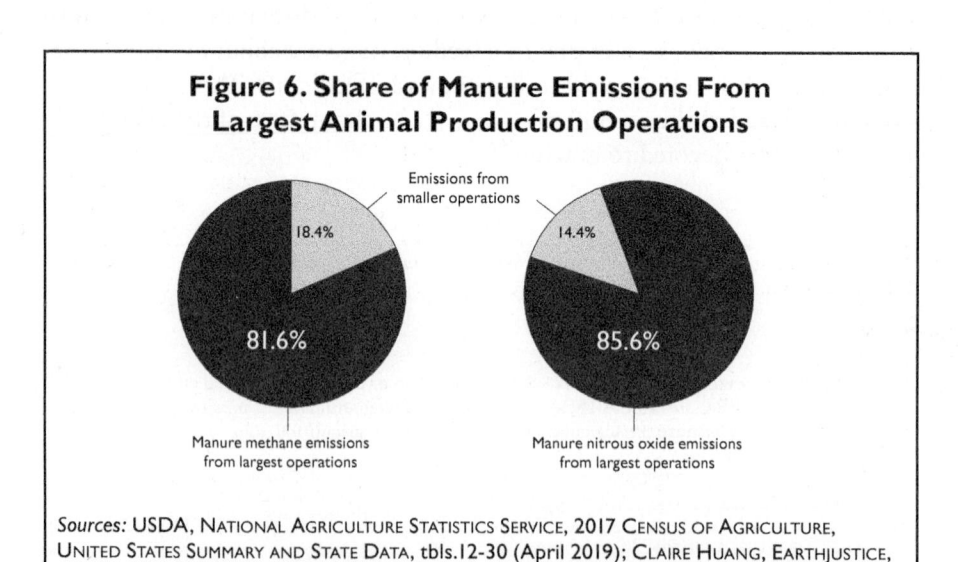

Figure 6. Share of Manure Emissions From Largest Animal Production Operations

- Manure methane emissions from largest operations: 81.6% (Emissions from smaller operations: 18.4%)
- Manure nitrous oxide emissions from largest operations: 85.6% (Emissions from smaller operations: 14.4%)

Sources: USDA, NATIONAL AGRICULTURE STATISTICS SERVICE, 2017 CENSUS OF AGRICULTURE, UNITED STATES SUMMARY AND STATE DATA, tbls.12-30 (April 2019); CLAIRE HUANG, EARTHJUSTICE, AIR AND GHG EMISSIONS CAFOS: QUANTIFYING A NATIONAL ESTIMATE 23 (Aug. 19, 2020).

Chapter III. The Climate Crisis and Agriculture   45

> **Figure 7. Annual Greenhouse Gas Emissions From Largest Animal Facilities** 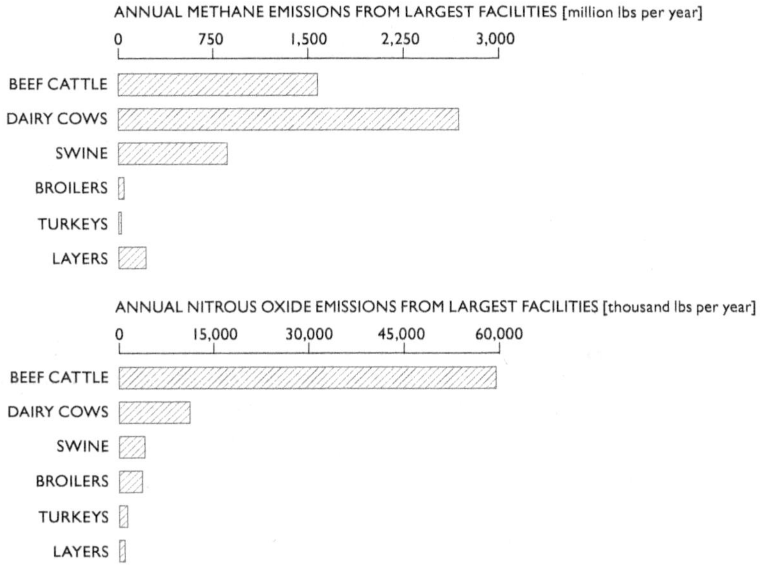
>
> ANNUAL METHANE EMISSIONS FROM LARGEST FACILITIES [million lbs per year]
>
> | Category | Value |
> |---|---|
> | BEEF CATTLE | ~1,500 |
> | DAIRY COWS | ~2,700 |
> | SWINE | ~500 |
> | BROILERS | ~50 |
> | TURKEYS | ~25 |
> | LAYERS | ~200 |
>
> ANNUAL NITROUS OXIDE EMISSIONS FROM LARGEST FACILITIES [thousand lbs per year]
>
> | Category | Value |
> |---|---|
> | BEEF CATTLE | ~60,000 |
> | DAIRY COWS | ~8,000 |
> | SWINE | ~3,000 |
> | BROILERS | ~3,500 |
> | TURKEYS | ~500 |
> | LAYERS | ~500 |
>
> *Source:* CLAIRE HUANG, EARTHJUSTICE, AIR AND GHG EMISSIONS CAFOS: QUANTIFYING A NATIONAL ESTIMATE 25 (Aug. 19, 2020).

In addition, approximately half of all harvested cropland is devoted to animal feed crop production, adding to animal agriculture's already capacious footprint.[32] This cropland is often cultivated more intensely than cropland growing human food and often emits more nitrous oxide per acre than the production of crops for human consumption.[33] However, only a fraction of those crop calories is delivered to humans because the feed-to-meat ratio is so inefficient. For example, the production of one pound of beef from feedlot

---

32. There were approximately 310 million acres of harvested cropland in 2007 according to the Census of Agriculture. NATIONAL AGRICULTURAL STATISTICS SERVICE, U.S. DEPARTMENT OF AGRICULTURE, 2007 CENSUS OF AGRICULTURE: U.S. NATIONAL LEVEL DATA 16 tbl.8 (2009). The U.S. Department of Agriculture (USDA) estimates that approximately 165 million of those acres were devoted to feed crops; however, up to 10% of the feed was diverted to biofuels. CYNTHIA NICKERSON ET AL., USDA, MAJOR USES OF LAND IN THE UNITED STATES, 2007, at 20 (2011) (EIB-89). This total does not include soybeans, which USDA considers a "food crop," despite the fact that soybean meal is typically used as animal feed. TANI LEE ET AL., USDA, MAJOR FACTORS AFFECTING GLOBAL SOYBEAN AND PRODUCTS TRADE PROJECTIONS (2016).
33. Conventionally grown feed crops, such as corn, soybean, and hay, generally result in high N_2O emissions. *See* EPA, *supra* note 15, at 5-23.

cattle requires 15 pounds of grain.[34] As a result, grazing and feed crop production contribute almost two-thirds of N_2O emissions from agricultural soils.[35]

Not only does animal agriculture overall have an outsized impact on climate change, but this impact is—under current production methods—particularly influenced by beef, dairy, and to a lesser extent, swine production. This impact is all the more striking given that Americans receive only 30% of their calories from animal products.[36]

3. State-Level Agricultural Greenhouse Gas Emissions

While virtually all U.S. states have important agricultural sectors, their agricultural emission rates vary significantly due to climate differences and the type and intensity of agriculture within the state. A major factor is the amount of livestock. The top 10 highest overall greenhouse gas emitters (including fossil fuel emissions) account for nearly 50% of national emissions, with Texas and California in the lead. (See total emissions listed below state names in Figures 8 and 9 on pages 47 and 48.) However, the states with the largest total agricultural greenhouse gas emissions are Iowa (63 MMT CO_2 eq./yr.) with more than 10% of all U.S. agricultural greenhouse gas emissions, Texas (38 MMT CO_2 eq./yr.), Nebraska (33 MMT CO_2 eq./yr.), and California (29 MMT CO_2 eq./yr.). All of these states have both very high numbers of livestock and extensive cropland.

States also vary in the significance of agricultural emissions, accounting for more than 50% of total state emissions in South Dakota, and more than 30% of state emissions in Idaho, Nebraska, and Iowa. Finally, agricultural greenhouse gas emissions also vary when scaled by total cropland acres, indicating the relative intensity of the agricultural practices and the portion of

---

34. The feed conversion ratio expresses the number of pounds of grain necessary to increase the "live weight" of a head of cattle by one pound. At industrial feedlots, a feed conversion ratio of 6:1 is common. DAN W. SHIKE, BEEF CATTLE FEED EFFICIENCY 3 (2013). About 40% of the live weight of a head of cattle is sold as beef, which means that 15 pounds of grain is necessary to yield one pound of beef. See ROB HOLLAND ET AL., UNIVERSITY OF TENNESSEE INSTITUTE OF AGRICULTURE, HOW MUCH MEAT TO EXPECT FROM A BEEF CARCASS 9 (2016) (PB-1822).
35. This includes grassland emissions, which account for 73.3 MMT CO_2 eq., as well as 48% of cropland emissions—the approximate percentage of harvested cropland devoted to feed crop production in 2007—which adds an additional 92.7 MMT CO_2 eq. *Compare* U.S. EPA, *supra* note 15, at 5-2 tbl.5-1, 5-25 tbl.5-15 (showing annual emissions from agriculture by source), *with supra* note 32(explaining how the percentage of harvested cropland devoted to feed crop production was calculated). Together, they were responsible for 166 MMT CO_2 eq. or 62% of all emissions from agricultural soils in 2016. This total does not include the approximately 16.5 million acres devoted to the production of biofuel feedstock. *See supra* note 32.
36. USDA Economic Research Service, *Seventy Percent of U.S. Calories Consumed in 2010 Were From Plant-Based Foods*, https://www.ers.usda.gov/data-products/chart-gallery/gallery/chart-detail/?chartId=81864 (last updated Jan. 6, 2017).

Chapter III. The Climate Crisis and Agriculture

Figure 8. Agricultural Emissions by State

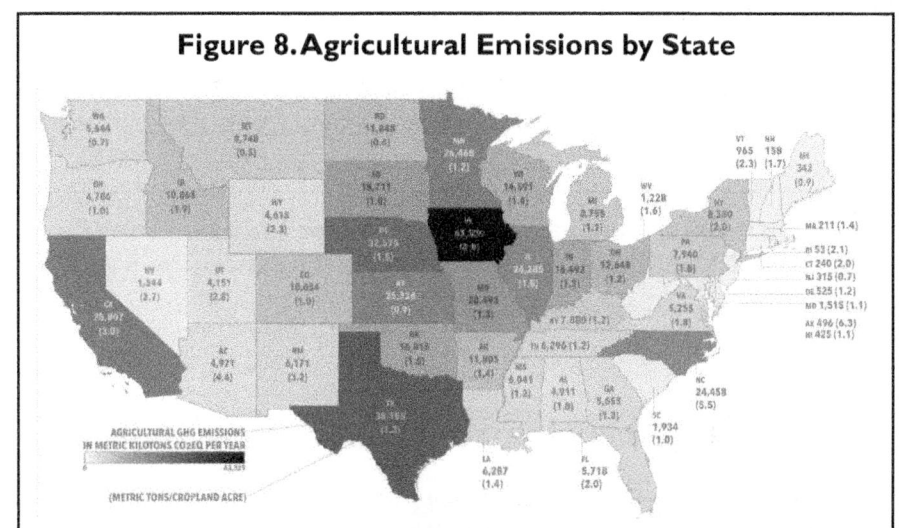

Notes: The 2014 state-specific agricultural greenhouse gas emissions are based on EPA's State Inventory Tool, available in the CAIT Climate Data Explorer (http://cait.wri.org/). States are shaded based on the amount of agricultural greenhouse gas emissions (between 0.05 MMT CO_2 eq. in Rhode Island, to 63.33 MMT CO_2 eq. in Iowa). State-specific agricultural greenhouse gas emission values are printed following state names, followed by emissions scaled by state-specific cropland acres from the 2012 Agricultural Census given in metric tons per cropland acre.

Source: Calculated by the authors using EPA, State Inventory Tool, and NATIONAL AGRICULTURAL STATISTICS SERVICE, USDA, 2012 CENSUS OF AGRICULTURE, STATE LEVEL DATA (2014).

livestock or poultry to cropland, with Arkansas, North Carolina, Arizona, New Mexico, and California in the lead under this metric. State-specific policies should take into account the particularities of each state's agricultural and total economic sectors.

C. Underestimates and Uncertainties

We must understand the limitations of current research and data in order to craft effective policies. This is particularly true in agriculture, which uses a vast amount of land—that could otherwise be used for different purposes—and where production cannot be standardized to the degree that it is in most other sectors of the economy. An apple tree in Washington will have radically different irrigation, nutrient, and anti-pest needs than an apple tree grown in New York state. It may also have substantially different needs than an apple tree down the road due to variations in microclimates and land use history. The production of widgets, or even energy, is much easier to standardize and thus analyze. We begin the section by examining the ways that EPA's official figures significantly underestimate agricultural emissions. We then examine the substantial uncertainties in those estimates. Both of these problems underscore the need to rapidly reduce net agricultural emissions.

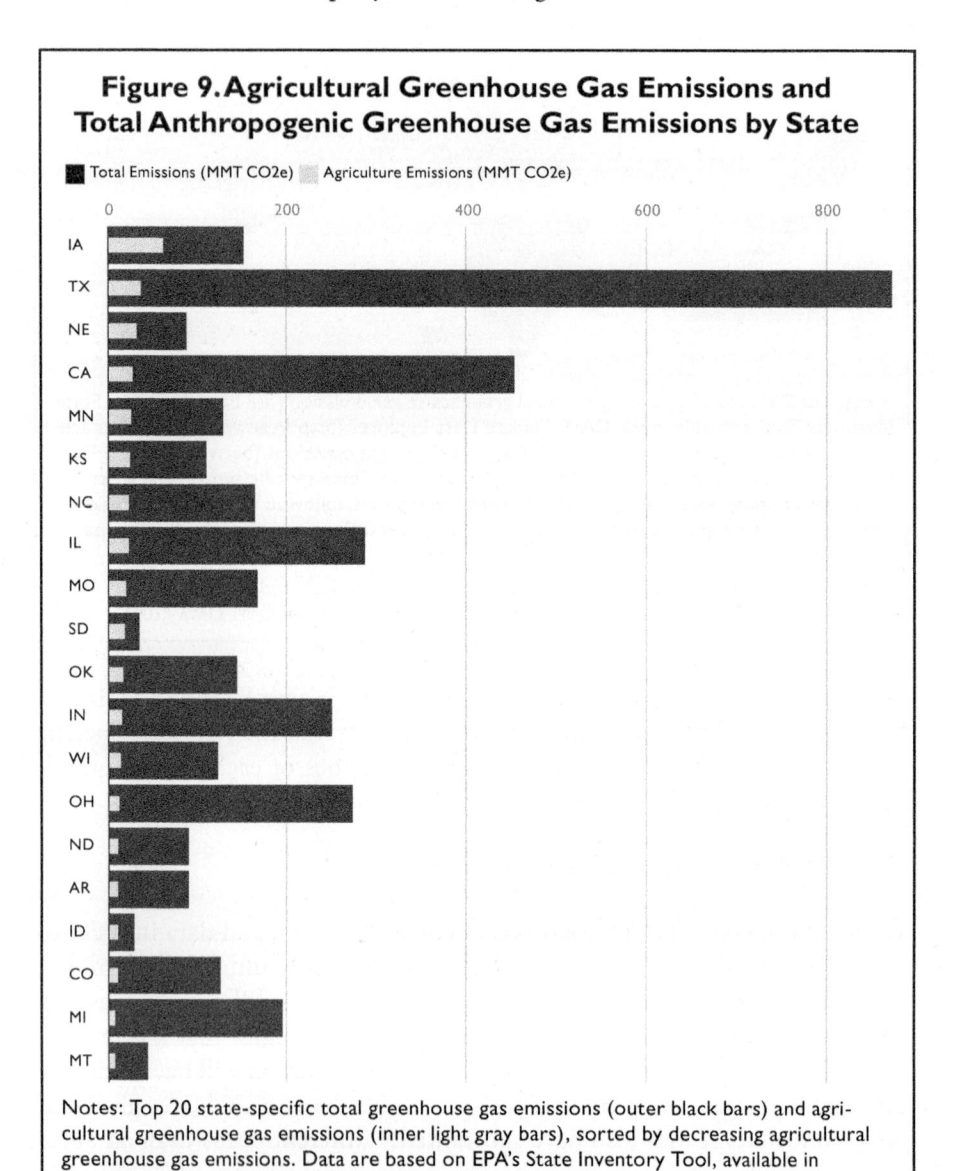

Figure 9. Agricultural Greenhouse Gas Emissions and Total Anthropogenic Greenhouse Gas Emissions by State

Notes: Top 20 state-specific total greenhouse gas emissions (outer black bars) and agricultural greenhouse gas emissions (inner light gray bars), sorted by decreasing agricultural greenhouse gas emissions. Data are based on EPA's State Inventory Tool, available in the CAIT Climate Data Explorer (http://cait.wri.org/).

Chapter III. The Climate Crisis and Agriculture

1. Underestimates

EPA's deceptively low number for agricultural greenhouse gas emissions often leads policymakers and journalists to focus on other sectors when considering climate change mitigation strategies. However, this estimate undercounts agriculture's actual emissions in at least three ways (as shown in Figure 10 on page 53). It does not consider the current climate change impacts of prior land conversion and the lost opportunity of that land to sequester more carbon. Nor does it include on-farm energy, annual land use conversion, agricultural inputs, and other components of the food system. Finally, EPA uses a method for calculating the impact of methane that does not reflect current policy discussions or the need for shorter-term action, reducing its estimate of agriculture's emissions by more than half. Accounting for all these adjustments brings the total to one-quarter to one-third of all U.S. emissions.

First, while the impact on climate change for most sectors of the economy stems almost entirely from their production-related greenhouse gas emissions, with agriculture one must also consider the impact of land use. The land footprint of other sectors is insignificant in relation to their emissions and therefore is not considered in EPA's greenhouse gas inventory. But agriculture's land footprint is the dominant part of the impact. The use of land for growing crops or raising livestock means that agricultural land—62% of the contiguous United States—cannot be used for other purposes, including those that could have a very different climate impact. Most agricultural land before development was grassland or forest land, which both stored and annually sequestered large amounts of carbon. This lost sequestration capacity of agricultural land is a very real climate impact of agriculture, although one rarely considered. If this impact is included, the total annualized climate change impact of agriculture is approximately 50% bigger than the total agriculture sector emissions in the EPA inventory.[37]

As one group of scholars explained: "Restoration of native ecosystems, including forests, is a land-based option for atmosphere carbon dioxide removal. Ecosystem restoration is constrained largely by land requirements of food production, the largest human use of land globally. Food production therefore incurs a 'carbon opportunity cost,' that is, the potential for natural carbon dioxide removal via ecosystem restoration on land."[38] These scholars calculate that "the cumulative potential of carbon dioxide removal on land

---

37. Matthew Hayek et al., *The Carbon Opportunity Cost of Animal-Sourced Food Production on Land*, 4 NATURE SUSTAINABILITY 21 (Jan. 2021) (annualized, U.S. carbon opportunity cost of approximately 264 MMT shared through personal communication with author and supplementary materials).
38. *Id.*

currently occupied by animal agriculture is comparable in order of magnitude to the past decade of global fossil fuel emissions."[39]

Similarly, other scholars have noted that "standard methods for evaluating the effect of land use on greenhouse gas emissions systematically underestimate the opportunity of land to store carbon if it is not used for agriculture."[40] They note that "typical lifecycle assessments, which estimate the [greenhouse gas] costs of a food's consumption, only estimate land use demand in hectares without translating them into carbon costs. Other [life cycle assessments] consider land use carbon costs only if a food is directly produced by clearing new land . . . ." A better approach would be to add to the production-related greenhouse gas emissions the "quantity of carbon that could be sequestered annually if [that land] were instead devoted to regenerating forest [or grassland]."[41]

Many already acknowledge this opportunity when they note the capacity of U.S. agricultural land to sequester carbon.[42] In many cases, the land has this capacity to increase carbon stored in vegetation and soils currently because earlier agricultural activities have significantly depleted what had been previously stored prior to cultivation.[43] Thus in reality, there is a need to restore land to its pre-agricultural condition to repay this debt before interpreting sequestration as an additional opportunity. While now generally discussed as a future sequestration opportunity (often in the context of proposed payment or offset schemes), this can also be seen as legacy harm in need of repair. Seeing it thus and recognizing this "carbon opportunity cost" of the land already in production for agriculture significantly increases agriculture's contribution to climate change. For example, in high income

---

39. *Id.*
40. Timothy Searchinger et al., *Assessing the Efficiency of Changes in Land Use for Mitigating Climate Change*, 564 NATURE 249 (Dec. 13, 2018).
41. *Id.* at 250.
42. *See, e.g.*, FOOD AND AGRICULTURE CLIMATE ALLIANCE, FOOD AND AGRICULTURE CLIMATE ALLIANCE PRESENTS JOINT POLICY RECOMMENDATIONS (2021), https://agclimatealliance.com/files/2020/11/faca_recommendations.pdf (recommendations of coalition led by American Farm Bureau Federation, Environmental Defense Fund, National Council of Farmer Cooperatives, and National Farmers Union to new administration and Congress including providing tools "to maximize the sequestration of carbon" to "achieve the highest number of appropriate soil health-focused practices on the highest number of acres in order to sequester carbon and reduce other GHGs").
43. Jonathan Sanderman et al., *Soil Carbon Debt of 12,000 Years of Human Land Use*, 114 PROC. NAT'L ACAD. SCI. 9575-9580 (Sept. 2017), https://doi.org/10.1073/pnas.1706103114 (modeling soil organic carbon indicates a global soil carbon debt due to agriculture of 133 billion metric tons of carbon, with the rate of loss increasing dramatically in the past 200 years). Note that approximately 440 billion metric tons of carbon have been released by fossil fuel burning since between 1850 and 2018. *See* Pierre Friedlingstein et al., *Global Carbon Budget 2019*, 11 EARTH SYST. SCI. DATA 1783-1838 (2019), https://doi.org/10.5194/essd-11-1783-2019.

Chapter III. The Climate Crisis and Agriculture

countries like the United States, the carbon opportunity cost contributes as much to climate change as all fossil fuel and cement emissions together.[44]

Second, EPA includes on-farm fuel combustion such as for tractors or direct heating in the industrial sector; on-farm electricity for irrigation pumps, cooling, heating, ventilation, and other needs in the electricity sector; and soil carbon lost from conversion of forest or other nonagricultural land to farmland in the consideration of land use. Thus, these emissions are not included in EPA's calculations for the "agriculture" sector. Nor does EPA's agricultural tally include emissions related to aquaculture and fisheries, which provide significant amounts of our food.[45] On-farm fuel combustion in 2018 contributed about 40 MMT CO_2 eq.,[46] as did the indirect emissions of on-farm electricity use, while land annually converted for agricultural use released 56 MMT CO_2 eq.[47] All told, these additional elements of agriculture's greenhouse gas emissions increase the sector's share to about 11%. This total does not include upstream and downstream food system emissions such as emissions associated with the manufacture of fertilizer (discussed below in Chapter VIII and itself adding at least one-half percent of total U.S. greenhouse gas emissions), refrigeration and transport of food, and managing food waste, which, if included, would bring the U.S. food system's total carbon footprint much higher.[48] At the global scale, as noted above, approximately one-third of all greenhouse gas emissions are attributed to the food system.[49]

Third, calculating agriculture's climate change contribution is also complicated by the fact that, unlike the energy and transportation sectors, which emit primarily CO_2 as fossil fuels are burned, crop and livestock greenhouse gas emissions consist largely of N_2O and CH_4. Comparing gases and their

---

44. Hayek, *supra* note 37, at 22, fig.2.
45. U.S. EPA, *supra* note 15, at 6-109 to 6-110.
46. *See id.* tbl.2-10.
47. *Id.* tbl.6-1, at 6-34. EPA uses land use history data from the USDA Natural Resources Conservation Service to determine the acreage of land that has been converted to cropland or has remained as cropland, and then models emissions. *Id.* at 6-54 to 6-72. Over the past several years, the conversion of forest to cropland has resulted in the largest land use-related annual emissions of CO_2. *Id.* at 6-34.
48. *See, e.g.*, Claudia Hitaj et al., *Greenhouse Gas Emissions in the United States Food System: Current and Healthy Diet Scenarios*, 53 Env't Sci. Tech. 5493–5503 (2019).
49. *See supra* notes 9–11. *See also* Sonja J. Vermeulen et al., *Climate Change and Food Systems*, 37 Ann. Rev. Env't Resources 195–222 (2012); Priyadarshi R. Shukla et al., Intergovernmental Panel on Climate Change, Climate Change and Land: An IPCC Special Report on Climate Change, Desertification, Land Degradation, Sustainable Land Management, Food Security, and Greenhouse Gas Fluxes in Terrestrial Ecosystems 476 tbl.5.4 (2019) (indicating 21-37% of anthropogenic emissions from food systems); Henning Steinfeld et al., Food and Agriculture Organization of the United Nations, Livestock's Long Shadow 113 tbl.3.12 (2006) (indicating ~18% of anthropogenic GHG emissions are attributed to livestock alone); Robert Goodland and Jeff Anhang, Worldwatch, Livestock and Climate Change (2009) (indicating 51% of anthropogenic GHG emissions are attributed to livestock alone).

climate impact implicates fundamental policy choices. N_2O, largely released as a result of fertilizer that is applied but not taken up by crops, is a particularly potent greenhouse gas, with an average global warming potential of 265-298 times that of CO_2 over 100 years.[50] Whether a calculation uses the lower or the higher number of that range for N_2O's global warming potential creates about a 10% variation in its relative contribution to climate change.[51]

Additionally, a 2016 study found that the Intergovernmental Panel on Climate Change's (IPCC's) Fifth Assessment Report underestimated CH_4's global warming potential by 20%-25% because its methods did not take into account the absorption of shortwave radiation by CH_4, among other factors.[52] The study's author estimates that the IPCC's Sixth Assessment Report may revise CH_4's 100-year global warming potential to 35 or higher.[53]

Calculating CH_4's global warming potential is further complicated by the fact that methane breaks down relatively quickly compared to N_2O or CO_2. The global warming potential of methane is about 84-86 times that of CO_2 over 20 years.[54] EPA, however, uses a longer time horizon for calculating the global warming potential of CH_4, reducing the relative impact of agriculture's total emissions by more than half. Instead of determining the CO_2 equivalent of CH_4 by comparing the two gases over a 20-year time span, EPA's report follows the IPCC's Fourth Assessment Report in using a 100-year time span. This significantly lowers CH_4's global warming potential since CH_4's potency declines relatively quickly. As a result, EPA's estimate assumes that CH_4 has only 25 times the radiative impact of CO_2.[55] The IPCC's Fifth Assessment Report, however, not only increased the 100-year global warming potential of CH_4 to 28-34 times that of CO_2,[56] but also supports the use of a 20-year timescale for measuring the impact of emissions from agriculture.[57] While a 100-year time period for CH_4 is still commonly used in scientific discussions, policy debates increasingly use a 20-year period

---

50. Intergovernmental Panel on Climate Change, Climate Change 2013: The Physical Science Basis Ch. 8, at 714 tbl.8-7 (2014).
51. N_2O emissions will also be the primary cause of stratospheric ozone destruction this century. A.R. Ravishankara et al., *Nitrous Oxide (N_2O): The Dominant Ozone-Depleting Substance Emitted in the 21st Century*, 326 Science 123, 123-25 (2009).
52. Maryam Etminan et al., *Radiative Forcing of Carbon Dioxide, Methane, and Nitrous Oxide: A Significant Revision of the Methane Radiative Forcing*, 43 Geophysical Res. Letters 12614 (2016).
53. Jessica McDonald, *How Potent Is Methane?*, FactCheck.Org, Sept. 24, 2018, https://www.factcheck.org/2018/09/how-potent-is-methane/.
54. Intergovernmental Panel on Climate Change, *supra* note 49.
55. U.S. EPA, *supra* note 15, at ES-3.
56. Intergovernmental Panel on Climate Change, *supra* note 49.
57. *Id.* at 720.

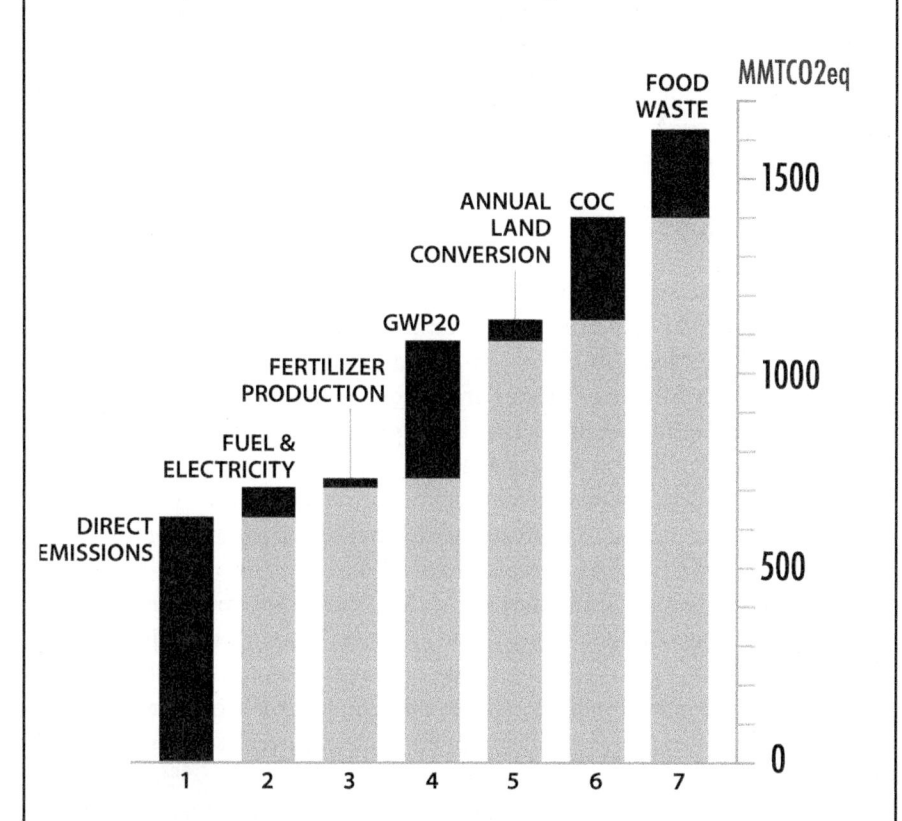

Chapter III. The Climate Crisis and Agriculture 53

Figure 10. More Complete Estimate of Agriculture's Climate Change Contribution

[Bar chart showing cumulative contributions in MMTCO2eq: (1) DIRECT EMISSIONS, (2) FUEL & ELECTRICITY, (3) FERTILIZER PRODUCTION, (4) GWP20, (5) ANNUAL LAND CONVERSION, (6) COC, (7) FOOD WASTE. Y-axis ranges from 0 to 1500 MMTCO2eq.]

Notes: Accounting for agriculture's broad climate footprint, cumulative factors: (1) EPA's estimate of direct emissions from agricultural activities as delineated by IPCC were 629 MMT in 2019; (2) EPA's estimate of emissions from on-farm fuel and electricity use; (3) Total emissions from nitric acid, ammonia and phosphoric acid production; (4) Scaling emission estimates from 100-year global warming potentials (GWP) to represent policy-relevant timescales using GWP20 values for methane and nitrous oxide; (5) Net carbon losses from annual land conversion to croplands; (6) Carbon opportunity cost (COC), converted to annualized greenhouse gas emission equivalent for the U.S.; (7) Emissions from 75% of total landfill methane emissions, anaerobic digestion at biogas facilities, and composting. References for figures in prior footnotes. For comparison, the combined total emissions (>1600 MMT) are equivalent to the annual carbon dioxide emissions from over 400 coal-fired power plants and slightly less than total U.S. transportation sector greenhouse gas emissions (1880 MMT).

due to the urgent need to reduce CH_4 emissions over the next 10-30 years.[58] For example, New York's Climate Leadership and Community Protection Act requires use of the 20-year time frame for analysis and policy development, which time frame increases CH_4 share of the state's total greenhouse gases by 3.4 times.[59] If EPA had calculated agricultural emissions using a 20-year time horizon, its estimate would nearly double, from 619 to 1,216 MMT CO_2 eq. each year,[60] shifting agricultural CH_4 emissions alone to contributing about 8% of total U.S. emissions.

2. Uncertainties

Agricultural emissions are much more difficult to calculate than those in other sectors and far less certain. Governments and the private sector keep precise data about the amount of coal, oil, and gas used, which can be used to accurately determine the amount of CO_2 entering the atmosphere. By contrast, EPA's methodologies for estimating agricultural greenhouse gas emissions are very different and far less exact.

All emission calculations involve some uncertainty due to challenges with collecting accurate and representative data, selecting appropriate model parameters, and simplifying complex natural processes into a series of equations. Experts calculating emissions can determine how model results vary according to a range of likely inputs, and thus can establish what is known as a "95% confidence interval"—the range of values surrounding the estimate for which there is a 95% likelihood that the true value lies between.

Many emission sources within the energy and industrial sectors are associated with precise mean or central estimates. For example, the 95% confidence interval for total CO_2 emissions from energy-related fossil fuel combustion narrows this estimate to within 2%-4% of the mean estimate.[61] In contrast, estimates for CH_4 and N_2O emissions from agriculture come with broader uncertainties across the board,[62] and several of the largest agricultural green-

---

58. *See, e.g.*, Robert W. Howarth, *A Bridge to Nowhere: Methane Emissions and the Greenhouse Gas Footprint of Natural Gas*, 2 ENERGY SCI. & ENGINEERING 47, 53-55 (2014).
59. *See* N.Y.S. Env't Conservation L. § 75-0101(2) ("'Carbon dioxide equivalent' means the amount of carbon dioxide by mass that would produce the same global warming impact as a given mass of another greenhouse gas over an integrated twenty-year timeframe after emission."); Robert Howarth, *Methane Emissions From Fossil Fuels: Exploring Recent Changes in Greenhouse Gas Reporting Requirements for the State of New York*, 17 J. INTEGRATIVE ENV'T SCI. 69 (2020), https://doi.org/10.1080/194381 5X.2020.1789666.
60. Calculated by the authors using a global warming potential of 84 instead of 25 for the CH_4 emission rates in U.S. EPA, *supra* note 15, at 5-2 tbl.5-1.
61. *See* U.S. EPA, *supra* note 15, at 3-37 tbl.3-17.
62. *Id.* at 1-26 tbl.1-6.

house gas emission sources, including soil N_2O emissions[63] and enteric fermentation,[64] have extremely wide confidence intervals. The confidence interval for the agricultural sector is between 451 and 847 MMT CO_2 eq., or from 7% to 13% of total U.S. greenhouse gas emissions, around the mean estimate of 10% of emissions. This broad range of uncertainty between the upper and lower bounds for U.S. agricultural emissions (396 million tons) is equivalent to the annual emissions from 102 coal-fired power plants.[65]

These wide uncertainties are partly attributed to fundamental differences in estimating agricultural emissions compared to other sectors. For example, to determine enteric emissions of methane from cattle, EPA uses U.S. Department of Agriculture (USDA) data on the age, weight, and location of different varieties of animals. Emissions from each subpopulation are then modeled based on parameters reflecting diet characteristics in the region and the CH_4 conversion rate, or fraction of calories converted to CH_4. A similar but coarser approach is used for non-cattle livestock. In addition to uncertainties associated with the demographic data on animal subpopulations, the cattle diet estimates are relatively speculative. EPA uses similarly complex models for manure emissions that incorporate the production rates of solid waste, CH_4 conversion factors, and N_2O emission factors, among other estimates, resulting in a 95% confidence interval from 18% below to 24% above the given figures.[66]

A recent paper suggests an additional substantial underestimate of modeled emissions from CAFOs.[67] The authors compared atmospheric measurements taken above and downwind of animal production regions to standard EPA and other models and found that the measurements showed animal CH_4 emissions 39%–90% higher than model estimates of animal CH_4. They note that "bottom up" models based on data on animal inventory and characteristics underpredict enteric CH_4 emissions for multiple animal species, potentially in part due to the prevalence of diseased animals with higher rates of enteric emissions than predicted from models with healthy herds. Additionally, they note that manure emission estimates from these bottom-up models, which use parameters based on laboratory experiments within controlled test chambers, "appear to routinely underpredict emissions from manure. . . . When methane

---

63. *Id.* at 5-44 tbl.5-20.
64. *Id.* at 5-8 tbl.5-6.
65. *See* U.S. EPA, *Greenhouse Gas Equivalencies Calculator*, https://www.epa.gov/energy/greenhouse-gas-equivalencies-calculator (last updated Mar. 2020).
66. U.S. EPA, *supra* note 15, at 5-16 tbl.5-9.
67. Matthew Hayek & Scot Miller, *Underestimates of Methane From Intensively Raised Animals Could Undermine Goals of Sustainable Development*, 16 Env't Res. Letters 063006 (June 2021), https://iopscience.iop.org/article/10.1088/1748-9326/ac02ef.

is measured outside of the lab, in the air directly above manure tanks, pits, and piles, emissions tend to be greater than models predict, sometimes by more than 300%." These findings suggest even greater attention must be paid to ways to reduce CH_4 emissions from CAFOs.

The greatest uncertainties in EPA's greenhouse gas inventory are attributed to estimating N_2O emissions from agricultural soils. The calculation must include each of the five different ways N_2O is released, including (1) emissions from the application of synthetic fertilizers and other inputs; (2) emissions following the breakdown of organic matter; (3) emissions following soil drainage; (4) emissions following livestock manure deposits; and (5) indirect emissions following leaching or volatilization. Even with reasonably good data on nitrogen application activities, there are many uncertainties since the model must use intricate biogeochemical interactions in soil that vary with the weather, inputs, and other environmental conditions. As a result of this complexity, EPA indicates that the true N_2O emissions from direct and indirect sources could be between 37% below to 50% above the given figure,[68] which encompasses a range of 292 MMT CO_2 eq., itself an amount equal to almost half of the given figure for total U.S. agricultural greenhouse gas emissions.

As a result of all these uncertainties, demonstrated in Figure 11, EPA's estimate for agricultural greenhouse gas emissions must be understood as simply one point in a wide range of possible figures. These uncertainties also point to a major challenge in developing policies to mitigate agricultural emissions and promote sequestration. When regulating emissions from other sectors, the government can identify emissions trends with minimal uncertainty, closely monitor emissions sources, and even compensate for emission reductions with precision. In contrast, agricultural emissions are diffuse. Monitoring and measuring emissions is often difficult or impossible, and model calculations are relatively uncertain. These factors make it challenging to disentangle trends or detect the impact of specific policies on total emissions relative to wide uncertainties. Fortunately, there is ample evidence that many climate-friendly practices do significantly reduce emissions or increase sequestration, and policymakers can craft programs that address the unavoidable uncertainty.

---

68. Calculated by the authors using the sum of direct and indirect sources in *id.* at 5-44 tbl.5-20.

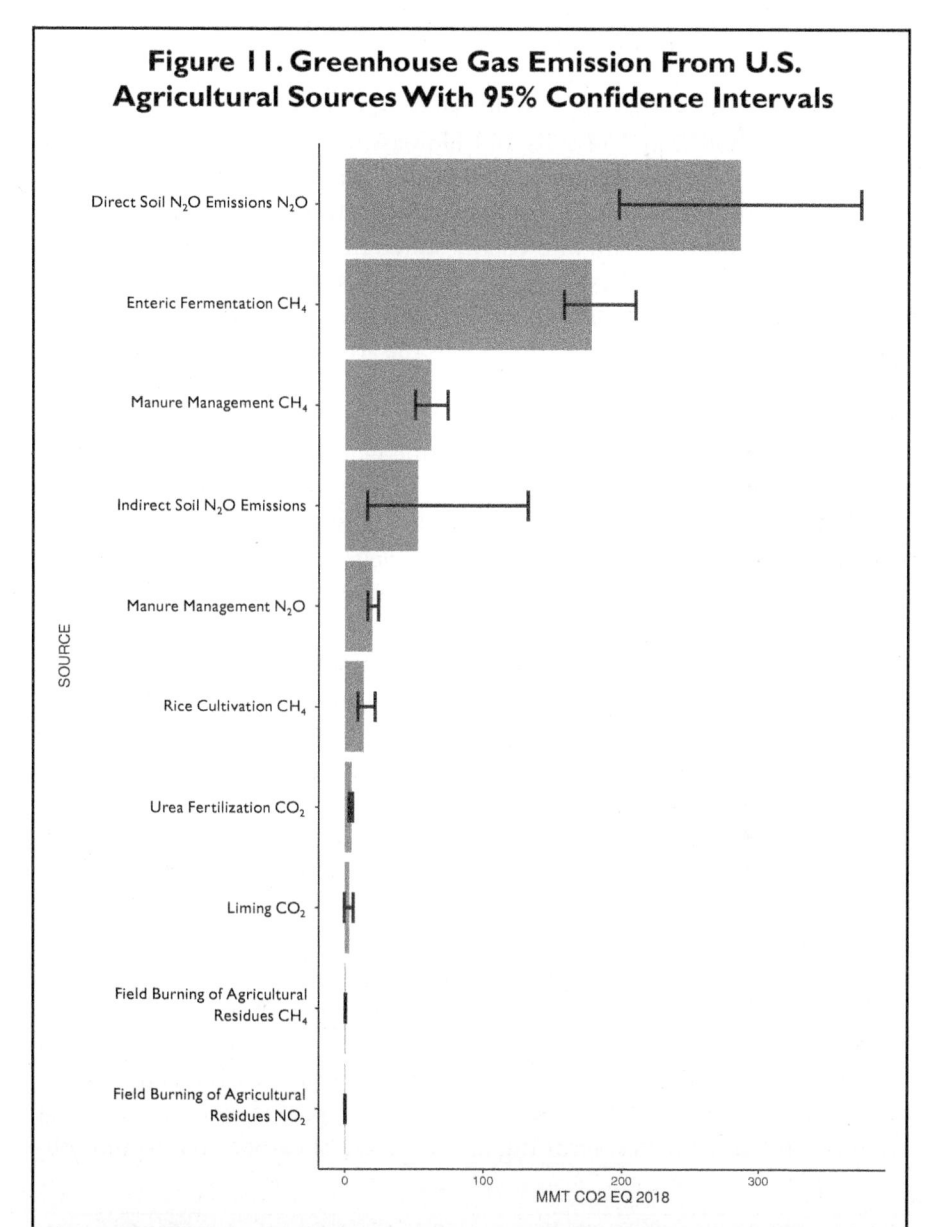

Chapter III. The Climate Crisis and Agriculture

Figure 11. Greenhouse Gas Emission From U.S. Agricultural Sources With 95% Confidence Intervals

Source	MMT CO2 EQ 2018
Direct Soil N_2O Emissions N_2O	~290
Enteric Fermentation CH_4	~180
Manure Management CH_4	~60
Indirect Soil N_2O Emissions	~50
Manure Management N_2O	~20
Rice Cultivation CH_4	~15
Urea Fertilization CO_2	~5
Liming CO_2	~3
Field Burning of Agricultural Residues CH_4	<5
Field Burning of Agricultural Residues NO_2	<5

Source: U.S. EPA, INVENTORY OF U.S. GREENHOUSE GAS EMISSIONS AND SINKS: 1990-2018, at 5-2 tbl.5-1 (2020) (EPA 430-R-20-002), https://www.epa.gov/sites/production/files/2020-04/documents/us-ghg-inventory-2020-main-text.pdf.

D. Agriculture's Dual Opportunity

Agricultural activities not only emit greenhouse gases but can change the amount of carbon stored in soils and biomass, thus effectively releasing or absorbing CO_2. Carbon storage is increased by plant growth, which removes CO_2 from the atmosphere during photosynthesis, the process by which plants convert energy from the sun into energy stored in the chemical bonds of carbohydrates, carbon-based molecules. Carbon storage is decreased when these bonds are broken by organisms to access the stored energy and the carbon contained in organic matter is returned to the atmosphere as CO_2. Thus, *net* carbon storage can be increased by increasing the amount of photosynthesis, such as by adding cover crops over bare ground or incorporating trees, or by slowing the decomposition of soil organic matter, such as through use of no-till practices.

As discussed in detail in Chapter IV, scientific studies have identified a number of agricultural practices that could help to slow climate change by reducing greenhouse gas emissions or capturing carbon—or both—while maintaining productivity. For example, in 2016, researchers concluded that the expansion of existing USDA conservation practices could lead to the sequestration of 277 MMT CO_2 eq. annually by 2050.[69] Capturing this volume of carbon in the soil would cut net agricultural greenhouse gas emissions in half. Similarly, agroforestry (incorporating trees and shrubs into cropland and pastureland) and perennial agriculture (plants that live year-round and do not need annual replanting, thus disturbing the soil less) offer significant climate benefits by locking carbon in the perennial biomass of the plant roots and shoots and stimulating a more biodiverse ecosystem that stores more carbon. According to a 2012 review, the widespread adoption of agroforestry practices in the United States could sequester 530 MMT carbon (or close to 2,000 MMT CO_2 eq.) each year, thereby transforming agricultural land into a carbon sink.[70]

Like cropland, rangeland used for livestock grazing can also sequester carbon. Overgrazing has damaged vegetation and degraded soil quality across the western United States, resulting in the release of carbon that would oth-

---

69. Adam Chambers et al., *Soil Carbon Sequestration Potential of U.S. Croplands and Grasslands: Implementing the 4 Per Thousand Initiative*, 71 J. Soil & Water Conservation 68A, 70A (2016). This total represents four times the carbon sequestration of forest soils. *See* Rattan Lal et al., *Achieving Soil Carbon Sequestration in the United States: A Challenge to the Policy Makers*, 168 Soil Sci. 827, 838 (2003) (finding that forest soils could sequester 63 MMT CO_2 eq. annually).
70. Ranjith P. Udawatta & Shibu Jose, *Agroforestry Strategies to Sequester Carbon in Temperate North America*, 86 Agroforestry Sys. 225, 239 (2012).

erwise remain locked in organic matter.[71] However, managing the location and intensity of grazing, while adjusting its timing to facilitate plant growth, can repair these landscapes[72] and restore their function as carbon sinks.[73]

As these examples demonstrate, methods already exist to mitigate agriculture's net contribution to climate change by reducing greenhouse gas emissions or increasing carbon sequestration. However, policies must recognize that while greenhouse gas emissions are permanent actions, biological sequestration is reversible and limited through the natural process of decomposition. Climatic events, such as droughts or wildfires, or human actions, such as resumed tillage, increased grazing, or deforestation, can quickly destroy biomass and disrupt soils, thereby releasing stored carbon.[74] In addition, gains in soil carbon slow as soils approach a new equilibrium under improved management practices.[75] (Additional research is needed to clarify how quickly this occurs, but location, prior soil quality, and land management practices all appear to be important factors.[76])

While sequestration alone cannot offset ever-increasing greenhouse gas emissions, it remains a necessary strategy for avoiding catastrophic climate change. Current levels of atmospheric carbon are so dangerously high that we cannot choose between reducing emissions on the one hand and sequestering carbon on the other.[77] We must do both.

---

71. *See* John Carter et al., *Moderating Livestock Grazing Effects on Plant Productivity, Nitrogen, and Carbon Storage*, 17 Nat. Resources & Env't Issues 191, 191-92 (2011).
72. Sherman Swanson et al., *Practical Grazing Management to Maintain or Restore Riparian Functions and Values on Rangelands*, 2 J. Rangeland Applications 1, 10-14 (2015).
73. David Lewis et al., University of California Cooperative Extension, Creek Carbon: Mitigating Greenhouse Gas Emissions Through Riparian Revegetation 22 (2015).
74. Uta Stockmann et al., *The Knowns, Known Unknowns, and Unknowns of Sequestration of Soil Organic Carbon*, 146 Agric. Ecosystems & Env't 80, 82 (2012).
75. Catherine Stewart et al., *Soil Carbon Saturation: Concept, Evidence, and Evaluation*, 86 Biogeochemistry 19, 25-28 (2007); Stockmann et al., *supra* note 74, at 94-95.
76. Stockmann et al., *supra* note 74, at 82.
77. For an informal discussion of carbon sequestration's potential to help address climate, see Marcia DeLonge, *Soil Carbon Can't Fix Climate Change by Itself—But It Needs to Be Part of the Solution*, Union Concerned Scientists, Sept. 26, 2016, http://blog.ucsusa.org/marcia-delonge/soil-carbon-cant-fix-climate-change-by-itself-but-it-needs-to-be-part-of-the-solution.

Key Recommendations

- Climate change will affect agriculture and the food system more than almost any other sector of the economy. Climate change induced weather changes already jeopardize agriculture with increased floods, droughts, pests, heat waves, wildfires, and more and will force disruptive dislocations as it shifts which crops are suitable for different regions. In addition, climate change threatens even our food itself as it is expected to reduce protein concentrations in staple crops, reduce micronutrients in vegetables, and more.

- Agriculture occupies 62% of the contiguous U.S. landmass.

- Global food systems contribute approximately one-third of total greenhouse gas emissions, mostly N_2O from soil management and CH_4 from cattle, dairy, and manure, as well as impacts of land use and soil carbon loss.

- In the United States, meat and dairy production, including emissions relating to production of their feed, grazing, enteric fermentation, and manure, accounts for about 80% of agriculture's greenhouse gas emissions. Yet, Americans receive only 30% of their calories from animal products.

- The vast majority of animal production greeenhouse gas emissions are produced by the very small number of the largest facilities that house almost all the animals produced. Overall, the largest animal production facilities—fewer than 6% of all facilities—produce 89% of the animals and about 85% of the greenhouse gas emissions of all animal production.

- Unless there are significant changes in the food system, agricultural emissions alone will make it impossible to achieve the climate stabilization goal of 2 degrees Celsius, let alone the safer target of 1.5 degrees Celsius. And if meat consumption continues to grow, climate change will be dramatically accelerated.

- Carbon sequestration should be an essential function of agriculture and be supported by federal agricultural programs and policies.

- Other countries are already investing significant sums into agroforestry research and production, yet the United States has lagged, despite robust research demonstrating its significant potential to sequester carbon while producing ample food.

Chapter III. The Climate Crisis and Agriculture

- Total agricultural greenhouse gas emissions and agricultural emission rates significantly vary among states due to climate differences and the type and intensity of agriculture within each state; state-based policies will likely also need to differ to best address each state's agriculture sector.

- We must understand the limitations of current research and data in order to craft effective policies.

- Agricultural emissions are much more difficult to calculate than those in other sectors and far less certain. EPA's estimates for agricultural greenhouse gas emissions must therefore be understood as simply one point in a wide range of possible figures.

- EPA estimates fail to consider impacts of prior land conversion and the lost opportunity to sequester more carbon or release less greenhouse gases from that land; the lost "carbon opportunity cost" contributes as much to climate change as the last decade of fossil fuel emissions.

- EPA analyses do not include in the "agriculture" sector the greenhouse gas emissions of on-farm energy and electricity, annual land use conversion, or production of agricultural inputs. Nor do they include other components of the food system, such as processing, distribution, preparation, and waste. Considering all these emissions together, the food system is responsible for over a third of all U.S. emissions.

- EPA analyses do not calculate the impact of methane in a way that reflects current policy discussions and the need for shorter-term action, reducing its estimate of agricultural emissions by more than half.

- There is ample evidence that many climate-friendly practices significantly reduce emissions or increase sequestration or do both.

- Methods exist to mitigate agriculture's net contribution to climate change by reducing greenhouse gas emissions or increasing carbon sequestration. However, policies must recognize that while greenhouse gas emissions are permanent actions, biological sequestration is reversible and limited through the natural process of decomposition.

Chapter IV.
Climate-Friendly Agricultural Systems and Practices

To implement sound policy and pursue effective legal strategies, decisionmakers and advocates must become familiar with the climate-friendly agricultural practices that constitute carbon farming.[1] To that end, this chapter briefly reviews the tools and technologies available to reduce agricultural greenhouse gas emissions and to sequester carbon.[2] The chapter begins by summarizing the practices and technologies applicable to cropland, before considering those available for grazing lands, animal feeding operations (AFOs), and on-farm fuel combustion and electricity.[3] It concludes by discussing several factors that make it difficult to achieve maximum decarbonization in the sector, including scientific uncertainties, the need to balance climate benefits against other environmental concerns, and the practical challenges of implementing carbon farming on a national scale. Subsequent chapters of the book will describe policy pathways to encourage adoption of these climate-friendly practices.

As can be seen in Table 1, there are many practices and systems that have great potential to reduce greenhouse gas emissions and increase carbon sequestration. Similarly, Figure 1 on page 65, reflecting an independent

---

1. Many climate-friendly agricultural practices are "regenerative," meaning that they regenerate healthy soil carbon levels as part of a holistic management system. *See* REGENERATION AGRICULTURE INITIATIVE & CARBON UNDERGROUND, WHAT IS REGENERATIVE AGRICULTURE? (2017).
2. This chapter does not provide an exhaustive literature review. However, we have briefly summarized the most commonly discussed and promising methods available to reduce agricultural emissions and increase carbon sequestration, and provided a rough estimate of their potential in the United States. Although further research and development is necessary—and, indeed, is one of this book's main recommendations for advancing carbon farming—most of the methods described in this chapter are currently in use and suitable for widespread adoption.
3. Estimated carbon sequestration rates and emissions reductions for each practice are included in Table 1 when possible. Most of the data are derived from COMET-Planner, an online tool developed by the U.S. Department of Agriculture (USDA) and Colorado State University that provides approximate net emissions reductions for a number of practices recognized by the USDA Natural Resources Conservation Service (NRCS). *See* AMY SWAN ET AL., USDA & COLORADO STATE UNIVERSITY, COMET-PLANNER: CARBON AND GREENHOUSE GAS EVALUATION FOR NRCS CONSERVATION PRACTICE PLANNING. Projections of the total amount of farmland where each practice is applicable are also included when possible. This is designed to allow readers to gauge not only how effective a practice might be on any given parcel of land, but also what its cumulative potential might be for the country as a whole.

Table 1. Average Annual Net Emissions Reductions of Select Agricultural Practices*

Practice	Maximum applicable area (million acres)	Average annual net emissions reductions (MT CO$_2$ eq. per acre)	Possible annual sequestration potential (MMT CO$_2$ eq.)
Cropland			
Improved synthetic fertilizer management	230	0.11	25
Reduced till	178	0.19	34
No-till	232	0.32	74
Cover cropping	126-245	0.32	40
Conservation crop rotations	310**	0.24	74
Organic amendments	Unknown	1.38***	Unknown
Biochar	306	0.26-7.90	80
Alley cropping	198	1.28	253
Windbreaks	11	1.59	19
Riparian buffers	2	1.78	4
Perennial biofuels and feedstock****	Unknown	1.74-2.43	Unknown
Grazing land			
Prescribed grazing	Unknown	0.19	Unknown
Organic amendments	Unknown	0.85-1.90	Unknown
Silvopasture	173	0.73	126

* Calculated by the authors using data derived from ALISON J. EAGLE ET AL., NICHOLAS INSTITUTE FOR ENVIRONMENTAL POLICY SOLUTIONS, GREENHOUSE GAS MITIGATION POTENTIAL OF AGRICULTURAL LAND MANAGEMENT IN THE UNITED STATES: A SYNTHESIS OF THE LITERATURE (2012); AMY SWAN ET AL., USDA & COLORADO STATE UNIVERSITY, COMET-PLANNER: CARBON AND GREENHOUSE GAS EVALUATION FOR NRCS CONSERVATION PRACTICE PLANNING [hereinafter COMET-PLANNER]; P.K. Ramachandran Nair & Vimala Nair, *Carbon Storage in North American Agroforestry Systems*, in THE POTENTIAL OF U.S. FOREST SOILS TO SEQUESTER CARBON AND MITIGATE THE GREENHOUSE EFFECT (John M. Kimble et al. eds., Lewis Publishers 2003).

** Calculated by the authors. *Compare* COMET-PLANNER at 8 (noting that conservation crop rotations are possible on all cropland where at least one annually planted crop is included in the crop rotation), *with* NATIONAL AGRICULTURAL STATISTICS SERVICE, USDA, 2012 CENSUS OF AGRICULTURE, U.S. NATIONAL LEVEL DATA 16 tbl.8, 27-32 tbl.37 (2014) (reporting the total number of harvested acres with annually planted crops).

*** This total does not account for nitrous oxide emissions. COMET-PLANNER at 7.

**** The perennials studied include poplar, willow, and switchgrass. R. Lemus & R. Lal, *Bioenergy Crops and Carbon Sequestration*, 24 CRITICAL REVS. PLANT SCI. 1, 15 (2005).

Chapter IV. Climate-Friendly Agricultural Systems and Practices

Figure 1. The Mitigation and Adaptation Potential of Select Food System Responses to Climate Change

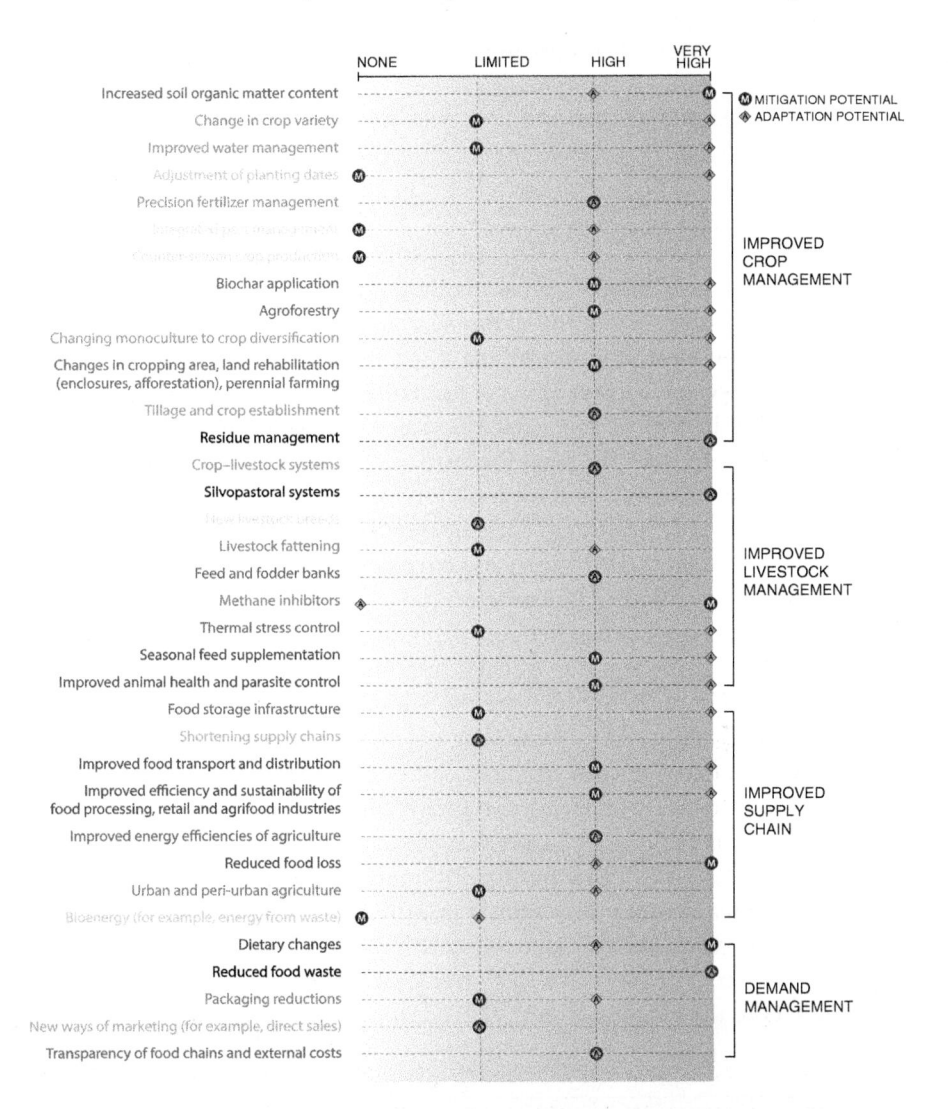

Response	Mitigation Potential	Adaptation Potential	Category
Increased soil organic matter content	Very High	High	Improved Crop Management
Change in crop variety	Limited	High	Improved Crop Management
Improved water management	Limited	High	Improved Crop Management
Adjustment of planting dates	None	High	Improved Crop Management
Precision fertilizer management	High	—	Improved Crop Management
Integrated pest management	None	High	Improved Crop Management
Greater emissions productivity	None	High	Improved Crop Management
Biochar application	High	High	Improved Crop Management
Agroforestry	High	High	Improved Crop Management
Changing monoculture to crop diversification	Limited	High	Improved Crop Management
Changes in cropping area, land rehabilitation (enclosures, afforestation), perennial farming	High	High	Improved Crop Management
Tillage and crop establishment	High	—	Improved Crop Management
Residue management	High	—	Improved Crop Management
Crop–livestock systems	High	—	Improved Livestock Management
Silvopastoral systems	High	—	Improved Livestock Management
New livestock breeds	Limited	—	Improved Livestock Management
Livestock fattening	Limited	High	Improved Livestock Management
Feed and fodder banks	High	—	Improved Livestock Management
Methane inhibitors	Very High	None	Improved Livestock Management
Thermal stress control	Limited	High	Improved Livestock Management
Seasonal feed supplementation	High	High	Improved Livestock Management
Improved animal health and parasite control	High	High	Improved Livestock Management
Food storage infrastructure	Limited	High	Improved Supply Chain
Shortening supply chains	Limited	—	Improved Supply Chain
Improved food transport and distribution	High	—	Improved Supply Chain
Improved efficiency and sustainability of food processing, retail and agrifood industries	High	—	Improved Supply Chain
Improved energy efficiencies of agriculture	High	—	Improved Supply Chain
Reduced food loss	Very High	High	Improved Supply Chain
Urban and peri-urban agriculture	Limited	High	Improved Supply Chain
Bioenergy (for example, energy from waste)	None	Limited	Improved Supply Chain
Dietary changes	Very High	High	Demand Management
Reduced food waste	High	—	Demand Management
Packaging reductions	Limited	High	Demand Management
New ways of marketing (for example, direct sales)	High	—	Demand Management
Transparency of food chains and external costs	High	—	Demand Management

Source: Adapted from Cynthia Rosenzweig et al., *Climate Change Responses Benefit From a Global Food System Approach*, 1 NATURE FOOD 94-97 fig.1 (2020). Responses with darkest text have the highest total combined adaptation and mitigation potential.

analysis that also quantified the adaptation or resilience benefit of a range of practices, comes to largely the same conclusion. That the climate-mitigation potential of these practices is so robustly demonstrated underscores the value of the policy changes we discuss in Chapters V-IX.

A. Agricultural Systems and Practices for Reducing Greenhouse Gas Emissions

1. Cropland

Responsible management of croplands should increase carbon sequestration and reduce greenhouse gas emissions. Carbon sequestration can be accomplished by growing additional biomass, particularly perennial plants, and by increasing the soil carbon levels of agricultural land. The most promising practices often do both. They also reduce soil disturbance, which releases carbon dioxide (CO_2), and the need for fertilizer, which releases nitrous oxide.[4]

We first describe four agricultural systems that offer a range of climate benefits: agroforestry, perennial agriculture, diversified farming, and organic agriculture. U.S. agricultural policy currently disfavors agroforestry and other forms of perennial agriculture, and these practices are unlikely to expand without significant changes in public law and policy. Nonetheless, this book highlights them due to their unrivaled capacity for long-term carbon sequestration. Diversified systems, which often incorporate trees and other perennial crops, have a number of climate advantages over conventional monocultures, including increased resilience and carbon sequestration. Although organic agriculture remains uncommon—less than 0.5% of agricultural land is devoted to organic production—it offers policymakers seeking to reduce emissions a number of important lessons.[5]

We next explain why policy should prioritize the production of crops that provide people with healthy food, rather than crops that become processed food, animal feed, or biofuels. The latter set of crops—namely corn, wheat, and soy—take up a huge amount of land and consume enormous amounts

---

4. As discussed in Chapter III, the application of excess fertilizer releases nitrous oxide and the manufacture of synthetic fertilizer is energy intensive and itself releases significant greenhouse gases. The use of manure or other water for fertilizer avoids the manufacturing emissions but can still release nitrous oxide.
5. *Compare* National Agricultural Statistics Service, USDA, 2014 Organic Survey 1 tbl.1 (2016), *and* National Agricultural Statistics Service, USDA, 2015 Certified Organic Survey 1 tbl.1 (2016), *with* National Agricultural Statistics Service, USDA, 2012 Census of Agriculture, U.S. National Level Data 16 tbl.8 (2014) (finding more than 914 million acres of farmland).

Chapter IV. Climate-Friendly Agricultural Systems and Practices

of energy,[6] though, as explained below, they offer little nutritional benefit. If humans consumed these crops or other whole foods directly, the food system would produce more nutritious food with much less land and energy.

The discussion then moves to three methods to reduce net CO_2 emissions by increasing the organic matter content of soil—reducing tillage, increasing carbon inputs from crops, and adding soil amendments. Soil organic matter, which consists primarily of decomposing plants and animals, is rich in carbon. Thus, practices that increase the organic matter content of soil generally also increase soil carbon sequestration and, thereby, reduce net emissions. Increasing soil organic matter is a particularly important method of sequestering carbon in temperate parts of the world, such as the United States, where soils contain vastly more carbon than plants (both above- and belowground).[7] Such healthier soils can also require less fertilizer, which decreases nitrous oxide emissions.

We next describe farming methods to reduce nitrous oxide emissions. And finally, we examine practices rice producers can adopt to reduce methane emissions.

❑ *Expand agroforestry.* "Agroforestry" is a collective name for agricultural systems that integrate management with woody perennials and agricultural crops or animals on the same piece of land.[8] While agroforestry is included in the broader category of perennial systems, discussed below, we focus on it here separately due to its unique potential to rapidly reduce emissions in the United States through already well-established crops and practices. The trees can be those producing food such as fruit or nut trees, those providing other services such as windbreaks, or those grown for wood products. Since trees substantially increase above- and below-ground biomass, agroforestry increases both the rate of sequestration and the total amount of carbon that a piece of agricultural land can store relative to annual cropping systems

---

6. Emily Cassidy et al., *Redefining Agricultural Yields: From Tonnes to People Nourished Per Hectare*, 8 Env't Res. Letters 1, 3-4 (2013).
7. In tropical forests, however, soil and vegetation sequester approximately the same amount of carbon. This has important land use implications. For example, conventional agriculture in tropical regions is generally worse for the climate than conventional agriculture in temperate ones. For more information, see Intergovernmental Panel on Climate Change, Land Use, Land-Use Change, and Forestry (2000).
8. Food and Agriculture Organization of the United Nations, *Agroforestry*, http://www.fao.org/forestry/agroforestry/80338/en/ (last updated Oct. 23, 2015). The *United States Mid-Century Strategy for Deep Decarbonization* recognized agroforestry as a promising strategy for mitigation and adaptation. *See* The White House, United States Mid-Century Strategy for Deep Decarbonization 78-79 (2016), http://unfccc.int/files/focus/long-term_strategies/application/pdf/mid_century_strategy_report-final_red.pdf.

(although less than unconverted forest).[9] As a result, agroforestry's per-acre sequestration potential is far higher than that found in annual crop systems. Over time, agroforestry can also reduce indirect emissions of nitrous oxide by reducing nitrogen runoff.[10]

In the United States, existing agroforestry systems typically use trees and shrubs as windbreaks, buffers, and hedges on otherwise conventionally managed cropland; however, agroforestry can also include alley cropping—the side-by-side planting of annual crops with trees in adjacent rows—and silvopasture, discussed below, which incorporates trees in pastures. A 2012 literature review estimated that agroforestry systems, implemented nationwide, could sequester 530 million metric tons (MMT) of carbon a year—an amount equivalent to one-third of all fossil fuel emissions in the United States.[11] (Since a CO_2 molecule is about 3.7 times heavier than a carbon atom, increases in carbon sequestered will lead to about 3.7 times that amount of reduction of atmospheric CO_2.) Alley cropping and silvopasture alone could sequester more than 516 MMT of carbon annually.[12] In contrast to other practices focused exclusively on soil carbon, these large gains from agroforestry reflect both aboveground and belowground sequestration.

In addition to its enormous potential for carbon sequestration, agroforestry also reduces environmental harms, although its impacts depend heavily on the selection of tree species and their management.[13] A 2019 review found that agroforestry systems reduce surface runoff, soil erosion, organic carbon losses, and related nutrient losses by an average of 58%, 65%, 9%, and 50%, respectively compared to conventional practices. Agroforestry practices also lower herbicide, pesticide, and other pollutant losses by 49% on average,

---

9. In tropical climates, well-established agroforestry systems have even been shown to sequester more carbon than natural forests in upper soil layers in some circumstances. P.K. Ramachandran Nair et al., *Carbon Sequestration in Agroforestry Systems*, 108 ADVANCES AGRONOMY 237, 272 (2010).
10. The loss of nitrogen as nitrate can result in indirect emissions of nitrous oxide when the nitrate is deposited in downstream ecosystems and converted to nitrous oxide by soil bacteria. The U.S. Environmental Protection Agency (EPA) estimates that indirect emissions of nitrous oxide accounted for 18% of nitrous oxide emissions from agricultural soils in 2015. U.S. EPA, INVENTORY OF U.S. GREENHOUSE GAS EMISSIONS AND SINKS: 1990-2018, at 5-29 tbls.5-17, 5-18 (2020) (EPA 430-R-20-002). Over time, agroforestry practices like riparian tree buffers can prevent the loss of nitrate and thereby prevent its downstream conversion to nitrous oxide. Ranjith P. Udawatta et al., *Agroforestry Practices, Runoff, and Nutrient Loss: A Paired Watershed Comparison*, 31 J. ENV'T QUALITY 1214, 1224-25 (2002).
11. Ranjith P. Udawatta & Shibu Jose, *Agroforestry Strategies to Sequester Carbon in Temperate North America*, 86 AGROFORESTRY SYS. 225 (2012).
12. *See id.* at 239.
13. Shibu Jose, *Agroforestry for Ecosystem Services and Environmental Benefits: An Overview*, 76 AGROFOREST SYS. 1–10 (2009), https://doi.org/10.1007/s10457-009-9229-7; ENIKOE BIHARI, EARTHJUSTICE, AGROFORESTRY: THE BENEFITS AND CHALLENGES OF PUTTING TREES ON FARMS (2021) (providing an updated literature review of the ecological benefits of agroforestry (on file with authors)).

decreasing overall nutrient and chemical runoff.[14] Agroforestry practices can also increase yields.[15] Windbreaks improve air quality and protect plants from wind-related damage, thereby enhancing wildlife and insect habitat, although they may reduce light infiltration very close to the trees, slightly reducing yields.[16] Finally, riparian forest buffers are effective at protecting rivers and streams from bacterial contamination,[17] surface runoff, and pesticide drift.[18]

❑ *Shift from annual crops to perennial crops.* Unlike annual crops, which only survive a single harvest, perennial crops do not need to be replanted each year. As a result, they substantially improve upon the carbon storage potential of annual crops: they reduce or eliminate the need for tillage, generally reduce irrigation and fertilizer needs, and sequester additional carbon through their considerable biomass and deep root systems. A 2016 study assessed the potential for increased root production—one of the mechanisms by which perennials reduce net emissions—to increase soil carbon sequestration in the United States. The study found that increasing root mass on all U.S. cropland with appropriate soil types, which includes 87% of the country's cropland, would sequester an additional 107 to 800 MMT CO_2 equivalent (eq.) each year.[19]

Perennial crops include a wide range of plants that are capable of fulfilling important ecological and human needs, including woody crops (used in agroforestry systems), herbaceous oilseed crops, grains, and grasses. In the United States, farmers grow several common perennial crops, mostly in monocultures, including grapes, apples, blueberries, stone fruits, citrus, and almonds and other nuts. Perennials can also be a significant source of

---

14. Xiai Zhu et al., *Reductions in Water, Soil and Nutrient Losses, and Pesticide Pollution in Agroforestry Practices: A Review of Evidence and Processes*, 444 PLANT SOIL 33 (2019).
15. Jo Smith et al., *Reconciling Productivity With Protection of the Environment: Is Temperate Agroforestry the Answer?*, 20 RENEWABLE AGRIC. & FOOD SYS. 80, 81-82 (2013); Matthew H. Wilson & Sarah T. Lovell, *Agroforestry—The Next Step in Sustainable and Resilient Agriculture*, 8 SUSTAINABILITY 574, 580 (2016).
16. SWAN ET AL., *supra* note 3, at 25.
17. *See, e.g.*, Rob Collins & Kit Rutherford, *Modelling Bacterial Water Quality in Streams Draining Pastoral Land*, 38 WATER RES. 700, 710-11 (2004).
18. SWAN ET AL., *supra* note 3, at 27.
19. The amount sequestered would depend on a number of factors, including variations in nitrous oxide fluxes, fertilizer emissions, and root depth and mass. KEITH PAUSTIAN ET AL., ASSESSMENT OF POTENTIAL GREENHOUSE GAS MITIGATION FROM CHANGES TO CROP ROOT MASS AND ARCHITECTURE 2 (2016). At the low end, a 25% increase in root production with no downward shift in root length would sequester 107 MMT CO_2 eq. annually. *Id.* at 26 tbl.13. At the high end, a doubling of root production accompanied by an extreme downward shift in root length could sequester up to 800 MMT CO_2 eq. annually. *Id.*

vegetables,[20] livestock forage,[21] biofuel feedstock,[22] and carbohydrate, protein, or oil constituents in food products that are now largely produced by annual crops.[23] Walnut and pistachio trees, for example, yield more oil per acre than either soy or corn.[24] Consumers, in response to dietary restrictions and health benefits,[25] now show a stronger demand for nut crops.[26] Since they can be processed into flour or meal, some nut flours are being used to replace wheat flour or corn meal on a commercial scale. Other tree crops, such as the leguminous pods produced by mesquite trees, also offer potential for commercial food production.[27]

While current cultivars of perennial starch, oil, and protein crops have lower per-hectare caloric yields than soybean or corn at present, this is due at least in part to disparities in breeding and agronomic research.[28] Yields for five major crops in the United States—corn, soybean, wheat, cotton, and rice—quadrupled between 1950 and 2010 as public and private research dollars flowed to these crops.[29] Similar investments in perennial crops are likely to produce significant gains in yield.[30] While there are now no perennial grains ready for widespread commercial use in the United States, the Land

---

20. Eric Toensmeier et al., *Perennial Vegetables: A Neglected Resource for Biodiversity, Carbon Sequestration, and Nutrition*, 15 PLoS ONE 1-19 (2020).
21. *See, e.g.*, Land Stewardship Project, *Farm Transitions: Valuing Sustainable Practices—Perennial Forages and Grazing*, https://landstewardshipproject.org/farmtransitionsvaluingsustainablepracticesperennialforagesandgrazing (last visited Oct. 30, 2020); Bill Jokela & Michael Russelle, *Benefits of Perennial Forages for Soils, Crops, and Water Quality*, Progressive Forage, Mar. 2, 2010, https://www.progressiveforage.com/forage-types/other-forage/benefits-of-perennial-forages-for-soils-crops-and-water-quality.
22. Approximately 40% of the corn grown in the United States is now devoted to ethanol production. *See* Peter Riley, *Interaction Between Ethanol, Crop, and Livestock Markets*, in U.S. Ethanol: An Examination of Policy, Production, Use, Distribution, and Market Interactions 27 (James A. Duffield et al. eds., USDA 2015). Soybean processing can produce soy oil for biofuels and protein for animal feed at the same time, so little to no soy is grown exclusively as a biofuel; however, approximately 30% of the soybean oil produced in 2013 was used for biodiesel. Jeremy Martin, *Biodiesel Update: Now With More Soy*, Union Concerned Scientists, Jan. 2, 2014, http://blog.ucsusa.org/jeremy-martin/biodiesel-update-now-with-more-soy-360.
23. Eric Toensmeier, The Carbon Farming Solution 129-38 (Brianne Goodspeed & Laura Jorstad eds., 2016); Kevin Wolz et al., *Frontiers in Alley Cropping: Transformative Solutions for Temperate Agriculture*, 24 Global Change Biology 6 (2018).
24. Wolz et al., *supra* note 23.
25. Emilio Ros, *Health Benefits of Nut Consumption*, 2 Nutrients 652 (2010); Ravila Graziany Machado de Souza, *Nuts and Human Health Outcomes: A Systematic Review*, 9 Nutrients 1311 (2017).
26. The amount of tree nuts consumed by Americans increased by more than 2.5 times between 1970 and 2016. USDA Economic Research Service, *Almonds Lead Increase in Tree Nut Consumption*, https://www.ers.usda.gov/data-products/chart-gallery/gallery/chart-detail/?chartId=93152 (last updated May 31, 2019).
27. *See, e.g.*, Toensmeier, *supra* note 23, at 166-70.
28. Wolz et al., *supra* note 23.
29. *Id.* at 6; *see* Sun Ling Wang et al., USDA, Agricultural Productivity Growth in the United States: Measurement, Trends, and Drivers 25 fig.17 (Economic Research Report No. 189, 2015) (showing yield growth in common crops).
30. Wolz et al., *supra* note 23.

Institute, a nonprofit research organization dedicated to developing perennial staple crops, has been making promising progress.[31]

Scientists at Iowa State University have also developed a system of row crop production integrated with strategically placed native perennial grasses, called prairie strips, modeled on agroforestry practices. The project, Science-based Trials of Rowcrops Integrated With Prairie Strips (STRIPS), is designed to create a scalable, resilient, and environmentally responsible system of agriculture in the Midwest.[32] Further research is needed to accurately measure its impact on net emissions, but scientists estimate that prairie strips sequester approximately one metric ton CO_2 eq. per acre, about three times the emissions reduction benefit of no-till farming.[33]

The ecosystem benefits of using perennial crops are well established.[34] Perennial crops generally have deeper rooting levels, reducing erosion risk and allowing them to conserve water more effectively.[35] Their extensive root systems also absorb nutrients more efficiently, reducing fertilizer runoff.[36] Additionally, perennial crops require less fertilizer and herbicide since the soil on which they sit is exposed and disturbed much less frequently than in annual systems.[37] Integrating livestock or additional crops into perennial systems can increase biodiversity, improve natural pest control, raise yields, and increase system resilience.[38]

❑ *Implement diversified farming systems.* Input-intensive crop monocultures and industrial feedlots dominate agricultural production in the United States, in large part due to government support in the form of research, subsidies, and lax regulations. These systems produce prodigious amounts of calories, but have a number of public health, environmental, and com-

---

31. *See, e.g.*, Pheonah Nabukalu & Thomas Cox, *Response to Selection in the Initial Stages of a Perennial Sorghum Breeding Program*, 209 EUPHYTICA 103, 108-10 (2016); Marisa Lanker et al., *Farmer Perspectives and Experiences Introducing the Novel Perennial Grain Kernza Intermediate Wheatgrass in the US Midwest*, 35 RENEWABLE AGRIC. FOOD SYSTEMS 653-62 (2020).
32. MEGHANN JARCHOW & MATT LIEBMAN, IOWA STATE UNIVERSITY EXTENSION, INCORPORATING PRAIRIES INTO MULTIFUNCTIONAL LANDSCAPES 14-15 (2011) (PMR 1007).
33. *Id.* at 20-21.
34. *See* J.D. Glover et al., *Increased Food and Ecosystem Security Via Perennial Grains*, 328 SCIENCE 1638, 1638 (2010); Ben Werling et al., *Perennial Grasslands Enhance Biodiversity and Multiple Ecosystem Services in Bioenergy Landscapes*, 111 PROC. NAT'L ACAD. SCI. U.S. AM. 1652, 1654-55 (2014) (demonstrating the ecosystem and biodiversity benefits of perennial biofuel feedstocks over annual ones).
35. Glover et al., *supra* note 34.
36. *Id.*
37. *See id.*
38. *See* Brenda Lin, *Resilience in Agriculture Through Crop Diversification: Adaptive Management for Environmental Change*, 61 BIOSCIENCE 183, 183-87 (2011).

munity externalities.³⁹ They also rely on emissions-intensive practices and inputs.⁴⁰ In response to the negative impacts of industrialized agriculture, many researchers and farmers have advanced an alternative approach known as agroecological intensification that uses ecological principles to minimize inputs and maximize production and environmental services.⁴¹ One of the key concepts of agroecological intensification is functional biodiversity, which seeks to diversify crop and animal production to minimize risk, fertilize soil, conserve resources, and intensify production.⁴² Many of the most promising carbon farming practices, such as alley cropping and crop-livestock systems, effectively utilize functional biodiversity.

Diversification on farms can reduce emissions from soil management and increase carbon sequestration in soil and biomass.⁴³ As discussed above, functional biodiversity is modeled on ecological principles and, thus, also has a number of environmental co-benefits, including improved resilience, soil health, wildlife habitat, natural pest control, and pollinator health, among others.⁴⁴

❏ *Employ organic farming and other more climate-friendly farming systems.*⁴⁵ There are several agricultural systems, including organic agriculture, permaculture, agroecology (which includes practices such as crop rotations, integration, and diversification), and regenerative agriculture, built on the fundamental premise that soil health and natural ecological systems, such as the nutrient cycle between livestock and crops, are paramount to long-

---

39. INTERNATIONAL PANEL OF EXPERTS ON SUSTAINABLE FOOD SYSTEMS, FROM UNIFORMITY TO DIVERSITY 15-27 (2016); Marcia DeLonge & Andrea D. Basche, *Leveraging Agroecology for Solutions in Food, Energy, and Water*, 5 ELEMENTA: SCI. ANTHROPOCENE 2 (2017); Claire Kremen et al., *Diversified Farming Systems: An Agroecological, Systems-Based Alternative to Modern Industrial Agriculture*, 17 ECOLOGY SOC'Y 10-11 (2012).
40. INTERNATIONAL PANEL OF EXPERTS ON SUSTAINABLE FOOD SYSTEMS, *supra* note 39, at 19; Timothy E. Crews, *Is the Future of Agriculture Perennial? Imperatives and Opportunities to Reinvent Agriculture by Shifting From Annual Monocultures to Perennial Polycultures*, 1 GLOBAL SUSTAINABILITY 9 (2018).
41. A number of other terms are also used to describe agroecological intensification and other similar approaches. *See* Manuel González de Molina & Gloria I. Guzmán Casado, *Agroecology and Ecological Intensification. A Discussion From a Metabolic Point of View*, 9 SUSTAINABILITY 87-88 (2017).
42. Teja Tscharntke et al., GLOBAL FOOD SECURITY, *Biodiversity Conservation and the Future of Agricultural Intensification*, 151 BIOLOGICAL CONSERVATION 53, 54 (2012); Miguel A. Altieri, *The Ecological Role of Biodiversity in Agroecosystems*, AGRIC. ECOSYSTEMS ENV'T 19, 19-29 (1999).
43. INTERNATIONAL PANEL OF EXPERTS ON SUSTAINABLE FOOD SYSTEMS, *supra* note 39, at 34; Timothy E. Crews & Brian E. Rumsey, *What Agriculture Can Learn From Native Ecosystems in Building Soil Organic Matter: A Review*, 9 SUSTAINABILITY 578, 589 (2017); Kevin J. Wolz et al., *Frontiers in Alley Cropping: Transformative Solutions for Temperate Agriculture*, GLOBAL CHANGE BIOLOGY 883, 886 (2017).
44. Tscharntke et al., *supra* note 42, at 56-57; INTERNATIONAL PANEL OF EXPERTS ON SUSTAINABLE FOOD SYSTEMS, *supra* note 39, at 34-36; Altieri, *supra* note 42.
45. The discussion here of organic and other climate-friendly farming systems also applies to animal agriculture. It is not repeated below.

Chapter IV. Climate-Friendly Agricultural Systems and Practices

term productivity. This subsection focuses on organic agriculture, since it is well-studied and the U.S. Department of Agriculture (USDA) already has national organic standards,[46] making it easier to classify. However, certified organic operations are not necessarily more climate-friendly than noncertified operations implementing these other models—all can have significant climate benefits.

Organic farming methods enhance production by supporting natural soil fertility and biological activity, and prohibiting the use of synthetic pesticides or fertilizers.[47] USDA, which sets standards for organic products in the United States, defines organic farming as a form of agriculture that uses methods designed to "support the cycling of on-farm resources, promote ecological balance, and conserve biodiversity."[48] Organic agriculture encourages many of the practices mentioned here, such as cover cropping, crop rotation, and the incorporation of diverse elements on cropland, including forestry and livestock. Its primary climate benefits are reduced nitrous oxide emissions, lower energy requirements, and increased soil carbon sequestration.[49] Some studies suggest that organic farming can obtain equivalent yields to conventional farming,[50] or come close in certain contexts,[51] while others suggest that the lower per-acre yields would reduce the climate benefits of the system by requiring more cropland.[52] Most productivity research to date has focused on conventional systems; increased research into organic farming would likely narrow or close any productivity gap.

Organic agriculture offers a wide range of environmental and social benefits in addition to its potential to reduce net agricultural emissions. As the

---

46. *See, e.g.*, 7 C.F.R. §205.203 (2016) (establishing the soil fertility and crop nutrient management standard).
47. Certified organic products in the United States, for example, must be "produced and handled without the use of synthetic chemicals." 7 U.S.C. §6504.
48. USDA, Introduction to Organic Practices (2015).
49. Tiziano Gomiero et al., *Environmental Impact of Different Agricultural Management Practices: Conventional vs. Organic Agriculture*, 30 Critical Revs. Plant Sci. 95, 101-04, 109-11 (2011) (summarizing research indicating that organic farming increases soil carbon levels and reduces energy requirements); Søren Petersen et al., *Nitrous Oxide Emissions From Organic and Conventional Crops in Five European Countries*, 112 Agric. Ecosystems & Env't 200, 203 (2006) (finding that nitrous oxide emissions from conventional crop rotations were higher than those in organic crop rotations in four out of five countries). *Contra* Hanna Tuomisto, *Does Organic Farming Reduce Environmental Impacts? A Meta-Analysis of European Research*, 15 J. Env't Mgmt. 309, 313 (2015) (concluding that nitrous oxide emissions are 31% lower in organic systems per unit of field area, but 8% higher per unit of product).
50. Rodale Institute, The Farming Systems Trial: Celebrating 30 Years 4, 9-10 (2012).
51. Verena Seufert et al., *Comparing the Yields of Organic and Conventional Agriculture*, 485 Nature 229, 231 (2012) (demonstrating that organic agriculture nearly matches conventional yields in certain environments); Lauren Ponisio et al., *Diversification Practices Reduce Organic to Conventional Yield Gap*, 282 Proc. Royal Soc'y B 1, 4 (2014) (finding that diversified organic systems were much closer to conventional yields than organic monocultures).
52. *See* Gomiero et al., *supra* note 49, at 111.

organic industry has grown, so too has the number of industrial-scale, capital-intensive organic operations, dampening these benefits.[53] Nonetheless, research consistently indicates that organic agriculture increases soil stability and fertility, on-farm biodiversity, and crop resilience to weather shocks, while reducing energy use (e.g., by reducing tractor usage) and the need for synthetic inputs.[54] Organic farms can also directly benefit people, especially in rural communities, who can enjoy better landscape preservation, less agricultural pollution,[55] reduced dietary exposure to pesticides,[56] and, according to some researchers, greater civic engagement.[57]

❏ *Shift to more ecologically efficient crop use.* Analyses of agricultural productivity generally focus on inputs, including labor, and crop yield. While these factors are important, they fail to provide an accurate account of whether a crop is a truly efficient use of land and energy from the perspective of meeting human needs. A crop with high yields and low labor requirements may be inefficient if it is integrated into an energy-intensive value chain, such as grain destined for a feedlot, or if it does not provide consumers with a nutritious end product, such as corn processed into high-fructose corn syrup. If the food system produced less animal feed, less feedstock for biofuels, and fewer processed foods—while producing more crops intended for human consumption as whole foods—it would be dramatically more efficient.

A 2013 study estimated that 67% of the calories and 80% of the protein in crops produced in the United States are diverted to animal feed.[58] This is an inefficient use of potential food. For example, approximately six pounds of grain are used for each pound of beef produced over the life-span of a cow.[59]

---

53. *See, e.g.*, Julie Guthman, Agrarian Dreams 1-22 (2004) (arguing that the organic industry has "replicated what it set out to oppose").
54. *See* Gomiero et al., *supra* note 49, at 100-13.
55. *Id.* at 106-08, 114.
56. Brian Baker et al., *Pesticide Residues in Conventional, IPM-Grown, and Organic Foods: Insights From Three U.S. Data Sets*, 19 Food Additives & Contaminants 427-46 (2002); Chengsheng Lu et al., *Organic Diets Significantly Lower Dietary Exposure to Organophosphorous Pesticides*, 114 Env't Health Persp. 260-63 (2006).
57. Jessica Goldberger, *Conventionalization, Civic Engagement, and the Sustainability of Organic Agriculture*, 27 J. Rural Stud. 288, 295 (2011).
58. Cassidy et al., *supra* note 6.
59. A total of six pounds of grain to one pound of beef was derived by dividing the number of pounds of grain during the finishing stage to one pound of live weight gain (2.4/0.4 = 6). Grain byproducts account for an increasing share of cattle feed. *Id.* at 7. Since byproducts are generally not fit for human consumption, it is sometimes argued that their contribution should be excluded when estimating the extent to which cattle feed displaces human-edible crops. This is misleading. Byproducts from the production of corn ethanol are the main source of industrial byproducts in cattle feed and, as discussed further below, corn raised for ethanol production displaces crops intended for human consumption. The use of corn ethanol byproducts in animal feed contributes to the profitability of the corn ethanol industry, effectively subsidizing this inefficient use of agricultural land and resources.

Chapter IV. Climate-Friendly Agricultural Systems and Practices 75

In the United States, approximately 70 million acres of cropland are used to produce corn and soybean for animal feed.[60] The same calories and protein now provided by animal products could be produced with a much smaller land footprint if crops were consumed directly by humans rather than fed to animals. Below, we discuss the concept of ecological leftovers, which would limit livestock production to grassland and food byproducts unsuited for human consumption, dramatically increasing the ecological efficiency of animal products.

The enactment of the renewable fuel standard (RFS) in 2005, and its subsequent strengthening in 2007 (RFS2), massively increased demand for biofuels in the United States.[61] Roughly 37% of the corn sold in the United State has been used to make ethanol since RFS2 went into effect in 2008.[62] While the biofuels mandate was justified (in part) as a way to reduce greenhouse gas emissions, studies examining the climate change impacts of corn ethanol show a wide range of results, reflecting the challenge of determining with precision or confidence the full life cycle impact. Most studies find a modest annual operational greenhouse gas benefit over gasoline, but they generally either ignore or minimize the impact of using land for fuel or the continuing impact of the prior conversion to cropland (referred to in Chapter III as the carbon opportunity cost).[63] However, other studies indicate that land use change impacts of current U.S. biofuels, and thus the carbon debt payback time from conversion of land to biomass feedstock production, are much more significant than conventional models indicate, resulting in a net harm to the climate.[64] Between March 2008 and May 2020, 75% of all corn sold was

---

60. Estimates of acres cultivated for corn and soybean used for animal feed in the United States were derived by multiplying total corn and soybean acreage in marketing year 2014/2015 (90.6 and 83.3 million acres planted, respectively) by the proportion of the corn supply used for animal feed (0.34) or the proportion of the soybean supply crushed (0.46), and multiplying this product by the proportion of the corn and soybean supply due to production in that year (0.92 and 0.97, respectively). For corn data, see USDA, FEED GRAINS: YEARBOOK TABLES (June 14, 2017), and for soybean, see USDA, OIL CROPS YEARBOOK (Mar. 29, 2017).
61. DAVID DEGENNARO, NAT'L WILDLIFE FED'N, FUELING DESTRUCTION: THE UNINTENDED CONSEQUENCES OF THE RENEWABLE FUEL STANDARD ON LAND, WATER, AND WILDLIFE 5-6 (2016).
62. Calculated by the authors. USDA ERS, *U.S. Bioenergy Statistics*, https://www.ers.usda.gov/data-products/us-bioenergy-statistics/ (last updated Oct. 27, 2020).
63. See e.g., Melissa Scully et al., *Carbon Intensity of Corn Ethanol in the United States: State of the Science*, 16 ENV'T RES. LETTERS 043001 (2021), https://iopscience.iop.org/article/10.1088/1748-9326/abde08 (finding "land use change a minor contributor" and corn ethanol to have a 46% lower greenhouse gas impact than gasoline); Jan Lewandrowski et al., *The Greenhouse Gas Benefits of Corn Ethanol—Assessing Recent Evidence*, 11 BIOFUELS (2020), https://doi.org/10.1080/17597269.2018.1546488 (finding the large increase in corn acreage for ethanol to cause a "relatively small increase in aggregate agricultural land" and thus finding corn ethanol's current greenhouse gas profile to be 39-43% lower than gasoline).
64. See Seth Spawn-Lee et al., *Comment on "Carbon Intensity of Corn Ethanol in the United States: State of the Science,"* EcoEvoRxiv Preprint (2021), https://ecoevorxiv.org/cxhz5/ (arguing that recent studies

used as animal feed or biofuel, a tremendously inefficient use of resources.[65] Moreover, a U.S. Environmental Protection Agency (EPA) study found that between four and almost eight million acres of grasslands were converted to cropland since RFS2, with nearly 20 million more acres being devoted to corn and soybean for fuel, so that now about 40% of the corn crop is used for ethanol production.

In addition, consumers now eat substantial amounts of processed and "ultra-processed" foods;[66] an estimated 75% of the average person's calories comes from such food.[67] In most wealthy countries, like the United States, most people have "commodity-based diets," where they eat heavily processed foods made largely from corn, wheat, and soy as well as some animal products.[68] These diets are deficient in nutrients and other beneficial compounds found in whole or minimally processed foods[69] and are associated with a higher risk of cancer.[70] It may not be feasible for farms to produce an adequate supply of nutritious foods if they do not reduce production

"grossly underestimate the [carbon] intensity of corn-grain ethanol"); Michael Abraha et al., *Carbon Debt of Field-Scale Conservation Reserve Program Grasslands Converted to Annual and Perennial Bioenergy Crops*, 14 ENV'T RES. LETTERS 024019 (2019), https://iopscience.iop.org/article/10.1088/1748-9326/aafc10 (finding the carbon debt payback term for no-till corn to be over 300 years); Ilya Gelfand et al., *Carbon Debt of Conservation Reserve Program (CRP) Grasslands Converted to Bioenergy Production*, 108(33) PNAS 13864 (2011), https://doi.org/10.1073/pnas.1017277108 (carbon debt payback of 29-40 years for no-till corn and 89-123 year for tilled corn). Studies that take land use change more fully into account find that the climate impact of corn ethanol is much great than that of gasoline. *See, e.g.*, Timothy Searchinger et al., *Assessing the Efficiency of Changes in Land Use for Mitigating Climate Change*, 564 NATURE 249 (Dec. 2018), https://doi.org/10.1038/s41586-018-0757-z. *See also* J.M. DeCicco et al., *Opinion: Reconsidering Bioenergy Given the Urgency of Climate Protection*, 39 PROC. NAT'L ACAD. SCI., 9642-45 (2018), https://doi.org/10.1073/pnas.1814120115; J.M. DeCicco et al., *Carbon Balance Effects of U.S. Biofuel Production and Use*, 138 CLIMATIC CHANGE 667-80 (2016), https://doi.org/10.1007/s10584-016-1764-4.

65. Calculated by the authors. USDA ERS, *supra* note 62.
66. The term was popularized by Carlos Monteiro, who argues, "The issue is not foods, nor nutrients, so much as processing." Carlos Monteiro, Commentary, *Increasing Consumption of Ultra-Processed Foods and Likely Impact on Human Health: Evidence From Brazil*, 12 PUB. HEALTH NUTRITION 729, 729 (2009). In a subsequent study, Monteiro and his collaborators divided food products into three groups: unprocessed or minimally processed, processed, and ultra-processed. Carlos Monteiro et al., *Increasing Consumption of Ultra-Processed Foods and Likely Impact on Human Health: Evidence From Brazil*, 14 PUB. HEALTH NUTRITION 5, 7 (2010). Ultra-processed foods are produced using industrial processes "designed to create durable, accessible, convenient, attractive ready-to-eat or ready-to-heat products." *Id.* Additionally, "they are formulated to reduce microbial deterioration ('long shelf-life'), to be transportable for long distances, to be extremely palatable ('high organoleptic quality') and often to be habit forming." *Id.* For a list of the industrial processes used in the production of ultra-processed foods, see *id.* at 7-8.
67. Jennifer Poti et al., *Is the Degree of Food Processing and Convenience Linked With the Nutritional Quality of Foods Purchased by US Households*, 101 AM. J. CLINICAL NUTRITION 1251, 1251 (2015).
68. David Ludwig, Commentary, *Technology, Diet, and the Burden of Chronic Disease*, 305 JAMA 1352, 1352 (2011).
69. *Id.*
70. Thibault Fiolet et al., *Consumption of Ultra-Processed Foods and Cancer Risk: Results From NutriNet-Santé Prospective Cohort*, 360 BMJ k322-k330 (2018).

of commodities used in processed foods. Research suggests that higher atmospheric levels of CO_2 will decrease the protein and mineral content of common staples such as wheat, rice, and soybeans, further increasing the need for a more ecologically efficient and nutritious food supply chain.[71] Shifting away from such high reliance on heavily processed foods would reduce inefficiencies in the food system, produce healthier food, and help mitigate climate change.[72]

❑ *Reduce or eliminate tillage.* To prepare for planting, farmers routinely till their land by plowing or otherwise breaking up the soil, and eliminating unwanted material. This process accelerates the breakdown of organic matter in the soil, increasing emissions of CO_2. Thus, researchers and farmers who want to reduce emissions are examining ways to prepare soil for planting with no, or reduced, tillage. No-till agriculture, which completely eliminates tillage, uses herbicides or other methods to control weeds instead of tillage, and leaves the soil physically undisturbed, protecting organic matter from soil microbes that could otherwise accelerate the carbon cycle by returning soil carbon to the atmosphere as CO_2.[73] Reduced tillage practices that integrate some amount of plant residue into soils can also reduce nitrous oxide emissions and further increase carbon sequestration in some circumstances.[74]

---

71. Samuel S. Myers et al., Letter, *Increasing CO_2 Threatens Human Nutrition*, 510 NATURE 139-42 (2014); Irakli Loladze, *Hidden Shift of the Ionome of Plants Exposed to Elevated CO_2 Depletes Minerals at the Base of Human Nutrition*, 3 ELIFE e02245 (2014). Climate change will also continue to negatively affect fruit and vegetable production. *See, e.g.*, Tapan B. Pathak et al., *Climate Change Trends and Impacts on California Agriculture: A Detailed Review*, 8 AGRONOMY 25 (2018). A 2016 Lancet study found that climate change is likely to decrease fruit and vegetable consumption in the United States as a result, leading to the premature deaths of millions. Marco Springmann et al., *Global and Regional Health Effects of Future Food Production Under Climate Change: A Modelling Study*, 387 LANCET 1937, 1942 (2016).
72. *See* Carlos Monteiro et al., *Dietary Guidelines to Nourish Humanity and the Planet in the Twenty-First Century. A Blueprint From Brazil*, 18 PUB. HEALTH NUTRITION 2311, 2317 (2015) (describing how dietary guidelines can enhance both human health and the environment by reducing the consumption of processed foods); Dariush Mozaffarian & David Ludwig, Commentary, *Dietary Guidelines in the 21st Century—A Time for Food*, 304 JAMA 681, 681-82 (2010) (emphasizing the importance of whole and minimally processed foods for human health); K.R. Siegel et al., *Association of Higher Consumption of Foods Derived From Subsidized Commodities With Adverse Cardiometabolic Risk Among US Adults*, 176 JAMA INTERNAL MED. 1124, 1124 (2016) (showing an association between consumption of subsidized food commodities and higher cardiometabolic risks). The Scientific Report of the 2015 Dietary Guidelines Advisory Committee also noted that diets with lower levels of animal products were associated with healthier outcomes and generally resulted in reduced greenhouse gas emissions. *See* DIETARY GUIDELINES ADVISORY COMMITTEE, SCIENTIFIC REPORT OF THE 2015 DIETARY GUIDELINES ADVISORY COMMITTEE pt. D ch. 5 (2015).
73. For an overview of this process, see DANIEL KANE, CARBON SEQUESTRATION POTENTIAL ON AGRICULTURAL LANDS: A REVIEW OF CURRENT SCIENCE AND AVAILABLE PRACTICES 5-11 (2015).
74. Cheryl Palm et al., *Conservation Agriculture and Ecosystem Services: An Overview*, 187 AGRIC. ECOSYSTEMS & ENV'T 87, 90 (2014).

No-till agriculture began to grow steadily in the United States after inexpensive herbicides and specialized equipment became widely available in the 1970s.[75] A 1972 survey of USDA conservationists found that there were an estimated 3.3 million acres of no-till cropland.[76] By 2017, farmers reported practicing no-till on 104 million acres and reduced tillage on another 98 million acres.[77] In contrast, intensive tillage was practiced on 80 million acres—down to 28% of the 282 million acres suitable for tilling according to the 2017 Census of Agriculture.[78] While no-till's impact on crop yields varies according to a number of factors, including soil conditions, management techniques, weather, and crop type, a 2016 meta-analysis found that no-till generally results in similar yields to conventional tillage after a transition period of five or more years.[79] Even with yield reductions during the transition phase, however, no-till may remain more profitable for farmers than conventional tillage due to its potential to reduce expenditures on labor, fuel, and, in some cases, fertilizer.[80] Although farmers initially adopted no-till to reduce their heavy reliance on tractors—and thus reduce costs—and to limit soil erosion by reducing the amount of bare soil, USDA,[81] industry groups,[82] and some scientists now promote no-till as a way to sequester carbon. Indeed, conservation tillage, which includes no-till farming and some methods of reduced tillage, is among the most widely studied agricultural practices with respect to climate change.

Despite this attention, however, there are questions about the potential of no-till practices to mitigate greenhouse gas emissions.[83] A 2007 review

75. David R. Huggins & John P. Reganold, *No-Till: The Quiet Revolution*, Sci. Am., July 2008, at 71, 73; Rattan Lal, Editorial, *Evolution of the Plow Over 10,000 Years and the Rationale for No-Till Farming*, 93 Soil & Tillage Res. 1, 6-7 (2007).
76. Frank Lessiter, *From 3.3 to 96.4 Million Acres*, No-Till Farmer, July 1, 2014, https://www.no-tillfarmer.com/articles/3918-from-33-to-964-million-acres?v=preview.
77. National Agricultural Statistics Service, USDA, 2017 Census of Agriculture, U.S. National Level Data 58 tbl.47 (2019) [hereinafter 2017 Census of Agriculture].
78. *Id.*
79. Unlike other crops, however, corn yields on no-till farms typically do not improve over time, resulting in lower yields than corn produced with conventional tillage. Cameron M. Pittelkow et al., *When Does No-Till Work? A Global Meta-Analysis*, 183 Field Crops Res. 156, 159 (2015).
80. Erica Goode, *Farmers Put Down Plow for More Productive Soil*, N.Y. Times, Mar. 9, 2015, at D1; Claire O'Connor, *Farmers Reap Benefits as No-Till Adoption Rises*, Nat. Resources Def. Council, Nov. 15, 2013, https://www.nrdc.org/experts/claire-oconnor/farmers-reap-benefits-no-till-adoption-rises.
81. USDA aims to increase no-till farming by 33 million acres as part of its goal to increase carbon sequestration by 120 MMT CO_2 eq. annually by 2025. USDA, Factsheet: USDA's Building Blocks for Climate Smart Agriculture and Forestry (2015).
82. *See, e.g.*, Bayer, *Farm Solutions to Address a Changing Climate*, https://www.cropscience.bayer.com/people-planet/climate-change/a/soil-below-and-satellites-above (last visited Jan. 11, 2021).
83. *See, e.g.*, A.J. VandenBygaart, Commentary, *The Myth That No-Till Can Mitigate Global Climate Change*, 216 Agric. Ecosystems & Env't 98 (2016); David S. Powlson et al., *Perspective Limited Potential of No-Till Agriculture for Climate Change Mitigation*, 4 Nature Climate Change 678 (2014). Contra Henry Neufeldt et al., Correspondence, *No-Till Agriculture and Climate Change*, 5 Nature Climate

noted flaws in how some of the more favorable studies measured soil organic carbon,[84] while a 2015 study found that increased earthworm activity can negate any carbon sequestration effects from no-till, at least in the short term.[85] Nonetheless, the evidence suggests that no-till agriculture can increase soil carbon stocks in many regions, although its effect varies considerably by soil type and location.[86] A 2013 meta-analysis also found that no-till significantly decreases nitrous oxide emissions after five years, especially in dry climates.[87]

Researchers have also expressed concerns that no-till farming as practiced by commercial farmers often differs considerably from how it is implemented on research fields.[88] The available data suggest that many farmers who consider their methods "no-till" actually till their fields periodically.[89] Since even a single tillage event can release carbon built up over years of no-tillage, these periodic tillings means many "no-till" farms sequester far less carbon than a naïve analysis suggests.[90] One expert estimates that less than a third of no-till farms in the United States are truly no-till, and that the num-

---

CHANGE 488 (2015) (responding to Powlson et al.'s argument that no-till's potential to mitigate climate change is overstated).

84. John M. Baker et al., Commentary, *Tillage and Soil Carbon Sequestration—What Do We Really Know?*, 118 AGRIC. ECOSYSTEMS & ENV'T 1, 2-3 (2007). *But see* A.N. Kravechenko & G.P. Robertson, *Whole-Profile Carbon Stocks: The Danger of Assuming Too Much From Analyses of Too Little*, 75 SOIL & WATER MGMT. & CONSERVATION 235, 240 (2011) (arguing that Baker et al. and similar analyses do not properly analyze carbon stock differences as a function of depth).

85. Ingrid Lubbers et al., *Reduced Greenhouse Gas Mitigation Potential of No-Tillage Soils Through Earthworm Activity*, SCI. REP., Sept. 2015, at 1.

86. Keith Paustian, *Carbon Sequestration in Agricultural Systems*, in ENCYCLOPEDIA OF AGRICULTURE AND FOOD SYSTEMS 140, 146 (Neal K. Van Alfen ed., Academic Press 2014).

87. Chris van Kessel, *Climate, Duration, and N Placement Determine N_2O Emissions in Reduced Tillage Systems: A Meta Analysis*, 19 GLOBAL CHANGE BIOLOGY 33, 33 (2013). *But see* Claudio Stöckle et al., *Carbon Storage and Nitrous Oxide Emissions of Cropping Systems in Eastern Washington: A Simulation Study*, 67 J. SOIL & WATER CONSERVATION 365, 376 (2012) (finding that increases in nitrous oxide offset gains in soil carbon sequestration at no-till sites in eastern Washington).

88. Bram Govaerts et al., *Conservation Agriculture and Soil Carbon Sequestration: Between Myth and Farmer Reality*, 28 CRITICAL REVS. PLANT SCI. 97, 111 (2009).

89. An extensive survey conducted from 1994-1999 found that no-till farms in Indiana and Illinois tilled their fields every 2.5 years on average, while no-till farms in Minnesota were tilled every 1.4 years on average. Peter R. Hill, *Use of Continuous No-Till and Rotational Tillage Systems in the Central and Northern Corn Belt*, 56 J. SOIL & WATER CONSERVATION 286, 289 (2001). Anecdotally, periodic tillage remains common on no-till farms throughout the United States. The writer and sustainable farmer Gene Logsdon, for example, wrote in 2011 that "[a]lmost all farmers, in my neck of the woods anyway, are finding it necessary to do quite a bit of soil tillage but because they use a 'no-till' planter, [the USDA NRCS] allows them to act out the farce of saying they are practicing no tillage." Gene Logsdon, *No Till Farming Not So Great After All*, CONTRARY FARMER, Dec. 28, 2011, https://thecontraryfarmer.wordpress.com/2011/12/28/no-till-farming-not-so-great-after-all/. *See also* TARA WADE ET AL., USDA, CONSERVATION-PRACTICE ADOPTION RATES VARY WIDELY BY CROP AND REGION 3 (2015) (EIB-147) (describing why some no-till farmers periodically till their fields).

90. Richard Conant et al., *Impacts of Periodic Tillage on Soil C Stocks: A Synthesis*, 95 SOIL & TILLAGE RES. 1, 4 (2007).

ber of these continuous no-till farms is likely decreasing.[91] Since estimates of carbon sequestration from no-tillage often assume continuous no-till,[92] the aggregate climate benefits of no-till agriculture as currently practiced are often overestimated.

While most no-till systems rely on herbicides to eliminate weeds, organic no-till systems could offer significantly higher levels of carbon sequestration. The Rodale Institute has developed a mechanical mounted roller that knocks down and kills cover crops, suppressing weed growth without herbicides.[93] Short-term studies of organic no-till systems indicate that they likely sequester more carbon than conventional no-till farming.[94] Although the Rodale Institute's field results are promising, it is still conducting trials, and commercial farmers have yet to adopt organic no-till.[95]

Given the uncertainties of the climate benefits of no-till as currently practiced, it may not deserve the attention it is getting as a strategy to fight climate change. Yet its many other well-documented environmental benefits suggest that researchers should continue to study, refine, and integrate it with other climate-friendly practices to optimize its climate impact. By leaving more plant residue and organic matter in and on the soil, it can improve soil quality, reduce erosion, provide food and cover for wildlife, and reduce dust and diesel pollution from tillage.[96] However, conservation tillage as commonly practiced in the United States often requires higher levels of herbicides than conventional tillage systems.[97]

❑ *Increase carbon inputs from plants through cover crops and crop rotations.* Farmers can also foster soil carbon by increasing carbon inputs from plants. Cover cropping and conservation crop rotations are among the most common practices designed to do this in annual crop systems. Cover crops are plants grown to enhance soil conditions rather than to produce an agricul-

---

91. Brad Reagan, *Plowing Through the Confusing Data on No-Till Farming*, WALL ST. J., Oct. 15, 2012, https://www.wsj.com/articles/SB10000872396390443855804577602931348705646.
92. *See* VandenBygaart, *supra* note 83, at 99.
93. Rodale Institute, *Organic No-Till*, http://rodaleinstitute.org/our-work/organic-no-till/ (last visited Oct. 30, 2020).
94. Patrick Carr et al., *Impacts of Organic Zero Tillage Systems on Crops, Weeds, and Soil Quality*, 5 SUSTAINABILITY 3172, 3184 (2013).
95. TOENSMEIER, *supra* note 23, at 69. Other researchers and practitioners are also working to develop functional and productive no-till systems.
96. SWAN ET AL., *supra* note 3, at 4-5. Conservation tillage can increase the number of small mammals in fields, resulting in crop loss; however, such damage is generally controllable. USDA, NRCS, *Conservation Tillage Systems and Wildlife*, FISH & WILDLIFE LITERATURE REV. SUMMARY, Sept. 1999, at https://www.nrcs.usda.gov/Internet/FSE_DOCUMENTS/nrcs143_022212.pdf.
97. Lionel Alletto et al., *Tillage Management Effects on Pesticide Fate in Soils: A Review*, 30 AGRONOMY SUSTAINABLE DEV. 367, 369 (2010).

tural product. Farmers generally grow them during the late fall and winter when common commodity crops such as corn, wheat, and soy are not in season. In addition to increasing soil organic carbon by increasing carbon inputs, cover crops have also been shown to significantly reduce nitrate loss, thereby indirectly reducing nitrous oxide emissions.[98] Cover cropping with legumes also increases biological nitrogen fixation, reducing the need for nitrogen fertilizers.[99]

Conservation crop rotations refer to planting systems designed to decrease the frequency at which fields are left fallow and to rotate between a diverse set of crops, thereby increasing carbon inputs.[100] Crop rotations that include perennial plants, such as alfalfa or grass hay, can be especially effective at sequestering carbon.[101] Iowa State University researchers have shown that three- and four-year rotations that include alfalfa increase yields and require less fertilizer and herbicides.[102] While farmers rotate most crops on a seasonal basis, producers with perennial crops in their rotation do not need to return to annual crops for one to three years.[103]

Although neither of these methods offers transformative climate benefits when practiced in isolation, both can play an important role in reducing net agricultural emissions when integrated into climate-friendly systems. Diversified crop rotations, for example, are even more effective at increasing soil carbon when combined with cover cropping,[104] although likely sequestration rates have not been established.[105] Cover cropping has also been shown to sequester carbon more quickly when used in conjunction with no-till agriculture and it likely complements other environmentally friendly practices as well.[106] As cover crops also use water, farmers who grow them may need

---

98. Andrea Basche et al., *Do Cover Crops Increase or Decrease Nitrous Oxide Emissions? A Meta-Analysis*, 69 J. SOIL & WATER CONSERVATION 471, 479-80 (2014).
99. *See* Seth M. Dabney et al., *Using Winter Cover Crops to Improve Soil and Water Quality*, 32 COMM. SOIL SCI. & PLANT ANALYSIS 1221, 1224, 1228 (2001).
100. Increasing crop diversity influences soil carbon and nitrogen concentrations, microbial communities, and soil ecosystem functions, often resulting in higher soil carbon levels. Marshall D. McDaniel et al., *Does Agricultural Crop Diversity Enhance Soil Microbial Biomass and Organic Matter Dynamics? A Meta-Analysis*, 24 ECOLOGICAL APPLICATIONS 560, 560 (2014).
101. ALISON J. EAGLE ET AL., NICHOLAS INSTITUTE FOR ENVIRONMENTAL POLICY SOLUTIONS, GREENHOUSE GAS MITIGATION POTENTIAL OF AGRICULTURAL LAND MANAGEMENT IN THE UNITED STATES: A SYNTHESIS OF THE LITERATURE 15 (2012). Perennial grasses grown for livestock may not be appropriate for water-scarce regions.
102. UNION OF CONCERNED SCIENTISTS, ROTATING CROPS, TURNING PROFITS 3-4 (2017).
103. EAGLE ET AL., *supra* note 101.
104. *See* McDaniel et al., *supra* note 100, at 560.
105. Telephone Interview with Amy Swan, Research Associate, Colorado State University, and Mark Easter, Senior Research Associate, Colorado State University (May 20, 2016).
106. *See* Humberto Blanco-Canqui, *Cover Crops and Ecosystem Services: Insights From Studies in Temperate Soils*, 107 AGRONOMY J. 2449, 2450 (2015); *see generally* Rattan Lal, *A System Approach to Conservation Agriculture*, 70 J. SOIL & WATER CONSERVATION 82A, 82A (2015) (arguing that basic components

more water for the cash crop. However, since cover crops reduce evaporation, they may also conserve water—the best practices for cover cropping will depend on the region. Cover crops that can be grazed or be marketed can also increase operational profitability.

Practices that increase carbon inputs from plants also offer a range of ecosystem benefits. Both cover cropping and diversified crop rotations have been shown to improve soil health, nutrient cycling, pest regulation, and crop productivity,[107] while reducing herbicide and fertilizer use.[108]

❑ *Add soil amendments.* Farmers who apply amendments such as manure or other organic fertilizers to their soil can lower emissions by decreasing manure waste, reducing emissions from the production of synthetic fertilizers,[109] and increasing soil carbon stocks.[110] While livestock manure remains the dominant source of organic fertilizer for agriculture, the United States has large amounts of compostable solid waste and solid residues from sewage treatment plants, called biosolids, which also can be, and often already are, used as soil amendments.[111] According to the *Guardian*, 60% of sewage sludge produced by treatment plants was sold as fertilizer in 2019.[112] Nonetheless, these biosolids are insufficiently regulated and pose serious health risks to farmworkers, local residents, and consumers.[113] In 2018, the EPA Office of Inspector General released a report on biosolids, warning that it had identified more than 350 pollutants, including 61 acutely hazardous, hazardous, or priority pollutants in biosolids samples. The report concluded that EPA "lacked the

---

of conservation agriculture, including cover cropping, must be implemented together in order to maximize their benefits).
107. *See* Meagan Schipanski et al., *A Framework for Evaluating Ecosystem Services Provided by Cover Crops in Agroecosystems*, 125 AGRIC. SYSTEMS 12, 13 (2014); UNION OF CONCERNED SCIENTISTS, *supra* note 102, at 5; Riccardo Bommarco et al., *Ecological Intensification: Harnessing Ecosystem Services for Food Security*, 28 TRENDS ECOLOGY & EVOLUTION 230, 233-34, 236 (2013).
108. *See, e.g.*, UNION OF CONCERNED SCIENTISTS, *supra* note 102; Raphaël A. Wittwer et al., *Cover Crops Support Ecological Intensification of Arable Cropping Systems*, SCI. REP., Feb. 3, 2017, at 1.
109. *See infra* Chapter VIII.A.1, for a discussion of upstream emissions from synthetic fertilizers.
110. *See, e.g.*, Maysoon M. Mikha & Charles W. Rice, *Tillage and Manure Effects on Soil and Aggregate-associated Carbon and Nitrogen*, 68 SOIL SCI. SOC'Y AM. J. 809, 809, 815 (2004) (discussing manure's impact on soil carbon content).
111. Half of all biosolids produced in the United States are applied to agricultural land, although this accounts for the nutrient needs of less than 1% of the country's farmland. EPA, *Frequent Questions About Biosolids*, https://web.archive.org/web/20200122134336/https://www.epa.gov/biosolids/frequent-questions-about-biosolids (last updated Feb. 22, 2018).
112. Tom Perkins, *Biosolids: Mix Human Waste With Toxic Chemicals, Then Spread on Crops*, GUARDIAN, Oct. 5, 2019, https://www.theguardian.com/environment/2019/oct/05/biosolids-toxic-chemicals-pollution.
113. *See, e.g., id.*; Amy Lowman et al., *Land Application of Treated Sewage Sludge: Community Health and Environmental Justice*, 121 ENV'T HEALTH PERSP. 537 (2013), *available at* https://ehp.niehs.nih.gov/doi/10.1289/ehp.1205470; Sharon Lerner, *Toxic PFAS Chemicals Found in Maine Farms Fertilized With Sewage Sludge*, INTERCEPT, June 7, 2019, https://theintercept.com/2019/06/07/pfas-chemicals-maine-sludge/.

Chapter IV. Climate-Friendly Agricultural Systems and Practices

data or risk assessment tools" needed to regulate biosolids effectively.[114] More recent reports have found that biosolids can be contaminated by per- and polyfluoroalkyl substances (PFAS) (a class of chemicals widely used in non-stick and other industrial uses that are extremely persistent), which can then be taken up by plants and grazing animals.[115] Until government agencies can ensure the safety of biosolids, they should not be used as agricultural soil amendments regardless of their potential to reduce net emissions.

A type of charcoal called biochar may be able to store even more carbon than traditional organic amendments.[116] Biochar is produced by pyrolysis: the thermal decomposition of organic material at high temperatures in the absence of oxygen. This process results in a carbon-rich char that is more stable than uncharred plant material, although local environmental conditions, such as climate and soil type, play an important role in determining how long it persists in soils.[117] Biochar primarily reduces emissions by stabilizing and adding to carbon stores in the soil;[118] however, it may also reduce nitrous oxide emissions and fertilizer requirements.[119]

Both organic fertilizer and biochar can increase agricultural productivity, particularly in degraded soils, and reduce irrigation and fertilizer requirements.[120] Organic soil amendments also have some potentially negative environmental impacts. If not managed well, they can result in odor and particulate pollution, nitrate runoff, and phosphorus loading.[121] As with

---

114. OFFICE OF INSPECTOR GENERAL, U.S. EPA, EPA UNABLE TO ASSESS THE IMPACT OF HUNDREDS OF UNREGULATED POLLUTANTS IN LAND-APPLIED BIOSOLIDS ON HUMAN HEALTH AND THE ENVIRONMENT 12-25 (2018) (Report 19-P-0002).
115. *See, e.g.,* Marc Mills, *PFAS Treatment in Biosolids—State of the Science,* PFAS Science Webinars for EPA Region 1 and State & Tribal Partners (Sept. 23, 2020).
116. Emissions from the production of biochar must be taken into account, however. Certain production methods negate some or all of its sequestration benefits. Dominic Woolf et al., *Sustainable Biochar to Mitigate Global Climate Change,* NATURE COMM., Aug. 10, 2010, at 1, 3.
117. Samuel Abiven et al., *Biochar by Design,* 7 NATURE 326, 326 (2014).
118. Woolf et al., *supra* note 116, at 2.
119. Lukas Van Zwieten et al., *The Effects on Nitrous Oxide and Methane Emissions From Soil, in* BIOCHAR FOR ENVIRONMENTAL MANAGEMENT: SCIENCE, TECHNOLOGY, AND IMPLEMENTATION 490-91 (Johannes Lehmann & Stephen Joseph eds., Routledge 2d ed. 2015); Saran P. Sohi et al., *A Review of Biochar and Its Use and Function in Soil,* 105 ADVANCES AGRONOMY 47, 70-72 (2010).
120. Melissa Leach et al., STEPS Centre, Working Paper No. 41, Biocharred Pathways to Sustainability? Triple Wins, Livelihoods, and Politics of Technological Promise 26-28 (2010) (discussing biochar's impact on productivity); Andrew Crane-Droesch, *Heterogeneous Global Crop Yield Response to Biochar: A Meta-Regression Analysis,* 8 ENV'T RES. LETTERS 044049 (2013) (finding that biochar's impact on yield varies considerably across different soil environments); Annette Cowie et al., *Biochar, Carbon Accounting, and Climate Change, in* BIOCHAR FOR ENVIRONMENTAL MANAGEMENT: SCIENCE, TECHNOLOGY, AND IMPLEMENTATION, *supra* note 119, at 767, 771, 774 (describing biochar's potential to reduce the need for irrigation and fertilizer inputs).
121. EAGLE ET AL., *supra* note 101, at 88.

synthetic fertilizers, farmers must manage application timing, methods, and rates with care to minimize nitrous oxide emissions.[122]

❑ *Improve management practices for synthetic fertilizers.* Because plants utilize nitrogen from the soil and crops carry it away from the field after harvest, farmers must replenish nitrogen in their fields to maintain yields. This is typically accomplished through the application of synthetic or organic (such as manure) nitrogen fertilizer. However, farmers routinely apply fertilizer at higher rates than crops require for a variety of reasons: as a form of insurance or risk avoidance, hope for a great year, over-focus on yield over return, habit, and misinformation.[123] On average, only 50% of the nitrogen applied as fertilizer to annual grains is removed at harvest.[124] Similarly, a 2011 study found that farmers applied at least 40% more nitrogen than the prior harvest removed on nearly one-third of acres planted with key commodity crops.[125] In addition, because farmers have applied excess fertilizer for so long there is often an excess in soil, so they can now apply fertilizer less frequently—and, when necessary, apply less fertilizer per acre—without reducing yield. When they do this, they will also reduce the amount by which the supply of nitrogen in the soil exceeds the demand for nitrogen by crops. This will limit excess nitrogen that is released into the environment, including as nitrous oxide.[126]

In general, best practices for fertilization include reducing the rate of application so that the amount of nitrogen is closer to what crops need; timing the application so that nitrogen is available when crops can best utilize it; and varying the placement of nitrogen within fields to account for spatial variability in utilization by crops. Fertilizer companies, industrial farmers, and many extension programs call these practices the "4Rs": apply the right fertilizer product at the right rate, right place, and right time.[127] These prac-

---

122. SWAN ET AL., *supra* note 3, at 7.
123. Farmers often apply excess fertilizer "in the hopes that 'this year will be the one in ten' when extra N will pay off." G. Philip Robertson & Peter M. Vitousek, *Nitrogen in Agriculture: Balancing the Cost of an Essential Resource*, 34 ANN. REV. ENV'T & RESOURCES 97, 117 (2009). As discussed *infra* Sections V.B.7 and V.D, both incentives, such as a payment-for-ecosystem-services program that rewarded farmers using best management practices, and disincentives, such as a tax on fertilizer, could be used to reduce overfertilization.
124. David S. Kanter & Timothy D. Searchinger, *A Technology-Forcing Approach to Reduce Nitrogen Pollution*, 1 NATURE SUSTAINABILITY 544-552 (Oct. 2018), https://www.nature.com/articles/s41893-018-0143-8. *See also* G. Philip Robertson, *Nitrogen Use Efficiency in Row-Crop Agriculture: Crop Nitrogen Use and Soil Nitrogen Loss*, *in* ECOLOGY IN AGRICULTURE 351 (Louise E. Jackson ed., Academic Press 1997).
125. MARC RIBAUDO ET AL., USDA, NITROGEN IN AGRICULTURAL SYSTEMS: IMPLICATIONS FOR CONSERVATION POLICY 11 (2011) (ERR-127).
126. Robertson & Vitousek, *supra* note 123, at 104.
127. *See* Terry L. Roberts, *Right Product, Right Rate, Right Time, and Right Place . . . the Foundation of Best Management Practices for Fertilizer*, *in* FERTILIZER BEST MANAGEMENT PRACTICES, GENERAL

tices are not mutually exclusive, and they will likely be most effective when combined in broader nutrient management plans.[128]

Even if the rate of fertilizer application matches crop needs, improper timing and placement can increase greenhouse gas emissions. One of the most important things a farm can do is apply fertilizer no earlier than the planting season.[129] Nonetheless, due to ease of application, soil and water conditions, the lower cost of fertilizer in the fall, availability of machinery, and other reasons, farmers now fertilize a significant portion of the nation's cropland each fall, even though those fertilized fields will not be seeded until the following spring.[130] Fertilizer left unutilized in the soil over winter is vulnerable to environmental loss, including as nitrous oxide.[131]

Some experts argue that farmers can increase efficiency by practicing "split application"—that is, applying small amounts of fertilizer early in the planting season and, again, when nitrogen demand is highest, typically after plants emerge from the ground.[132] Studies have found that split application may reduce emissions by a significant amount. In one study on potatoes, an especially nitrogen-intensive crop, split application resulted in a 30% reduction in cumulative emissions compared to a single application.[133] Slow-release fertilizer formulations can also improve efficiency. For instance, polymer-coated urea fertilizes crops continuously as soil temperature, moisture, and other factors break down its coating over the course of the growing season.[134]

Nitrogen availability can vary within fields, as factors like prior yields (and thus nitrogen removal at harvest) affect its distribution. Precision agriculture, also called satellite or soil-specific farming, allows farmers to optimize placement via the Global Positioning System (GPS) and other forms of tech-

---

Principles, Strategy for Their Adoption and Voluntary Initiatives vs Regulations 29-32, Proceedings of the IFA International Workshop on Fertilizer Best Management Practices, March 7-9 2007, Brussels, Belgium (International Fertilizer Industry Ass'n 2007).

128. G. Philip Robertson et al., *Nitrogen-Climate Interactions in U.S. Agriculture*, 114 Biogeochemistry 41, 55-56 (2013).
129. Ribaudo et al., *supra* note 125, at 6.
130. According to a USDA study, farmers applied fertilizer unnecessarily early on nearly one-quarter of acres planted with key commodity crops. Ribaudo et al., *supra* note 125.
131. Ribaudo et al., *supra* note 125, at 75; X. Hao et al., *Nitrous Oxide Emissions From an Irrigated Soil as Affected by Fertilizer and Straw Management*, 60 Nutrient Cycling Agroecosystems 1, 5 (2001); C. Wagner-Riddle & G.W. Thurtell, *Nitrous Oxide Emissions From Agricultural Fields During Winter and Spring Thaw as Affected by Management Practices*, 52 Nutrient Cycling Agroecosystems 151, 162 (1998).
132. Bijesh Maharjan et al., *Fertilizer and Irrigation Management Effects on Nitrous Oxide Emissions and Nitrate Leaching*, 106 Agronomy J. 703, 712 (2014).
133. David L. Burton et al., *Effect of Split Application of Fertilizer Nitrogen on N_2O Emissions From Potatoes*, 88 Canadian J. Soil Sci. 229, 233 tbl.3 (2008).
134. Maharjan et al., *supra* note 132, at 711.

nology that use spatial and temporal data about fields.[135] Precise harvesting machines can track the yield in each small section of each row; improved satellite imagery can accurately estimate plant nitrogen and soil moisture levels in each area; and soil and plant samples can determine soil type and plant needs. These data then inform how and when fields are fertilized as well as irrigated, sprayed with pesticides, and harvested, leading to productivity gains and reduced pollution. The market for precision agriculture technology is developing rapidly, but nonetheless such tools may still be prohibitively expensive for smaller operations.[136]

Farmers can also improve nitrogen placement by applying fertilizer in irrigation water via subsurface drip irrigation (SDI) systems. These systems deliver nitrogen precisely and in proximity to plant roots, increasing plant uptake and limiting excess nitrogen in the soil.[137] SDI is also less likely to fill soil pore space with water, avoiding the anaerobic conditions that are especially conducive to the generation of nitrous oxide.[138] At present, SDI systems have been studied only on fruit and vegetable crops.[139] However, some evidence indicates that SDI systems would be cost effective for corn in the Great Plains.[140] Because corn cultivation uses almost half of the nitrogen fertilizer applied in the United States,[141] SDI systems could substantially reduce nitrous oxide emissions.

Some studies have suggested that nitrification inhibitors, chemicals that delay the conversion of ammonium to nitrate, may reduce nitrous oxide emissions by allowing plants to absorb a larger share of nitrogen.[142] How-

---

135. Rattan Lal, *Preface* to SOIL-SPECIFIC FARMING: PRECISION AGRICULTURE vii (Rattan Lal & B.A. Stewart eds., CRC Press 2015).
136. *See* MICHAEL MCLEOD ET AL., COST-EFFECTIVENESS OF GREENHOUSE GAS MITIGATION MEASURES FOR AGRICULTURE: A LITERATURE REVIEW 26 (Organisation for Economic Co-operation and Development Food, Agriculture, and Fisheries Papers No. 89, 2015). *See also* FACT AND FACTORS, PRECISION FARMING MARKET BY COMPONENT (HARDWARE, SOFTWARE, AND SERVICES), BY TECHNOLOGY (GEOMAPPING, INTEGRATED ELECTRONIC COMMUNICATION, REMOTE SENSING, AND VARIABLE RATE TECHNOLOGY (VRT)), AND BY APPLICATION (WEATHER MONITORING, FIELD MAPPING, YIELD MONITORING, IRRIGATION MANAGEMENT, WASTE MANAGEMENT, AND OTHERS): GLOBAL INDUSTRY PERSPECTIVE, COMPREHENSIVE ANALYSIS AND FORECAST, 2020–2026 (2021) (the global precision farming market is expected to grow at a compound annual growth rate of 12.7% from 2019 to 2026).
137. Diego Abalos et al., *Management of Irrigation Frequency and Nitrogen Fertilization Mitigate GHG and NO Emissions From Drip-Fertigated Crops*, 490 SCI. TOTAL ENV'T 880, 880 (2014).
138. *Id.*
139. *See generally* Taryn L. Kennedy et al., *Reduced Nitrous Oxide Emissions and Increased Yields in California Tomato Cropping Systems Under Drip Irrigation and Fertigation*, 170 AGRIC. ECOSYSTEMS & ENV'T 16-27 (2013) (discussing studies within single cropping systems).
140. Freddie R. Lamm & Todd P. Trooien, *Subsurface Drip Irrigation for Corn Production: A Review of 10 Years of Research in Kansas*, 22 IRRIGATION SCI. 195, 198 (2003).
141. ECONOMIC RESEARCH SERVICE, USDA, FERTILIZER USE AND PRICE (last updated Oct. 30, 2019).
142. Maharjan et al., *supra* note 132, at 712.

ever, reductions may be modest compared to split application.[143] Moreover, nitrification inhibitors are antimicrobial pesticides that kill or inhibit the soil microbes involved in nitrification. The broader impact of these inhibitors on soil microbial communities, and soil health and fertility, requires further study.[144] Growers can also reduce net emissions by replacing synthetic nutrients with manure or other organic soil amendments, discussed above.

In addition to climate benefits, reducing excess fertilizer and improving fertilizer management would reduce surface and subsurface runoff of nitrogen, a major source of contamination of rivers, lakes, and drinking water supplies.[145] It can also save farmers money, as fertilizer purchase and application are often a significant expense.

❑ *Optimize flood irrigation and drainage in rice cultivation.* Flood irrigation of rice fields, a standard part of rice cultivation, causes methane emissions because it creates anaerobic conditions in which methane-producing bacteria thrive.[146] While rice cultivation is a relatively small source of national greenhouse gas emissions, accounting for about 0.2% of all emissions and 2% of agricultural emissions in 2016,[147] the concentration of rice production in two regions, the lower Mississippi River basin and California, makes it an important consideration for policymakers in these areas.[148] Furthermore, increased atmospheric CO_2 concentrations, temperatures, and natural flood risks may increase methane emissions from rice cultivation over time—one study estimated that emissions per ton of rice may double by 2100.[149]

---

143. *Id.*
144. A single gram of soil contains between 10,000 and 50,000 species of bacteria. Amber Dance, *Soil Ecology: What Lies Beneath*, 455 NATURE 724 (2008). *Nitrosomonas* bacteria are primarily responsible for the conversion of ammonium to nitrite, which is subsequently converted to nitrate. Darrell W. Nelson & Don Huber, *Nitrification Inhibitors for Corn Production*, *in* NATIONAL CORN HANDBOOK 1 (Iowa State University Extension 1992) (NCH-55). While *Nitrosomonas* are the targets of nitrification inhibitors, the impact of nitrification inhibitors on other soil microorganisms needs to be characterized as well.
145. SWAN ET AL., *supra* note 3, at 6.
146. EPA, *supra* note 10, at 5-19.
147. *Id.* at 2-3 to 2-4 tbl.2-1, 5-2 tbl.5.1.
148. *See id.* at 5-18 to 5-19. Arkansas, Louisiana, Mississippi, and Missouri accounted for 75% of methane emissions from rice cultivation in 2012. California accounted for 17%, and Texas accounted for the remaining 8%. *See id.* at 5-18 tbl.5-11.
149. TAPAN K. ADHYA ET AL., WORLD RESOURCES INSTITUTE, WORKING PAPER INSTALLMENT NO. 8 OF CREATING A SUSTAINABLE FOOD FUTURE, WETTING AND DRYING: REDUCING GREENHOUSE GAS EMISSIONS AND SAVING WATER FROM RICE PRODUCTION 5 (2014). The increase in the greenhouse gas intensity of rice cultivation would be due both to the direct effects of increasing atmospheric CO_2 concentrations, which increases the availability of carbon used by methanogens to generate methane, and to declines in yields due to increasing temperatures and natural flood risks, which would necessitate the cultivation of additional land for rice production. Flood irrigation and the resulting anaerobic conditions would increase methane emissions from the cultivated land. *Id.*

Rice farmers can lower methane emissions by reducing the continuous flooding during the growing season and instead using alternate wetting and drying. Periodic drainage temporarily restores aerobic conditions, which rapidly diminishes the amount of methane-producing bacteria and stimulates other bacteria that metabolize methane for energy.[150] The Intergovernmental Panel on Climate Change estimated that, on average, draining once per season reduces emissions by 40% while draining multiple times reduces emissions by 48%.[151] In 2016, California approved a protocol for rice farmers to quantify reductions at the farm level as the basis for generating credits under the state's cap-and-trade program; this system may incentivize rice growers to adopt mitigation practices.[152] Periodic drainage, which requires farmers to suspend irrigation, uses less water and can help farmers and communities in areas that experience water shortages.[153]

2. Grazing Land

Grazing lands cover almost one-third of the contiguous United States.[154] More than 80% of this land is rangeland, uncultivated land with minimal inputs, while the remainder is cultivated and more intensively managed grazing land, or pasture.[155] Pasture has greater potential for carbon sequestration as a result of its higher biomass unit production, but it requires irrigation or high precipitation levels, making it impractical in much of the arid West.[156]

---

150. *Id.* at 6.
151. 4 Intergovernmental Panel on Climate Change, 2006 IPCC Guidelines for National Greenhouse Gas Inventories: Agriculture, Forestry, and Other Land Use 5.44-5.53 (Simon Eggleston et al. eds., 2006).
152. *See* California Environmental Protection Agency, Air Resources Board, *Potential New Compliance Offset Protocol Rice Cultivation Projects*, https://ww2.arb.ca.gov/sites/default/files/classic//cc/capandtrade/protocols/rice/riceprotocol2015.pdf (last reviewed Dec. 2, 2014). Microsoft just purchased some such offsets. USDA NRCS, *Nature's Stewards: U.S. Rice Farmers Embrace Sustainable Agriculture and Earn First-Ever Carbon Credits for Rice Production*, http://nrcs.maps.arcgis.com/apps/Cascade/index.html?appid=c00a7710dbe04790823c4133777e49c0 (last visited Oct. 30, 2020).
153. In addition to reducing irrigation requirements, periodic drainage can increase water savings by decreasing the amount of water lost to percolation and, in some cases, evaporation. Adhya et al., *supra* note 149, at 8.
154. Of the 1,937.7 million acres of nonfederal land in the contiguous United States, 130.9 million are pastureland, 417.9 are rangeland, and 56.1 are grazed forestland. Cynthia Nickerson et al., USDA, Major Uses of Land in the United States, 2007, at 7 (2011) (EIB-89). USDA's data for the 48 contiguous states do not include federal lands, however, which account for a significant proportion of national grazing lands. *Id.* at 6. *See also* T.M. Sobecki et al., *A Broad-Scale Perspective on the Extent, Distribution, and Characteristics of U.S. Grazing Lands*, *in* The Potential of U.S. Grazing Lands to Sequester Carbon and Mitigate the Greenhouse Effect 21, 29 (Ronald F. Follett et al. eds., CRC Press 2001).
155. Eagle et al., *supra* note 101, at 36.
156. The majority of pasture is east of the Missouri River, where precipitation levels are higher. R.R. Schnabel et al., *The Effects of Pasture Management Practices*, *in* The Potential of U.S. Grazing Lands

(Degraded pasture has higher potential for new sequestration since it can be improved more.) Regardless of whether ranchers use pasture or rangeland, well-managed silvopasture systems—those that integrate the production of woody perennials and livestock on the same land—offer substantially more climate benefits than conventional grazing systems. Researchers have also proposed limiting livestock feed and forage to grass on marginal lands and food byproducts, which would substantially reduce the carbon intensity of animal products. Ranchers using conventional systems can generally reduce emissions and increase soil carbon sequestration through better grazing management, and by optimizing feed, breed, and herd health. Emerging research indicates that new practices, such as spreading organic soil amendments, may be able to further improve carbon sequestration on grazing lands.

❏ *Expand silvopasture.* Silvopasture (often included within the broader term "agroforestry") is the practice of planting woody perennials on grazing lands. As with agroforestry on cropland, silvopasture offers significant greenhouse gas mitigation potential. Adding trees to grazing lands provides a substantial new source of carbon storage, while also increasing livestock productivity (due to reduced heat-stress loss) and introducing new revenue streams for farmers whose trees produce food or forestry products. Silvopasture systems have the potential to sequester more carbon than either plantation forests or grasslands, because they can integrate perennial grasses and trees, each of which offers distinct sequestration avenues, as discussed in previous sections.[157] A 2012 literature review estimated that silvopasture systems would sequester an average of 2.5 metric tons of carbon per acre annually in the United States through both additional biomass and increased soil carbon storage.[158] USDA's estimated range for sequestration rates for silvopasture systems, while substantially lower, still markedly outperforms conventional grazing.[159]

Co-benefits of silvopasture systems include improved water quality, reduced erosion, and additional habitat for wildlife.[160]

❏ *Limit livestock production to grazing and byproducts.* As discussed above, approximately half of all harvested cropland in the United States is devoted to feed crop production.[161] Devoting such a high share of cropland to feed production substantially increases greenhouse gas emissions relative to other

TO SEQUESTER CARBON AND MITIGATE THE GREENHOUSE EFFECT, *supra* note 154, at 291, 293.
157. Udawatta & Jose, *supra* note 11, at 227.
158. *Id.* at 230.
159. SWAN ET AL., *supra* note 3, at 33.
160. *Id.*
161. *See* Chapter III, note 31.

agricultural practices since animal products have a poor feed-to-food conversion rate.[162] A hundred calories of feed, for example, only produces three calories of edible beef.[163] Eggs and dairy products have the highest conversion rate at 17%, meaning that 100 calories of feed results in 17 calories of eggs or dairy products.

Animal products are not inherently inefficient, however. Animals can be fed with resources that do not contribute to human food consumption, namely grassland unsuited for crop cultivation and byproducts from crop production and food processing.[164] As a result, researchers have proposed limiting livestock production to so-called "ecological leftovers," or "low-opportunity-cost feed," in order to render animal products more sustainable.[165] This would likely decrease the amount of animal products available—one study found that such limitations would reduce the amount of animal protein available per capita in the European Union by 40%[166]—but it would provide a number of climate benefits. Researchers estimate that diets with animal products limited to ecological leftovers would not only reduce the land use of the food system compared to both vegan diets and contemporary standard diets, allowing for reforestation or other climate-friendly land uses, but it would also directly reduce agricultural emissions by a substantial amount.[167]

When produced using ecological leftovers, livestock production can sustain biodiversity,[168] promote the use of perennials,[169] and recycle plant nutrients.[170]

❑ *Improve grazing management.* Grazing practices not only affect methane emissions from the grazing animals themselves but also soil emissions of greenhouse gases. The impacts of grazing on soil carbon storage depend on interactions between precipitation, soil texture, and grassland plant community composition, among other factors,[171] but grazing itself generally

---

162. *See* Alon Shepon et al., *Energy and Protein Feed-to-Food Conversion Efficiencies in the US and Potential Food Security Gains From Dietary Changes*, 11 Env't Res. Letters 3 fig.2 (2016).
163. *Id.* at 2 fig.1.
164. *See* Elin Röös et al., *Limiting Livestock Production to Pasture and By-Products in a Search for Sustainable Diets*, 58 Food Pol'y 2 (2016); Elin Röös et al., *Greedy or Needy? Land Use and Climate Impacts of Food in 2050 Under Different Livestock Futures*, 47 Global Env't Change 2 (2017).
165. Röös et al., *Limiting Livestock*, *supra* note 164; Röös et al., *Greedy or Needy?*, *supra* note 164.
166. Ollie van Hal et al., *Upcycling Food Leftovers and Grass Resources Through Livestock: Impact of Livestock System and Productivity*, 219 J. Cleaner Production 485, 494 (2019).
167. *Id.*
168. Henry H. Janzen, *What Place for Livestock on a Re-Greening Earth?*, 166/167 Animal Feed Sci. & Tech. 783, 787 (2011).
169. *Id.* at 787-88.
170. *Id.* at 787.
171. Megan McSherry & Mark Ritchie, *Effects of Grazing on Grassland Soil Carbon: A Global Review*, 19 Global Change Biology 1347, 1347 (2013).

decreases grassland soil, root, and microbial carbon pools.[172] Heavy rates of grazing can even reduce soil greenhouse gas emissions compared to grasslands without grazing.[173] Due to these interactions, managing rotation durations, stocking rates, and grazing pattern complexity can influence carbon sequestration on grazing lands.

Several factors, including climate, precipitation, topography, local plant communities, soil type, and ranch size, influence the types of practices appropriate for any given location and the magnitude of their impacts on carbon and nitrogen cycling. Management systems that rotate livestock through a series of pastures, if implemented well, may improve grassland productivity, increase soil organic carbon, and reduce greenhouse gas emissions.[174] At the same time, continuous systems, which allow unrestricted grazing, are more likely to lead to soil carbon losses.[175]

The USDA Natural Resources Conservation Service (NRCS) includes rotational systems that rotate livestock in order to foster optimal plant and animal health as a component of "prescribed grazing." There are different types of prescribed grazing systems, such as management-intensive grazing, adaptive multi-paddock grazing, and less intensive forms of rotational and planned grazing. While not widely adopted, there are numerous such operations that produce livestock while restoring rangelands, increasing soil carbon, reducing emissions of nitrous oxide and methane, and enhancing other ecological benefits.[176] These can be viewed as models for other farms, education programs, and government incentives.

---

172. *See* Shiming Tang et al., *Heavy Grazing Reduces Grassland Soil Greenhouse Gas Fluxes: A Global Meta-Analysis*, 654 Sci. Total Env't 1218-24 (2019).
173. *See id.*
174. *See* Richard T. Conant et al., *Grassland Management Impacts on Soil Carbon Stocks: A New Synthesis*, 27 Ecological Applications 662 (2017); Benjamin B. Henderson, *Greenhouse Gas Mitigation Potential of the World's Grazing Lands: Modeling Soil Carbon and Nitrogen Fluxes of Mitigation Practices*, 207 Agric. Ecosystems Env't 91 (2015); McSherry & Ritchie, *supra* note 171; Richard T. Conant et al., *Land Use Effects on Soil Carbon Fractions in the Southeastern United States. I. Management-Intensive Versus Extensive Grazing*, 38 Biology & Fertility Soils 386, 391 (2003); Chad Hellwinckel & Jennifer Phillips, *Land Use Carbon Implications of a Reduction in Ethanol Production and an Increase in Well-Managed Pastures*, 3 Carbon Mgmt. 27, 28 (2012). *Contra* David D. Briske et al., *Rotational Grazing on Rangelands: Reconciliation of Perception and Experimental Evidence*, 61 Rangeland Ecology & Mgmt. 3, 11 (2008) (arguing that rotational grazing offers few, if any, benefits over other systems of grazing according to experimental evidence).
175. *See, e.g.*, John Carter et al., *Moderating Livestock Grazing Effects on Plant Productivity, Nitrogen, and Carbon Storage*, 17 Nat. Resources & Env't Issues 191, 202 (2011). *See also* Paige L. Stanley et al., *Impacts of Soil Carbon Sequestration on Life Cycle Greenhouse Gas Emissions in Midwestern USA Beef Finishing Systems*, 162 Agric. Systems 249 (2018).
176. *See, e.g.*, Brown's Ranch, *Home Page*, http://brownsranch.us/ (last visited Oct. 30, 2020); Pinhook Farm, *Home Page*, http://pinhookfarm.blogspot.com/ (last visited Oct. 30, 2020); LeftCoast Grassfed, *Home Page*, http://www.leftcoastgrassfed.com/ (last visited Oct. 30, 2020). *See generally* Regeneration International, *Home Page*, http://regenerationinternational.org/ (last visited Oct. 30, 2020); Savory Institute, *Home Page*, http://www.savory.global/ (last visited Oct. 30, 2020).

The ability of individual systems to sequester carbon has been vigorously debated,[177] varies by region and land use history,[178] and hits an upper limit when soils become saturated.[179] Environmental factors beyond the control of ranchers, such as drought conditions, can also overshadow and overwhelm the impact of even the most effective management practices, particularly in arid rangelands.[180] The net impact of grazing practices on the climate also depends on the balance of specific greenhouse gas emissions and sequestration rates, which may not shift in a consistent direction in response to changes in management. For example, while adaptive multi-paddock grazing reduces soil emissions of nitrous oxide and methane, it may also result in higher CO_2 emissions from soil respiration.[181]

Advocates of adaptive multi-paddock grazing and other forms of intensive rotational grazing note that current conventional extensive operations can transition to managed grazing in a few years, depending on the region's climate and that various cost savings, such as reduced forage costs, can occur even before all the carbon benefits start to accrue, thus facilitating the transition.[182] Using a meta-analysis including hundreds of observations, a group of researchers in 2017 found that improved grazing practices (including lower stocking rates, removal of grazing livestock, several types of rotational or short-duration grazing, and seasonal grazing) generally (but not always) increased annual carbon sequestration rates on average by 0.28 metric ton carbon per hectare per year, a significant amount especially if applicable to all or most U.S. grazing land.[183] A small number of studies indicate that adaptive multi-paddock grazing, at least in some regions, can increase sequestration rates over five times that amount, which would mean that the sequestration could offset the associated bovine methane emissions, leading to a net drawdown of atmospheric carbon.[184] As managed grazing is still not widely

---

177. *See, e.g.*, John Carter et al., *Holistic Management: Misinformation on the Science of Grazed Ecosystems*, 2014 INT'L J. BIODIVERSITY 1, 5-7 (2014).
178. McSherry & Ritchie, *supra* note 171.
179. Catherine Stewart et al., *Soil Carbon Saturation: Concept, Evidence, and Evaluation*, 86 BIOGEOCHEMISTRY 19, 25-28 (2007); Uta Stockmann et al., *The Knowns, Known Unknowns, and Unknowns of Sequestration of Soil Organic Carbon*, 146 AGRIC. ECOSYSTEMS & ENV'T 80, 94-95 (2012).
180. Kayje Booker et al., *What Can Ecological Science Tell Us About Opportunities for Carbon Sequestration on Arid Rangelands in the United States?*, 23 GLOBAL ENV'T CHANGE 240, 240-44 (2013).
181. Steven L. Dowhower et al., *Soil Greenhouse Gas Emissions as Impacted by Soil Moisture and Temperature Under Continuous and Holistic Planned Grazing in Native Tallgrass Prairie*, 287 AGRIC. ECOSYSTEMS ENV'T 106647 (2020).
182. *See, e.g.*, CARBON COWBOYS (Carbon Nation 2020) (depicting the grazing transition in one to eight years), https://www.carboncowboys.org/.
183. Conant et al., *supra* note 174.
184. Stanley et al., *supra* note 175. *See* W.R. Teague et al., *Grazing Management Impacts on Vegetation, Soil Biota and Soil Chemical, Physical, and Hydrological Properties in Tall Grass Prairie*, 141 AGRIC. ECOSYSTEMS ENV'T 310 (2011); *see also* W.R. Teague et al., *The Role of Ruminants in Reducing Agriculture's*

used, further examples and additional research will likely lead to improved performance and a better understanding of the impact and the regions and land types on which it may be most effective.[185]

Rotational grazing co-benefits include increased species diversity, decreased erosion, improved soil quality, better quantity and quality of wildlife habitat, improved water quality, and improved riparian ecosystem health and watershed quality.[186]

❑ *Optimize feed, breed, and herd health.* Grazing practices have been the subject of significant attention and debate; however, ranchers can also take important steps to reduce net emissions through improved feed, breed, and animal health management. By carefully managing their herds' feed and forage options, operators may be able to decrease enteric emissions.[187] Operators can also reduce emissions by maintaining herd health and choosing or developing breeds best adapted to the local environment.[188] The capacity of different breeds to thrive in local conditions, such as weather and native plant communities, affects how quickly they mature. Breeds optimized for local conditions will therefore reach slaughter weight more quickly, reducing their emissions.

❑ *Add soil amendments.* New research has demonstrated that organic soil amendments like compost may be able to boost carbon sequestration on grazing land. Over the course of three years, researchers found that a single application of composted organic matter to rangeland increased net carbon

---

Carbon Footprint in North America, 71 J. SOIL WATER CONSERVATION 156, 160 (2016) (showing a three ton of carbon per-hectare per-year emission reduction from adaptive multi-paddock practices versus conventional practices). *See also* SWAN ET AL., *supra* note 3, at 34 (estimating 0.08-0.41 megagrams (Mg) CO_2 eq. per acre per year); Rattan Lal, *Soil Carbon Sequestration Impacts on Global Climate Change and Food Security*, 304 SCIENCE 1623-27 (2004) (showing estimate of 0.18 and 0.55 Mg CO_2 per hectare per year); Pete Smith et al., *Agriculture, in* CLIMATE CHANGE 2007: MITIGATION 499-540 (Bert Metz et al. eds., Intergovernmental Panel on Climate Change 2007) (showing estimate of 0.11-0.81 Mg CO_2 per hectare per year).

185. *See* Henderson, *supra* note 174, which estimates 0.15 gigatons (Gt) CO_2 per year global sequestration from optimizing grazing pressures, compared to earlier estimates of 1.4 Gt CO_2 per year in Pete Smith et al., *Greenhouse Gas Mitigation in Agriculture*, 363 PHIL. TRANSACTIONS ROYAL SOC'Y B 789, 789-813 (2008).
186. SWAN ET AL., *supra* note 3, at 34.
187. DOUG GURIAN-SHERMAN, UNION OF CONCERNED SCIENTISTS, RAISING THE STEAKS: GLOBAL WARMING AND PASTURE-RAISED BEEF PRODUCTION IN THE UNITED STATES 13-19 (2011) (summarizing practices to reduce methane emissions through improved feed and forage); Karen A. Beauchemin et al., *Mitigation of Greenhouse Gas Emissions From Beef Production in Western Canada—Evaluation Using Farm-Based Life Cycle Assessment*, 166/167 ANIMAL FEED SCI. & TECH. 663, 674-75 (2011).
188. GLOBAL RESEARCH ALLIANCE ON AGRICULTURAL GREENHOUSE GASES ET AL., REDUCING GREENHOUSE GAS EMISSIONS FROM LIVESTOCK: BEST PRACTICE AND EMERGING OPTIONS 12-14, 20-23 (Karin Andeweg & Andy Reisinger eds., 2015).

storage by 25%-70%,[189] while also increasing the production of grass for feed and thereby making rangelands more productive.[190] Some scientists have expressed concern that applying compost on natural grasslands could negatively alter soil chemistry and water quality, favor invasive plants species, and decrease native plant diversity.[191] A 2019 meta-analysis of 92 studies found that "organic amendments, on average, provide some environmental benefits (increased soil carbon, soil water holding capacity, aboveground net primary productivity, and plant tissue nitrogen; decreased runoff quantity), as well as some environmental harms (increased concentrations of soil lead, runoff nitrate, and runoff phosphorus; increased soil CO_2 emissions)."[192] Further study and field trials will need to confirm these results and measure results in different ecosystems.[193]

3. Animal Feeding Operations

Animal feeding operations (AFOs) are lots or facilities in which confined animals are fed, raised, and maintained.[194] Unlike farms that allow livestock to graze or be integrated into crop production, AFOs are focused on one task: maximizing the production of meat, dairy, or eggs. EPA classifies AFOs as concentrated animal feeding operations (CAFOs) if they exceed a

---

189. Rebecca Ryals & Whendee Silver, *Effects of Organic Matter Amendments on Net Primary Productivity and Greenhouse Gas Emissions in Annual Grasslands*, 23 ECOLOGICAL APPLICATIONS 46, 56 (2013). This total does not include the carbon directly added to the soil from the compost. *Id.* at 46.
190. *Id.* at 51. Biodegradable waste appropriate for compost includes animal manure, crop residues, composted urban waste, and sewage sludge. *Id.* at 46.
191. KELLY GRAVUER, AGRONOMIC RATES OF COMPOST APPLICATION FOR CALIFORNIA CROPLANDS AND RANGELANDS TO SUPPORT A CDFA HEALTHY SOILS INITIATIVE PROGRAM, VERSION 1.0, at 10-11 (2016).
192. Kelly Gravuer et al., *Organic Amendment Additions to Rangelands: A Meta-Analysis of Multiple Ecosystem Outcomes*, 25 GLOBAL CHANGE BIO. 1152–70 (2019), https://doi.org/10.1111/gcb.14535. *See also* Rebecca Ryals et al., *Grassland Compost Amendments Increase Plant Production Without Changing Plant Communities*, 7 ECOSPHERE 1, 7-8 (2016).
193. The original study was followed by a modeling study demonstrating the possibility for long-term effect. Rebecca Ryals et al., *Long-Term Climate Change Mitigation Potential With Organic Matter Management on Grasslands*, 25 ECOLOGICAL APPLICATIONS 531, 531 (2015). A 2019 meta-analysis found that the "efficacy and outcomes" of soil amendment application on rangelands "are relatively poorly studied, and the potential for negative environmental consequences are higher in rangelands than croplands due to their starkly different ecology and management context." Kelly Gravuer et al., *Organic Amendment Additions to Rangelands: A Meta-Analysis of Multiple Ecosystem Outcomes*, 25 GLOBAL CHANGE BIOLOGY 1152, 1153 (2019), *available at* https://onlinelibrary.wiley.com/doi/pdf/10.1111/gcb.14535.
194. U.S. EPA, *National Pollutant Discharge Elimination System (NPDES): Animal Feeding Operations (AFOs)*, https://www.epa.gov/npdes/animal-feeding-operations-afos (last updated Aug. 3, 2020).

Chapter IV. Climate-Friendly Agricultural Systems and Practices 95

certain size threshold or, in some circumstances, if they discharge waste into surface waters.[195]

There are roughly 450,000 AFOs[196] in the United States, including 20,000 CAFOs.[197] CAFOs alone hold the majority of the country's food-producing animals.[198] While AFOs are credited with lowering consumer costs for animal products, they have considerable externalities. They can harm animal welfare, increase antibiotic resistance due to the routine use of antibiotics,[199] emit air and water pollution,[200] depress property values,[201] hurt small-scale farms and businesses,[202] and diminish quality of life in rural communities.[203]

There are AFO systems for production of all types of meat—beef, pork, and poultry—as well as production of eggs and dairy products. While the details vary, in general, swine and dairy AFOs often rely on liquid manure systems, poultry and egg AFOs produce a dry litter, and cattle feedlots leave the animal waste on the open ground. In liquid systems, workers wash the manure from the animal pens to a storage lagoon, usually uncovered, where it is eventually pumped out and spread onto fields.

---

195. 40 C.F.R. §122.23(b)-(c) (2016). "Large CAFOs" are defined as CAFOs by EPA solely due to the number of animals they hold, "medium CAFOs" are operations that exceed a smaller size threshold, but discharge waste into surface water, and "small CAFOs" are facilities that do not meet any size threshold, but have been designated as "significant contributor[s] of pollutants to waters" by regulatory authorities. *Id.*
196. USDA NRCS, *Animal Feeding Operations*, https://www.nrcs.usda.gov/wps/portal/nrcs/main/national/plantsanimals/livestock/afo/ (last visited Oct. 30, 2020).
197. U.S. EPA, NATIONAL POLLUTANT DISCHARGE ELIMINATION SYSTEM, 2017 CAFO PERMITTING STATUS REPORT (2017).
198. *See* Chapter III.B.2, figs. 5 & 6; MARC RIBAUDO ET AL., USDA, MANURE MANAGEMENT FOR WATER QUALITY: COSTS TO ANIMAL FEEDING OPERATIONS OF APPLYING MANURE NUTRIENTS TO LAND iii (2003) (noting that while CAFOs make up less than 5% of AFOs, they contain 50% of all animals and produce more than 65% of all manure).
199. Joan A. Casey et al., *High-Density Livestock Operations, Crop Field Application of Manure, and Risk of Community-Associated Methicillin-Resistant Staphylococcus aureus Infection in Pennsylvania*, 173 JAMA INTERNAL MED. 1980, 1981 (2013); David Tillman et al., *Agricultural Sustainability and Intensive Production Practices*, 418 NATURE 671, 674 (2002); Ellen Silbergeld et al., *Industrial Food Animal Production, Antimicrobial Resistance, and Human Health*, 29 ANN. REV. PUB. HEALTH 151, 162-63 (2008).
200. *See* Virginia T. Guidry et al., *Hydrogen Sulfide Concentrations at Three Middle Schools Near Industrial Livestock Facilities*, 27 J. EXPOSURE SCI. & ENV'T EPIDEMIOLOGY 167, 167 (2017); Michael A. Mallin et al., *INDUSTRIAL SWINE AND POULTRY PRODUCTION CAUSES CHRONIC NUTRIENT AND FECAL MICROBIAL STREAM POLLUTION*, 226 WATER, AIR, SOIL & POLLUTION 407, 407 (2015).
201. Kelley Donham et al., *Community Health and Socioeconomic Issues Surrounding Concentrated Animal Feeding Operations*, 115 ENV'T HEALTH PERSP. 317, 319 (2007).
202. *Id.* at 317.
203. *See id.*; M. Tajik et al., *Impact of Odor From Industrial Hog Operations on Daily Living Activities*, 18 NEW SOLUTIONS 193, 201 (2008); Steve Wing & Susanne Wolf, *Intensive Livestock Operations, Health, and Quality of Life Among Eastern North Carolina Residents*, 108 ENV'T HEALTH PERSP. 233, 235-37 (2000).

AFO manure management systems produce much more methane than pasture-based livestock operations. When manure is left as a solid (as naturally happens on grazing lands and pasturelands), it typically decomposes aerobically and produces little to no methane. However, when it is stored or handled in a system that creates an anaerobic environment, such as a lagoon, it releases large amounts of methane.[204] Storage in uncovered anaerobic lagoons can result in methane conversion rates over 100 times as high as those in pasture and range.[205]

AFOs produce an enormous amount of waste and emit tremendous amounts of greenhouse gases. Iowa's 4,000 hog AFOs generated more than 68 billion pounds of manure in 2019, 68 times the amount of human excreta produced by the state's residents, according to a conservative estimate.[206] Meanwhile, Wisconsin's 8,600 dairy farms are estimated to have generated almost 28 million tons of manure in 2018[207]—more than 50 times the amount of human excreta produced within the state.[208] And since human waste must be treated before being released into the environment, reducing both the threat of pathogens[209] and volume of organic material discharged, the relative impact from CAFO waste is even more stark—a CAFO with 250 dairy cows produces more organic waste every day than does a city the size of Albany.[210] As a result of their reliance on anaerobic

---

204. While dry management can reduce methane emissions, switching to dry management can increase nitrous oxide emissions. Smith et al., *supra* note 185, at 794. Dry management does not always increase nitrous oxide emissions, however, and increases in nitrous oxide emissions resulting from dry management are likely to be exceeded by decreases in methane emissions. *See, e.g.*, JUSTINE J. OWEN ET AL., NICHOLAS INSTITUTE, GREENHOUSE GAS MITIGATION OPPORTUNITIES IN CALIFORNIA AGRICULTURE: REVIEW OF EMISSIONS AND MITIGATION POTENTIAL OF ANIMAL MANURE MANAGEMENT AND LAND APPLICATION OF MANURE 7 tbl.4 (2014) (showing emission estimates of cows in California by manure management system).
205. Olga Gavrilova et al., *Emissions From Livestock and Manure Management, in* 2019 REFINEMENT TO THE 2006 IPCC GUIDELINES FOR NATIONAL GREENHOUSE GAS INVENTORIES 67 tbl.10.17 (2019), https://www.ipcc-nggip.iges.or.jp/public/2019rf/pdf/4_Volume4/19R_V4_Ch10_Livestock.pdf.
206. JAMIE KONOPACKY & SOREN RUNDQUIST, ENVIRONMENTAL WORKING GROUP, EWG STUDY AND MAPPING SHOW LARGE CAFOS IN IOWA UP FIVEFOLD SINCE 1990 (2020).
207. Calculated by the authors. *Compare* NATIONAL AGRICULTURAL STATISTICS SERVICE, USDA, MILK COW NUMBERS, WISCONSIN (2020) (estimating Wisconsin's dairy cow population to be 1,274,000 in 2018), https://www.nass.usda.gov/Statistics_by_State/Wisconsin/Publications/Dairy/Historical_Data_Series/milkcowno.pdf, *with* Riverkeeper, Inc. v. Seggos, 75 N.Y.S. 3d 854, 858 n.5 (N.Y. Sup. Ct. 2018) (noting that cows produce 120 pounds of manure each day).
208. Wisconsin residents produced an estimated 528,837 tons of human waste in 2018 (calculated by the authors). *Compare* U.S. EPA, RISK ASSESSMENT EVALUATION FOR CONCENTRATED ANIMAL FEEDING OPERATIONS 9 tbl.3.3 (2004) (estimating that the average 150 pound person produces 182.5 pounds of excreta each year), *with* U.S. Census Bureau, *QuickFacts: Wisconsin* (listing Wisconsin's population as 5,795,483 on July 1, 2017), https://www.census.gov/quickfacts/WI (last visited Oct. 20, 2020).
209. U.S. EPA, REPORT TO CONGRESS: IMPACTS AND CONTROL OF CSOS AND SSOS 4-3 (2004) (EPA 833-R-04-001).
210. Calculated by the authors. *Compare* Livestock & Poultry Environmental Learning Center, *Liquid Manure Storage Ponds, Pits, and Tanks* (Mar. 2019), https://lpelc.org/liquid-manure-storage-ponds-pits-

storage practices, dairy and hog operations are responsible for almost 90% of methane emissions from manure management.[211] When comparing net greenhouse gas emissions from AFO systems, however, it is important to take into account other factors as well, including enteric emissions and the greater length of time that may be needed for animals to reach market weight in pasture-based systems.

This section evaluates four different strategies for reducing emissions from AFOs. The first strategy—transitioning to integrated crop-livestock systems—offers the most significant co-benefits, although its impact on greenhouse gas emissions will vary considerably by animal type and breed, the local environment, and other factors. The second strategy focuses on the benefits of eliminating concentrated liquid manure, which, as discussed above, is the dominant source of emissions from manure management. The third strategy considers methods for reducing emissions from liquid manure. The final strategy focuses on feed additives and vaccines designed to reduce enteric emissions from ruminants.

❏ *Reincorporate animals into croplands.* The most effective way to reduce emissions from AFOs would be to replace them with well-managed integrated crop-livestock systems. Traditionally, most farms incorporated animals into cropping systems by allowing them to forage on plant residues after harvest, but early agricultural scientists and extension agents discouraged this practice, perceiving it as archaic and inefficient. As late as 1974, 19% of farms in the United States used a crop-grazing rotation, accounting for 52% of cropland.[212] By 2012, only 7% of farms used livestock in their rotations on less than 2% of all cropland.[213] As scientists have begun to understand the ecology of agriculture better, however, they have started to encourage it as an environmentally friendly way to intensify agricultural

---

and-tanks/ (noting that cows produce 120 pounds of manure each day), *with* National Research Council, Use of Reclaimed Water and Sludge in Food Crop Production 46-47 (1996) (discussing research showing that the typical person produces about .30 pound of post-treatment sludge each day), *and* U.S. Census Bureau, *QuickFacts: Albany City, New York* (listing Albany's population as 98,111 on July 1, 2017), https://www.census.gov/quickfacts/fact/table/albanycitynewyork/PST045216 (last visited Oct. 30, 2020). *See* Eve C. Gartner, Letter to the Editor, *Environment Group to Cornell: We Stand by Our Numbers on Animal v. Human Waste*, Post-Standard, Jan. 4, 2019, http://www.syracuse.com/opinion/index.ssf/2017/07/environment_group_to_cornell_our_numbers_on_animal_v_human_waste_are_right.html.
211. U.S. EPA, *supra* note 10, at 5-12 tbl.5-7.
212. Rachael D. Garrett et al., *Policies for Reintegrating Crop and Livestock Systems: A Comparative Analysis*, 9 Sustainability 473, 485 (2017).
213. *Id.*

production.[214] Some even argue that crop-livestock farms are economically and environmentally optimal, creating an efficient nutrient cycle between plants and animals.[215]

Mixed crop-livestock systems increase inputs of plant litter and manure into soil along with the impacts of living plant roots and livestock, which can increase soil health and carbon sequestration. They can substantially reduce methane emissions from manure management because manure in integrated systems is typically left to decompose aerobically.[216] As noted above, however, both animal growth rates and enteric emissions must be taken into account when comparing net emissions from different systems of animal agriculture.

In addition to their climate benefits, properly managed crop-livestock systems naturally control pest and weed populations[217] and can improve soil structure, animal health, water quality, and biodiversity.[218]

❑ *Eliminate concentrated liquid manure.* Liquid manure is typically stored in lagoons and then spread or sprayed on fields. Measures can be taken to reduce emissions from both stages. Anaerobic digesters, which work by converting volatile solids in organic matter to biogas and a material called digestate, can be added to manure lagoons. The biogas, which is predominantly methane and CO_2, releases CO_2 when burned for energy. The digestate can be applied to fields as a fertilizer, which lowers net emissions by offsetting synthetic fertilizers and increasing carbon sequestration, or can be composted and used as bedding.[219]

---

214. *See, e.g.*, Michael Russelle et al., *Reconsidering Integrated Crop-Livestock Systems in North America*, 99 AGRONOMY J. 325, 325 (2007); Gilles Lemaire et al., *Integrated Crop-Livestock Systems: Strategies to Achieve Synergy Between Agricultural Production and Environmental Quality*, 190 AGRIC. ECOSYSTEMS & ENV'T 4, 4 (2014); Paulo César de Faccio Carvalho et al., *Managing Grazing Animals to Achieve Nutrient Cycling and Soil Improvement in No-Till Integrated Systems*, 88 NUTRIENT CYCLING AGROECOSYSTEMS 259, 271 (2010) (examining the environmental and productivity benefits of integrating livestock into a no-tillage crop system in southern Brazil).
215. Patrick Veysset et al., *Mixed Crop-Livestock Farming Systems: A Sustainable Way to Produce Beef? Commercial Farms Results, Questions, and Perspectives*, 8 ANIMAL 1218, 1218 (2014). The authors acknowledge that the current policy and market environment disincentivize crop-livestock systems, making it less than optimal in the real world. *Id.* at 1225-26. *See* V. Gupta et al., *Integrated Crop-Livestock Farming Systems: A Strategy for Resource Conservation and Environmental Sustainability*, 2 INDIAN RES. J. EXTENSION EDUC. 1 (2012) (arguing that integrating crops and livestock "create[s] a synergy, with recycling allowing the maximum use of available resources").
216. Unmanaged manure deposited on grassland by grazing animals still emits significant amounts of nitrous oxide, however. *See* EPA, *supra* note 10, at 5-26, 5-29 tbl.5-18.
217. Lemaire et al., *supra* note 214, at 4-5.
218. Bertrand Dumont et al., *Prospects From Agroecology and Industrial Ecology for Animal Production in the 21st Century*, 7 ANIMAL 1028, 1030-38 (2013).
219. It is not clear yet whether nitrous oxide emission rates differ for synthetic or organic fertilizers; however, organic fertilizers can offset emissions from nitrogen-based fertilizer manufacturing plants, which are a significant source of CO_2 as discussed in Chapter VIII.A.1.

Industry advocates claim that biogas reduces greenhouse gas emissions, but these assertions rely on unfounded assumptions and overly optimistic projections, while ignoring methane leakage and more effective forms of manure management. Digesters reduce methane emissions when compared to unregulated liquid manure management systems, but liquid manure management systems have the highest per-head methane emission rates among all methods of manure management.[220] Instead of comparing biogas to the most carbon-intensive system—and ignoring other methods, as anaerobic digestion proponents often do—we should compare it to other systems currently in use. When we do this, we find that is not only the most expensive method for reducing manure emissions, but it is also among the least effective.

Dairy and swine industrial facilities produce 90% of manure management emissions because they typically rely on liquid management systems, where manure is washed from animal pens to storage lagoons. This produces emissions rates as much as 90 times higher than those found in grazing-based and other dry manure systems.[221] Anaerobic digestion can reduce the scale of these emissions, but digesters still release a considerable amount of methane. Large-scale, well-managed digesters can achieve leakage rates around 3% in optimal conditions[222]—much less than some studies have found[223]—but still enough to make digestion an inferior method of reducing emissions.[224] In addition, the natural gas supply chain has a leakage rate of 2.3%,[225] if not higher,[226] adding to the amount of methane emitted by digesters.

While anaerobic digestion may reduce emissions relative to conventional liquid manure management, it is less beneficial to the climate than dry management and pastured-based alternatives—and, in contrast to those alternatives—it reinforces highly polluting practices. As discussed *infra* in Chapter VI.D, liquid manure systems also decrease soil, air, and water quality in rural

---

220. Justine J. Owen & Whendee L. Silver, *Greenhouse Gas Emissions From Dairy Manure Management: A Review of Fieldbased Studies*, 21 Global Change Biology 550, 558 fig.3 (2014).
221. Gavrilova et al., *supra* note 205.
222. Thomas K. Flesch et al., *Fugitive Methane Emissions From an Agricultural Biodigester*, 35 Biomass & Bioenergy 3927, 3933 (2011) (finding that leakage increased by a factor of 10 during feed loading and flaring). This does not include leakage associated with the transportation and storage of biogas.
223. *E.g.*, Jan Liebetrau et al., *Methane Emissions From Biogas-Producing Facilities Within the Agricultural Sector*, 10 Engineering Life Sci. 595, 599 (2010) (finding an average fugitive emission rate of 6%).
224. *See* William H. Schlesinger, *Natural Gas or Coal: It's All About the Leak Rate*, Cool Green Sci., June 24, 2016 (explaining that a leakage rate above "1 percent of gross production negates the advantages of natural gas with respect to mitigating climate change"), https://blog.nature.org/science/2016/06/24/natural-gas-coal-leak-rate-energy-climate/.
225. Ramón A. Alvarez et al., *Assessment of Methane Emissions From the U.S. Oil and Gas Supply Chain*, 361 Science 186, 186 (2018).
226. Warren Cornwall, *Natural Gas Could Warm the Planet as Much as Coal in the Short Term*, Sci. Mag., June 21, 2018, https://www.sciencemag.org/news/2018/06/natural-gas-could-warm-planet-much-coal-short-term.

communities, while lowering property values, regardless of whether anaerobic digestion is used.

❑ *Improve spreading of concentrated liquid manure.* The Clean Water Act requires that the manure be spread at "agronomic rates"—that is in quantities that the plants need and can use.[227] However, that provision is often ignored, with the result that manure can pollute nearby waters and release greenhouse gases.[228] There is some evidence that specific practices relating to manure spreading can also affect emissions and soil carbon sequestration levels. Spreading on frozen or saturated soils, for example, tends to lead to water pollution and higher nitrous oxide emissions because the manure is more likely to enter waterways instead of being incorporated into the soil.[229]

❑ *Develop methane inhibitors and vaccines.* A number of feed additives have been demonstrated to decrease methane emissions from livestock in short-term experiments.[230] When studied over the long term, however, these effects disappear or decrease significantly as the microflora in livestock's rumen adapt to the new diet.[231] Nonetheless, scientists are studying novel approaches that they hope will remain effective throughout a ruminant's life-span. One study documented a 30% decrease in enteric methane emissions over 12 weeks with the addition of 3-nitrooxypropanol, a chemical compound that blocks an enzyme critical to methane formation.[232] Another promising study found that red seawood supplementation reduced enteric emissions from cattle by 80% over 5 months.[233]

---

227. 40 C.F.R. §503.14 (2016).
228. *See, e.g.,* OLGA NAIDENKO ET AL., ENVIRONMENTAL WORKING GROUP, TROUBLED WATERS: FARM POLLUTION THREATENS DRINKING WATER 7, 11, 14 (2012) (explaining that the overapplication of manure is one of the primary sources of nutrient pollution); Michael Mallin & Lawrence Cahoon, *Industrialized Animal Production—A Major Source of Nutrient and Microbial Pollution to Aquatic Ecosystems,* 24 POPULATION & ENV'T 369, 377-78 (2003) (discussing runoff from manure spreading).
229. Andrew C. VanderZaag et al., *Strategies to Mitigate Nitrous Oxide Emissions From Land Applied Manure,* 166/167 ANIMAL FEED SCI. & TECH. 464, 469-70 (2011).
230. Mario Herrero et al., *Greenhouse Gas Mitigation Potentials in the Livestock Sector,* 6 NATURE CLIMATE CHANGE 452, 454 (2016).
231. *Id.*
232. Alexander Hristov et al., *An Inhibitor Persistently Decreased Enteric Methane Emission From Dairy Cows With No Negative Effect on Milk Production,* 112 PROC. NAT'L ACAD. SCI. U.S. AM. 10663, 10663 (2015) (finding a 30% decrease in enteric methane emissions over 12 weeks with the addition of 3-nitrooxypropanol). A subsequent study is J. Dijkstra et al., Short Communication, *Antimethanogenic Effects of 3-Nitrooxypropanol Depend on Supplementation Dose, Dietary Fiber Content, and Cattle Type,* 101 J. DAIRY SCI. 9041 (2018), *available at* https://www.journalofdairyscience.org/article/S0022-0302(18)30673-8/fulltext.
233. As of January 2021, the study is still in preprint and has not undergone peer review. Breanna M. Roque et al., *Red Seaweed* (Asparagopsis Taxiformis) *Supplementation Reduces Enteric Methane*

Chapter IV. Climate-Friendly Agricultural Systems and Practices 101

Other researchers have developed vaccines designed to reduce methane emissions. A New Zealand team is researching the viability of a vaccine that would target methanogenic archaea, reducing their prevalence in the rumen.[234] Industry officials estimate that the vaccine could reduce enteric emissions by 25%-30%;[235] however, as with methane inhibitors, the vaccine has yet to be proven safe, effective, or financially feasible.

Finally, research indicates that eliminating antibiotic use in livestock may reduce the prevalence of methane-producing archaea. As a result, eliminating the nontherapeutic use of antibiotics in animals could have an effect on emissions similar to that of methane inhibitors and vaccines. Animals in confined production facilities routinely receive antibiotics to increase their growth rates and to prevent disease.[236] This alters the microbiota of confined animals, affecting their health and physiology,[237] and may increase the amount of methane-producing microflora. For example, a 2016 study showed that tetracycline, a common antibiotic used in livestock production, nearly doubled methane emissions from cow dung.[238] Future research will be necessary to test additional classes of antibiotics and confirm that the effect will hold for enteric emissions.[239] Nonetheless, these initial results are promising.

Eliminating nontherapeutic antibiotic use in livestock would reduce the development of new resistance genes and the transmission of antibiotic resistance from animals to humans.[240] While new chemical inhibitors and vaccines may prove to be an effective and important pathway for reducing

---

by Over 80 Percent in Beef Steers (bioRxiv Preprint), https://www.biorxiv.org/content/biorxiv/early/2020/07/16/2020.07.15.204958.full.pdf.

234. D. Neil Wedlock et al., *Progress in the Development of Vaccines Against Rumen Methanogens*, 7 ANIMAL 244, 244 (2015).

235. Lucie Bell, *New Zealand Vaccine to Reduce Cattle Methane Emissions for Dairy and Beef Industry Reaches Testing Stage*, ABC RURAL, Nov. 9, 2015, http://www.abc.net.au/news/rural/2015-11-10/mitigating-methane-emissions-from-cattle-via-vaccine/6925676.

236. In 2012, the Food and Drug Administration released a guidance calling for the voluntary phaseout of antibiotic use in animals for growth promotion. However, livestock antibiotic use has increased by nearly 5% since the start of the phaseout program. FOOD AND DRUG ADMINISTRATION, 2014 SUMMARY REPORT ON ANTIMICROBIALS SOLD OR DISTRIBUTED FOR USE IN FOOD-PRODUCING ANIMALS 40 (2015). The agency is unlikely to realize lower usage rates without more active regulation and enforcement. *See* Frank Aaerestrup, Comment, *Get Pigs Off Antibiotics*, 486 NATURE 465, 465-66 (2012) (on the inadequacy of bans that fail to set and enforce reduction goals).

237. Nadia Gaci et al., *Archaea and the Human Gut: New Beginning of an Old Story*, 20 WORLD J. GASTROENTEROLOGY 16062, 16071 (2014).

238. Tobin Hammer et al., *Treating Cattle With Antibiotics Affects Greenhouse Gas Emissions, and Microbiota in Dung and Dung Beetles*, 283 PROC. ROYAL SOC'Y B 1, 5 (2016).

239. The effect may not hold for other forms of manure management due to a variety of factors including the timing of manure collection and aeration, which inhibits methanogenesis. E-mail From Tobin Hammer, Ph.D. Candidate, University of Colorado, Boulder, to Nathan Rosenberg, Visiting Assistant Professor, University of Arkansas School of Law (June 3, 2016).

240. Bonnie Marshall & Stuart Levy, *Food Animals and Antimicrobials: Impacts on Human Health*, 24 CLINICAL MICROBIOLOGY REV. 718, 729 (2011).

greenhouse gas emissions, they are unlikely to have significant social or environmental co-benefits. As with any feed additive or animal drug, their effects on animal welfare and human food safety should be rigorously assessed prior to marketing.

4. On-Farm Fuel Combustion and Electricity

While generally not included in discussions of agriculture's climate change footprint, on-farm fuel combustion, largely for vehicles and facility heating, and electricity for air and water pumps, cooling, and other actions, cause significant greenhouse gas emissions. These areas also offer many opportunities for emissions reductions. It is beyond the scope of the book to address all the opportunities for energy and electrical efficiency, such as more efficient vehicles or electric motors, better thermal insulation, and process improvements, but we do highlight a few key opportunities for agriculture. Similarly, while there are many other excellent discussions of shifting our electricity grid to clean and renewable energy, below we outline a few of the issues around the siting of renewable energy on farmland.

❑ *Improve on-farm machinery and vehicle efficiency.* The NRCS notes that CAFOs "require a great deal of energy for lighting; heating of barns and brooders; fans for ventilation and cooling of facilities; pumps for moving water, waste, or milk; electric motors to run feeders; and electricity for cooling milk and eggs." Because CAFOs must be intensively managed, it is relatively easy to make energy-saving changes to the operation. Efficiency measures such as maintaining or upgrading electric motors, switching to more efficient lighting, and using heat exchangers to capture excess heating when cooling milk can reduce energy use and thus emissions significantly.[241]

Farm vehicles tend to turn over slowly—the average life is more than 20 years—and there has been little pressure or market interest in accelerating the shift to greater efficiency. One analysis found that adopting zero-emission farm vehicles would reduce global emissions by about 537 MMT CO_2 eq. by 2050 at cost savings of about $229 per ton.[242] This possibility, while holding great promise, is still far off with only several pilots and no commercial products in place. However, even before a complete switch to zero-emission

---

241. *See* NRCS, USDA, Conservation Programs That Save: Energy Conservation in Confined Animal Operations (2006), https://www.nrcs.usda.gov/Internet/FSE_DOCUMENTS/nrcs143_023625.pdf.
242. Justin Ahmed et al., McKinsey & Co., Agriculture and Climate Change: Reducing Emissions Through Improved Farming Practices 15-16 (2020).

vehicles powered by clean electricity and batteries, significant improvements in farm vehicle fuel efficiency are possible.

❑ *Increase on-farm renewable energy production.* The co-location of photovoltaics (PV) and crops could improve outcomes across many sectors, increasing crop production, reducing water loss, and improving the efficiency of PV arrays. Adopting such synergistic paths forward can help build resilient food-production and energy-generation systems. A 2019 U.S. Department of Energy National Renewable Energy Laboratory study found that placing solar panels among plants can have benefits for both.[243] The plants cool the ground, increasing the efficiency of the panel, while the panels can improve water retention, reduce temperature fluctuations, and increase yields.[244] On the other hand, care must be taken to preserve the country's best farmland as food production is a necessity, so solar development should be directed toward marginal or sub-optimal farmland.[245]

B. Agriculture's Maximum Contribution to Curbing Climate Change

This book lays out the pathways necessary for agriculture to achieve climate neutrality. Even greater reductions in net greenhouse gas emissions may be technologically feasible. Nevertheless, net climate neutrality is a much more ambitious target than those set by the Deep Decarbonization Pathways Project and the United States Mid-century Strategy for Deep Decarbonization. The Deep Decarbonization Pathways Project proposes an 8% cut in nitrous oxide emissions and a 6% decrease in methane emissions from the agricultural sector and does not address agricultural carbon emissions or carbon sequestration.[246] The United States Mid-century Strategy for Deep Decarbonization is slightly more aggressive, calling for a 25% reduction in non-CO_2 emissions from agriculture.[247] It also highlights soil

---

243. Greg A. Barron-Gafford, *Agrivoltaics Provide Mutual Benefits Across the Food-Energy-Water Nexus in Drylands*, 2 Nature Sustainability 848 (2019).
244. *Id.* at 848-49.
245. *See, e.g.,* American Farmland Trust, Solar Siting Guidelines for Farmland (March 2020); American Farmland Trust, Smart Solar Siting in New York (2019).
246. James H. Williams et al., Pathways to Deep Decarbonization in the United States, U.S. 2050 Report, Volume 1: Technical Report 52 tbl.9 (2015), *available at* http://usddpp.org/downloads/2014-technical-report.pdf. The Deep Decarbonization Pathways Project specifically calls for a 9% decrease in nitrous oxide emissions from agricultural soils, which account for 93% of nitrous oxide emissions from agriculture, and a 9% decrease in methane emissions from enteric fermentation, which is responsible for 68% of methane emissions from agriculture. U.S. EPA, *supra* note 10, at 5-2 tbl.5-1 (showing annual emissions from agriculture by source).
247. The White House, *supra* note 8, at 91.

carbon sequestration on agricultural soils as a promising method for reducing net emissions, although it does not include soil carbon sequestration in its modeling results.[248]

The maximum possible contribution of agriculture to deep decarbonization is difficult to estimate. While researchers are rapidly advancing our understanding of the chemical and biological processes that result in agricultural emissions and sinks, there is still much to learn. Additionally, greenhouse gas emissions and sequestration rates vary significantly according to a number of local conditions, including climate, historical land use, and the composition of microbes in the soil. Finally, high rates of soil carbon sequestration cannot continue indefinitely; soil eventually becomes saturated with carbon, eliminating its ability to provide further decarbonization.

In addition, there are often trade offs to actions taken to reduce net agricultural emissions. Manure digesters capture methane but may increase incentives for concentration in livestock production; organic approaches may lower productivity, necessitating the use of more land; and no-till and cover cropping usually require greater use of herbicides. Despite these trade offs, it is clear that climate neutrality in agriculture is both a technologically and economically feasible goal, if an ambitious one.

The vast majority of nitrous oxide emissions result from the application of fertilizers, which climate-friendly practices can reduce. Additionally, manure can be used to fertilize fields or produce energy in ways that dramatically decrease methane emissions and the need for synthetic fertilizers. There are innumerable other strategies, practices, and tools available to cut agricultural emissions, many of which increase soil carbon and make farms or ranches better able to handle changing weather patterns.

Not all of these practices can be used together, and among those that can, it is not always clear how their interactions will affect net emissions. Since not all practices can be adopted in all geographies and their impact will vary according to local conditions, it is not possible to simply subtract the sum of aggregate soil carbon sequestration possibilities from total emissions. Yet the potential emissions reductions from these practices are so large, and their potential to increase soil carbon storage so substantial, that they make climate neutrality a realistic goal.

---

248. *Id.* at 77-79.

Chapter IV. Climate-Friendly Agricultural Systems and Practices

Figure 2. Emission Reductions From National Resources Conservation Service COMET-Planner Versus Maximum Applicable Land Area

Points are scaled by magnitude of impact (average emission reduction estimate multiplied by maximum applicable land area).

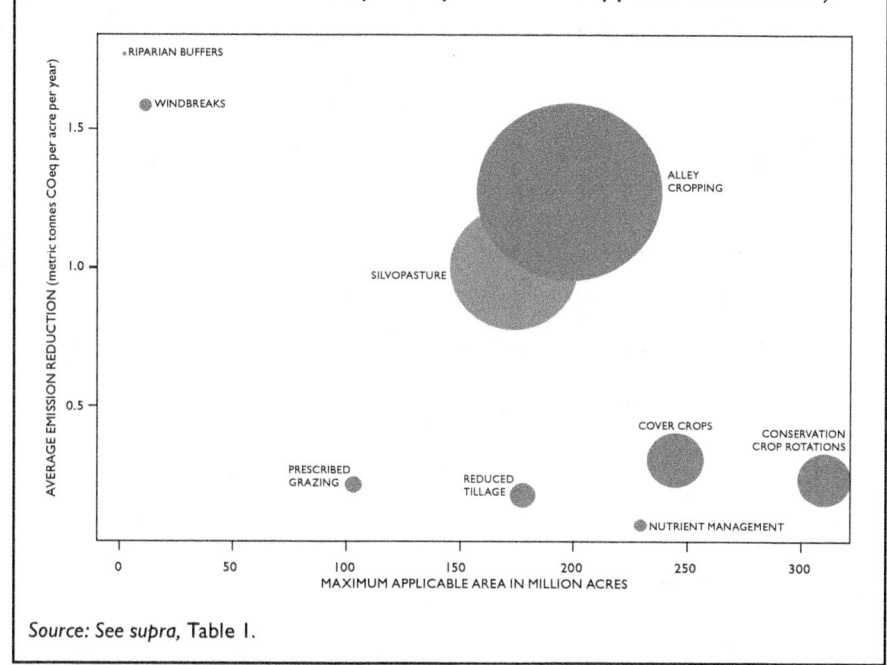

Source: See supra, Table 1.

Table 1 at the beginning of this chapter and Figure 2, above, provide the average annual net emissions reductions of the practices discussed in this book for which quantitative data are available. The table and figure offer a range of net emission reductions potentially achievable through the adoption of carbon sequestering practices on existing crop and grazing acres. Given the diversity of geographies and uncertainties of these practices, these totals are only illustrative. Figure 3, on the next page, provides a visual illustration of the data, showing that the sequestration capacity of these practices far exceeds agriculture's current emission levels.

While these practices can be cost beneficial for farmers or ranchers and have important additional benefits, uptake of new approaches can be slow and may require significant incentives, outreach and education, and more robust regulatory requirements. Whether agriculture will achieve climate neutrality will depend on whether our federal and state governments enact and implement policies with that goal—and that is ultimately a question of political will, not of science. Next, we outline legal pathways for reaching this objective.

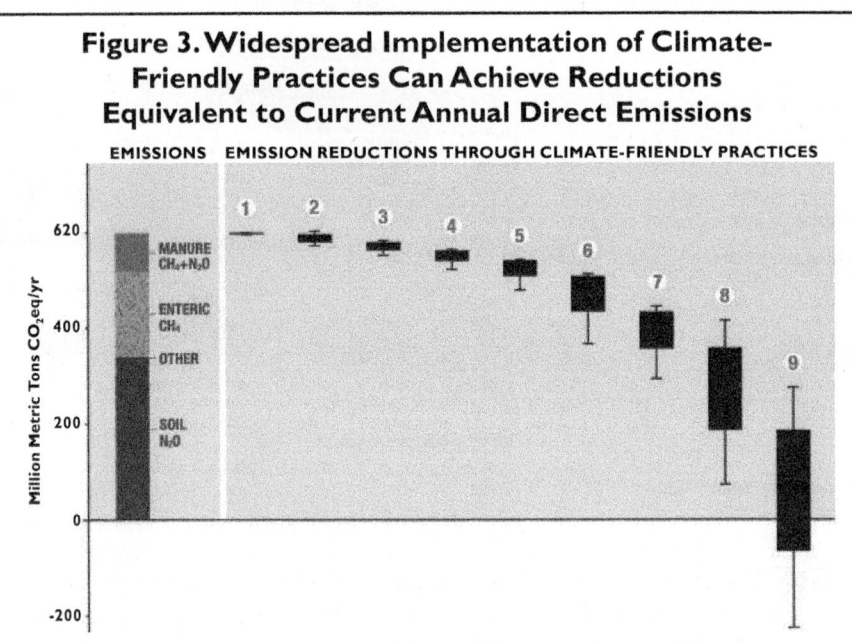

Figure 3. Widespread Implementation of Climate-Friendly Practices Can Achieve Reductions Equivalent to Current Annual Direct Emissions

Notes: Emission reduction estimates for adoption of (1) riparian buffers, (2) nutrient management practices, (3) windbreak establishment, (4) prescribed grazing, (5) reducing tillage, (6) conservation crop rotations, (7) cover crops, (8) silvopasture, and (9) alley cropping on maximum applicable area in the U.S. Estimates are derived by scaling emission reduction estimates from the National Resources Conservation Service COMET Planner by maximum applicable area. The net impact is determined by the estimated effect on carbon sequestration as well as the potential applicability of the practice across national agricultural acreage. The COMET Planner provides estimates of emission reductions for each practice under dry and moist ecotypes separately. Upper whiskers show emission reduction based on lowest average emission reduction estimate. Lower whiskers show emission reduction based on highest average emission reduction estimate. Thick bars show average emission reduction estimate. Fertilizer management efficiency only accounts for reductions in N_2O emissions from improvements in timing and volume (~15% reduction) of N applications.

Source: See supra, Table 1.

Chapter IV. Climate-Friendly Agricultural Systems and Practices

Key Recommendations

- Practices that increase the organic matter content of soil generally also increase soil carbon sequestration and, thereby, reduce net emissions.

- Increasing soil organic matter is a particularly important method of sequestering carbon in temperate parts of the world, such as the United States, where soils contain vastly more carbon than plants.

- Agroforestry, perennial agriculture, diversified farming, and organic agriculture offer a range of climate benefits, but current U.S. agricultural policy disfavors these systems.

- Since trees substantially increase above- and below-ground biomass, agroforestry increases both the rate of sequestration and the total amount of carbon that a piece of agricultural land can store relative to annual cropping systems. As a result, agroforestry's per-acre sequestration potential is far higher than that found in annual crop systems. Many tree crops are also very productive or provide other benefits to grazing animals and annual crops.

- Perennial crops substantially improve upon the carbon storage potential of annual crops because they reduce or eliminate the need for tillage, generally reduce irrigation and fertilizer needs, and sequester additional carbon through their considerable biomass and deep root systems.

- Diversified farming systems can reduce emissions from soil management and increase carbon sequestration in soil and biomass.

- Organic and other more climate-friendly farming systems can increase soil stability and fertility, on-farm biodiversity, and crop resiliency while reducing energy use and the need for synthetic inputs.

- Our food system would be dramatically more efficient, and thus have a lower climate change impact, if it produced more crops intended for human consumption as whole foods, and if it produced less animal feed, less feedstock for biofuels, and fewer processed foods.

- By leaving more plant residue and organic matter in and on the soil, reducing or eliminating tillage can improve soil quality, reduce erosion, provide food and cover for wildlife, and reduce dust and diesel pollution. However, the climate change benefits of no-till practices are somewhat uncertain and depend significantly on region, soil depth, and how the no-till system is implemented.

- Cover crops and crop rotations can play an important role in reducing net agricultural emissions when integrated into climate-friendly systems, although neither offers transformative climate benefits when practiced in isolation.

- Adding manure or other organic fertilizers to soil can decrease manure waste (and thus methane emissions), reduce emissions from the production of synthetic fertilizers, and increase soil carbon stock.

- Best practices for fertilization include reducing the rate of application so that the amount of nitrogen is closer to what crops need, timing the application so that nitrogen is available when crops can best utilize it, and varying the placement of nitrogen within fields to account for spatial variability in utilization by crops. Key opportunities include prohibiting or reducing fall and winter fertilization or manure spreading since crops are not taking up nutrients in those seasons.

- Rice farmers can lower methane emissions by reducing the continuous flooding during the growing season and instead using alternate wetting and drying.

- Regardless of whether ranchers use pasture or rangeland, well-managed silvopasture systems—those that integrate the production of woody perennials and livestock on the same land—offer substantially more climate benefits than conventional grazing systems.

- Livestock production should be limited to grazing and byproducts. Given that it takes many pounds of grain feed to add a pound of meat on livestock, the concentrated animal feeding operation (CAFO) system must be closely re-assessed.

- Grazing practices not only affect methane emissions from the grazing animals themselves but also the amount of carbon in the soil. Managing rotation durations, stocking rates, and grazing pattern complexity can influence carbon sequestration on grazing lands and multi-paddock intensive rotational grazing appears to hold great promise.
- By carefully managing their herds' feed and forage options, operators should be able to decrease enteric emissions while improving herd health.
- Organic soil amendments like compost and biochar may be able to boost carbon sequestration on grazing land.
- Transitioning to integrated crop-livestock systems offers significant co-benefits, although its impact on greenhouse gas emissions will vary considerably by animal type and breed, the local environment, and other factors.
- Concentrated liquid manure systems generate large amounts of methane while dry management and pastured-based alternatives generate far less methane.
- Anaerobic digesters may reduce emissions when compared to conventional liquid manure management methods, but they are more expensive and less climate-friendly than dry manure and well-managed pasture-based systems.
- On-farm fuel combustion, largely for vehicles and facility heating, and electricity for air and water pumps, cooling, and other actions, cause significant greenhouse gas emissions. These areas offer many opportunities for emissions reductions. More research is needed to harness greater efficiencies.
- Co-location of solar panels or wind turbines on farmland, if directed to marginal farmland and implemented in ways that allow continued agricultural production, holds great promise.

Chapter V.
Transforming Farm Policy Toward Climate-Neutral Agriculture

At first glance, reducing net agricultural greenhouse gas emissions through public law poses a considerable challenge. Agriculture operates on a "parallel regulatory framework," in which farms are provided safe harbors from regulations in a number of areas, including labor, antitrust, and the environment.[1] Indeed, the agriculture industry's exemptions from environmental regulations are so ubiquitous that the environmental regulation of agriculture has been referred to as a body of "anti-law."[2]

While the federal government may do little to regulate agriculture's negative externalities, it is heavily involved in the industry as a financer, insurer, and funder of research. Agriculture has its own cabinet position and an agency charged with ensuring the sector's financial well-being, which it does through funding for research, training, crop insurance, loans, direct subsidies, and numerous other programs. The industry has sought to "privatize profits and socialize losses" through these subsidies, often with great success.[3] Although the government has by and large served industrial agriculture, there are a number of ways government support for the sector could be used to reduce net agricultural greenhouse gas emissions. In this chapter we examine how the U.S. Department of Agriculture (USDA) can be transformed to accelerate the shift to climate-neutral agriculture.

Many of the proposals to reduce emissions in this chapter benefit a wide range of constituencies, increasing their political viability. Climate-friendly practices can increase the supply of healthy food,[4] reduce pollution in rural

---
1. Susan Schneider, *A Reconsideration of Agricultural Law: A Call for the Law of Food, Farming, and Sustainability*, 34 WM. & MARY ENV'T L. & POL'Y REV. 935, 937 (2010).
2. J.B. Ruhl, *Farms, Their Environmental Harms, and Environmental Law*, 27 ECOLOGY L.Q. 263, 263 (2000). *See* Margot Pollans, *Drinking Water Protection and Agricultural Exceptionalism*, 77 OHIO ST. L.J. 1195, 1213-24 (2016), for a discussion of agricultural exceptionalism in the context of drinking water contamination.
3. James C. Scott, *Forward* to BILL WINDERS, THE POLITICS OF FOOD SUPPLY: U.S. AGRICULTURAL POLICY IN THE WORLD ECONOMY xi (2009).
4. ERIC TOENSMEIER, THE CARBON FARMING SOLUTION 58-59 (Brianne Goodspeed & Laura Jorstad eds., 2016).

communities,[5] and improve farmworker health outcomes.[6] These practices can also reduce costs, increase soil health and fertility, and make farms and grazing lands more resilient to climate change, all of which can increase producer profitability, as has already been demonstrated by thousands of farmers and ranchers around the country. They can also help revitalize rural communities. Thus, politicians can support climate-friendly practices for their co-benefits along with their effects on net emissions. Indeed, advocacy should emphasize the quintuple benefits of climate mitigation, soil health, climate resilience, improved public health, and producer benefits.

The federal government spends almost $3 billion annually on agricultural research, development, and extension programs, much of which can be used to support climate mitigation efforts. Congress should increase funding for the department to develop and disseminate information on and assistance for climate-friendly practices and crop varieties. The department should also strengthen its capacity to collect, report, and analyze data on climate-friendly practices and systems. Since 2018, the federal government has subsidized agriculture with about $20 billion annually through farm bill commodity, conservation, and insurance programs, while also providing the sector with tens of billions of dollars through short-term subsidy, lending, and trade programs. These subsidies should prioritize programs and practices with the greatest climate benefits and exclude environmentally harmful operations and products. And agricultural operations that do not follow basic conservation practices should not be eligible to receive funds through USDA. Congress should create, or USDA should develop, a system that pays for ecosystem services in place of the current farm safety net. Such a system would be independent of the volatility of commodity markets and thus would provide greater and more stable benefits to farmers, rural communities, and the environment.

After examining traditional farm programs, we turn to grazing on public land and perennial agriculture. Each of these two types of production

---

5. We provide a brief summary of the environmental co-benefits associated with each climate-friendly practice examined in Chapter IV.
6. Farmworkers are at much greater risk of pesticide poisoning than both the general population and people in other agricultural occupations, such as farmers, significantly affecting farmworker health. Geoffrey M. Calvert et al., *Acute Pesticide Poisoning Among Agricultural Workers in the United States, 1998-2005*, 51 Am. J. Indus. Med. 883, 889 tbl.3 (2008), found that 71% of all acute pesticide poisoning cases in the agricultural industry were among farmworkers. There are several long-term health effects associated with both acute and chronic pesticide exposure, including increased rates of cancer, neurobehavioral deficits, depression, Parkinson's disease, Alzheimer's disease, hearing loss, diabetes, obesity, and respiratory disease. Aaron Blair et al., *Pesticides and Human Health*, 72 Occupational & Env't Med. 81 (2015). The most effective carbon-farming practices substantially reduce pesticide use. *See supra* Chapter IV.

Chapter V. Transforming Farm Policy Toward Climate-Neutral Agriculture

must play an important role in decarbonizing agriculture and, thus, merits their own set of recommendations. The Bureau of Land Management (BLM) (within the U.S. Department of the Interior) and U.S. Forest Service (USFS) (within USDA), which together oversee more than one-third of all grazing lands in the United States, can regulate grazing on those lands, but have so far failed to sufficiently limit practices that increase emissions. Addressing this regulatory gap will reduce emissions while potentially fostering carbon sequestration on hundreds of millions of acres. Federal policy should also be redirected to support perennial agriculture, which provides a powerful means of natural carbon sequestration. We conclude the chapter with a set of recommendations designed to reduce disparities in funding between annual and perennial crop production, and quickly expand perennial systems.

A. Research, Extension, and Technical Assistance Programs

Congress' expressed purpose for supporting agricultural research and extension is not only to increase the productivity of agriculture,[7] but also to "[maintain and enhance] the natural resource base on which rural America and the United States agricultural economy depend."[8] Many existing USDA programs already focus on conservation, through which the department has significant leeway to increase funding for climate-friendly practices. Doing so, whether through congressional or agency action, will be needed to decarbonize agriculture. State governments and land-grant institutions should also provide funding for research on climate-friendly practices, particularly in the absence of strong federal research support.

Congress should couple increased financial support for climate-related agricultural research with generous funding to disseminate climate-friendly practices and research. By creating a nationwide network of climate extension professionals, while significantly increasing funding for climate-related outreach, education, and technical assistance, Congress can provide carbon farming with the support it needs to rapidly expand.

1. Federal Research Programs

Congress appropriated more than $2 billion to agricultural research and development in fiscal year (FY) 2018, making up less than 2% of the federal

---

7. Agricultural research includes basic, applied, and developmental research. *See* 7 U.S.C. §3101. The term "research" in this book thus encompasses the development of new technologies, practices, and products in addition to basic and applied research.
8. *Id.*

research expenditures.[9] The overwhelming majority of these funds go to two USDA agencies: the National Institute of Food and Agriculture (NIFA) and the Agricultural Research Service (ARS). USDA's in-house research agency, ARS, received roughly $1.1 billion for research in FY 2018, while NIFA, which primarily funds research at land-grant universities and administers grants to organizations outside of USDA, received $824 million.[10]

About 20% of ARS' FY 2020 research budget is allocated to environmental research, which includes research on climate change.[11] ARS' climate change research is focused on adaptation, however, with relatively few resources allocated toward mitigation.[12] The majority of ARS' funding should be reoriented to research projects that include mitigation components; however, this does not mean that ARS will not be able to meet other research priorities at the same time. For example, 27% of ARS' 2020 budget is dedicated to livestock and crop production.[13] The agency could increase its support for research that advances production and mitigation simultaneously, such as projects to develop productive livestock breeds, better plant materials for cover crops, and high-yielding crops that facilitate lower emissions and sequester more carbon. This research would help farmers prepare for a decarbonized economy, while helping the United States meet its emissions goals.

NIFA administers dozens of programs authorized through the Farm Bill and other pieces of legislation. While NIFA does not release total expenditures on climate-related research, all available evidence indicates that relatively little NIFA funding goes to climate mitigation.[14] In response to a Freedom of Information Act request we submitted, we found that NIFA's Division of Global Climate Change had only six full-time employees in 2016—well below the agency average of 11 full-time employees per division.[15] Only one of NIFA's 12 research divisions had fewer employees.[16] President Obama's 2017 budget for NIFA's main competitive research program, the Agriculture and Food Research Institute (AFRI), included $15 million for a program area focused on "climate variability and change."[17] The Trump

---

9. NATIONAL SCIENCE FOUNDATION, FEDERAL R&D FUNDING, BY BUDGET FUNCTION: FISCAL YEARS 2018-20, tbl.1 (NSF 20-305) (2019), https://ncses.nsf.gov/pubs/nsf20305.
10. Id.
11. USDA, FY 2020 BUDGET SUMMARY 67-69 (2019).
12. Id.
13. See id. at 67.
14. See Marcia DeLonge et al., Investing in the Transition to Sustainable Agriculture, 55 ENV'T SCI. & POL'Y 266, 269 (2016).
15. Freedom of Information Act Response No. 2020-REE-04663-F From USDA to Authors (July 31, 2020) (on file with authors).
16. Id.
17. NIFA, USDA, FY 2017 PRESIDENT'S BUDGET PROPOSAL 8 (2016).

Administration, however, ended five different AFRI program areas in FY 2018, folding all them into a single program called "Sustainable Agricultural Systems." The new program received less funding than its predecessor programs, falling from $144 million for the five programs in FY 2017—the last year that they were fully funded—to a proposed $134 million for the Sustainable Agricultural Systems Program in FY 2020.[18] Of the Sustainable Agricultural Systems Program's five goals, only one—land stewardship—is focused on environmental concerns.[19] On its own or with direction from Congress, NIFA, like ARS, should steadily increase funding used for climate mitigation and adaptation, shifting research funding to projects designed to reduce greenhouse gas emissions or increase carbon sequestration while improving soil health and resilience.

Both agencies should also prioritize funding for research into agroecology, which has a much greater potential to mitigate climate change than conventional systems.[20] Research into agroforestry and perennial agriculture in particular is severely underfunded.[21] Because research into these systems is unlikely to develop highly profitable products for agrochemical and seed corporations—agroforestry and perennial farmers do not need new seeds each year and require much lower rates of chemical inputs—privately funded agricultural research in this area is likely to remain minimal.[22] A USDA report on agricultural research spending, for example, found that agricultural research in several critical areas—including the environment and nutrition—relied entirely on public funding in 2013, the most recent year analyzed in the report.[23] Publicly funded research into these practices will be critical for kick-starting these systems, ultimately bringing farmers, consumers, and local communities significant social, environmental, and economic returns.[24]

---

18. OFFICE OF BUDGET AND PROGRAM ANALYSIS (OBPA), USDA, FY 2020 USDA EXPLANATORY NOTES—NATIONAL INSTITUTE OF FOOD AND AGRICULTURE 19-53 (2019).
19. *Id.* at 19-51 to 19-52.
20. *See, e.g.,* Marcia DeLonge, *Soil Carbon Can't Fix Climate Change by Itself—But It Needs to Be Part of the Solution*, UNION CONCERNED SCIENTISTS, Sept. 26, 2016, http://blog.ucsusa.org/marcia-delonge/soil-carbon-cant-fix-climate-change-by-itself-but-it-needs-to-be-part-of-the-solution.
21. *Id.*
22. The public sector remains responsible for "much of the fundamental research that creates the building blocks for major agricultural innovations," in large part because private research "[gravitates] toward technologies that are easy to patent or otherwise protect intellectual property rights." Matthew Clancy et al., *U.S. Agricultural R&D in an Era of Falling Public Funding*, AMBER WAVES, Nov. 10, 2016, https://www.ers.usda.gov/amber-waves/2016/november/us-agricultural-r-d-in-an-era-of-falling-public-funding/.
23. *Id.*
24. Peter Lehner & Nathan Rosenberg, *Promoting Climate-Friendly Agriculture for the Benefit of Farmers, Rural Communities, and the Environment*, 33 NAT. RESOURCES & ENV'T 7 (2018).

More agricultural research will also be critical for maintaining agricultural productivity as weather patterns become more extreme and unpredictable due to climate change. As farming is perhaps the sector hardest hit by changing weather patterns, farmers have a particularly strong interest in adaptation to, and mitigation of, climate change. Many adaptation measures work in concert with mitigation strategies, though that is not always the case.[25] For example, adaptation strategies for livestock often include the expansion of cooling and ventilation systems, which increase energy use and may therefore result in higher emissions, depending on the energy source.[26] By contrast, adding trees to pasture land in silvopasture systems helps to both cool livestock and reduce climate change. Government-funded research into adaptation practices should be increased and priority should be given to those practices that reinforce mitigation strategies. Similarly, USDA research designed to advance objectives other than mitigation, such as crop productivity or food safety, should be designed to bolster, and work in conjunction with, climate-friendly systems. Climate-friendly practices will be more readily adopted—and ultimately more sustainable—if they meet other human needs in addition to climate stability.

Congress will need to significantly expand funding for agricultural research in order to achieve climate neutrality while maintaining crop and livestock productivity. Indeed, agricultural productivity is expected to stagnate without significant increases in public funds for research even without taking climate change into account.[27] In 2018, despite this urgent need, relative funding for research was at a historical low, with 2% of USDA's total budget devoted to agricultural research.[28] Between 1940 and 1980, a period when agricultural productivity rose dramatically—in large part due to research funded by USDA—USDA directed about 4% of its budget to research.[29]

---

25. *See supra* Chapter IV fig.1; Cynthia Rosenzweig et al., *Climate Change Responses Benefit From a Global Food Systems Approach*, 1 NATURE FOOD 94-97 (2020); Cynthia Rosenzweig & Francesco Nicola Tubiello, *Adaptation and Mitigation Strategies in Agriculture: An Analysis of Potential Synergies*, 12 MITIGATION & ADAPTATION STRATEGIES FOR GLOBAL CHANGE 855, 866-67 (2007); Pete Smith & Jørgen E. Olesen, *Synergies Between the Mitigation of, and Adaptation to, Climate Change in Agriculture*, 148 J. AGRIC. SCI. 543, 550 (2010).
26. *See* Rosenzweig & Tubiello, *supra* note 25, at 866; *see* DANIEL TOBIN ET AL., USDA, NORTHEAST AND NORTHERN FORESTS REGIONAL CLIMATE HUB ASSESSMENT OF CLIMATE CHANGE VULNERABILITY AND ADAPTATION AND MITIGATION STRATEGIES 23 (Terry Anderson ed., 2015) (recommending the enhancement of cooling and ventilation systems as an adaptation strategy).
27. Paul W. Heisey et al., USDA, Economic Brief No. 17, Public Agricultural Research Spending and Future U.S. Agricultural Productivity Growth: Scenarios for 2010-2050 (2011).
28. Calculated by the authors using USDA, *supra* note 11, at 86.
29. *The Role and Development of Public Agricultural Research*, *in* AN ASSESSMENT OF THE U.S. FOOD AND AGRICULTURAL RESEARCH SYSTEM 9 (Congress Office of Technology Assessment 1981) (providing historical data on USDA's research expenditures); JIM MONKE, CONGRESSIONAL RESEARCH SERVICE,

Public funding for agricultural research is also declining in real terms: between 2003 and 2013, public funding fell from $6 billion to $4.5 billion after adjusting for inflation.[30] (See Figure 1.) Prior to this period, the public sector enjoyed much larger investments in agricultural research than the private sector.[31] Private-sector investment in agricultural research increased rapidly as public funding began to decrease, however, and by 2010 private expenditures had surpassed public expenditures for the first time since the USDA Economic Research Service (ERS) began tracking funding in the 1970s.[32] In 2014, the most recent year for which data are available, private-

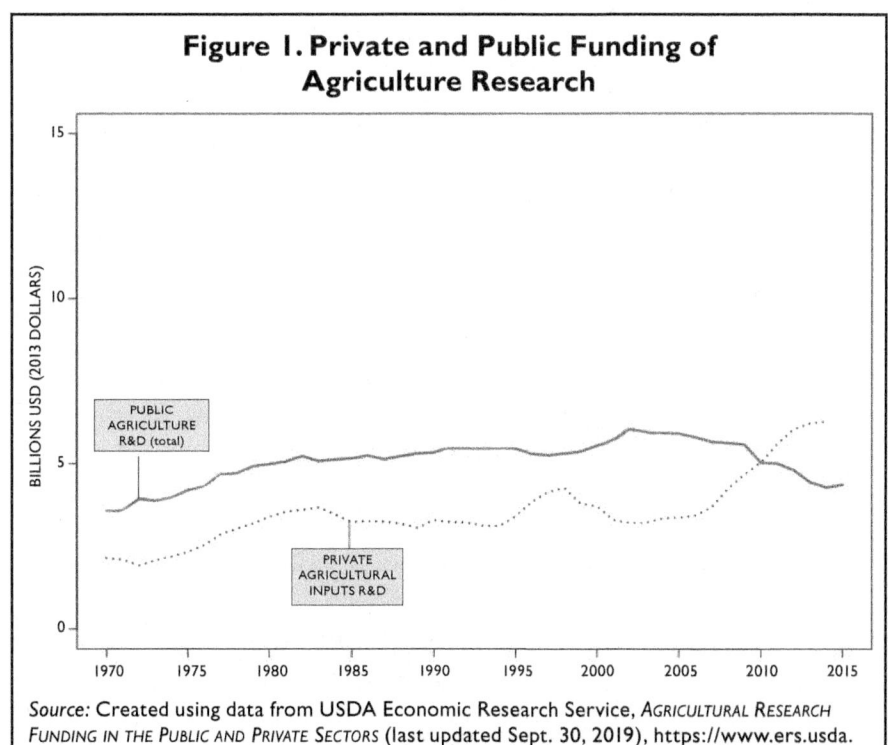

Figure 1. Private and Public Funding of Agriculture Research

Source: Created using data from USDA Economic Research Service, AGRICULTURAL RESEARCH FUNDING IN THE PUBLIC AND PRIVATE SECTORS (last updated Sept. 30, 2019), https://www.ers.usda.gov/webdocs/DataFiles/47755/agresearchfunding2015.xls?v=1064.8.

AGRICULTURAL RESEARCH: BACKGROUND AND ISSUES (2016) (describing the role of publicly funded agricultural research in productivity gains).
30. Clancy et al., *supra* note 22.
31. *Id.*
32. *Id.*

sector funding was almost 50% higher than public funding.[33] As a result, North America is the only region in the world where private agricultural research spending is higher than public spending,[34] and the United States now devotes only a tiny share of its total public research budget to agriculture—much less than almost every other advanced economy in the world.[35] This has important consequences because, as discussed above, private-sector research primarily supports practices that require the purchase of agrochemicals or other patentable products.

Several surveys of publicly funded agricultural research have concluded that public research into sustainable systems is, as one such survey put it, "woefully under-resourced."[36] In light of the challenge presented by climate change—and the current dearth of funding for sustainable farming systems—Congress should at a minimum restore the research budget to at least its prior level within USDA. Devoting 4% of USDA's budget to research in 2018 would have resulted in almost $3 billion in additional funding for agricultural research.[37] Public funding would once again exceed private funding, if only slightly, and agriculture's share of total public research spending in the United States would be closer to the level of investment enjoyed by other advanced economies. Such an investment would have significant economic benefits: every dollar spent on publicly funded agricultural research yields roughly $20 in benefits.[38] USDA should allocate these funds to develop the tools, measurement protocols, crops, and practices necessary to achieve climate neutrality in agriculture. While significant, this is only a fraction of the amount spent annually on crop insurance and other subsidies, which has ranged from $17 to $55 billion in recent years, [39] and which could, over time,

---

33. Calculated by the authors using ERS, USDA, AGRICULTURAL RESEARCH FUNDING IN THE PUBLIC AND PRIVATE SECTORS, 1970-2015, https://www.ers.usda.gov/webdocs/DataFiles/47755/agresearchfunding2015.xls?v=455.4 (last updated Feb. 14, 2019).
34. PAUL W. HEISEY & KEITH O. FUGLIE, USDA, AGRICULTURAL RESEARCH INVESTMENT AND POLICY REFORM IN HIGH-INCOME COUNTRIES 14 tbl.3.2 (Economic Research Report No. 249, 2018).
35. Among the 31 advanced economies included in USDA's survey of public agricultural research expenditures between 2009 and 2013, only Greece and Luxembourg devoted a smaller share of their public research budgets to agriculture. *Id.* at 23 fig.4.1.
36. Liz Carlisle & Albie Miles, *Closing the Knowledge Gap: How the USDA Could Tap the Potential of Biologically Diversified Farming Systems*, 4 J. AGRIC. FOOD SYSTEMS & COMMUNITY DEV. 219, 221 (2013) (arguing that a lack of research has limited organic agriculture's development); *see also* URS NIGGLI ET AL., RESEARCH INSTITUTE OF ORGANIC AGRICULTURE, A GLOBAL VISION AND STRATEGY FOR ORGANIC FARMING RESEARCH 19 (2016) (same).
37. Calculated by the authors. *Compare* USDA, *supra* note 11, at 2 fig. 1 (giving USDA's total outlays in FY 2018 as $137 billion), *with* USDA, FY 2018 BUDGET SUMMARY fig. REE-1 (2017) (listing USDA's research and development budget at $2.5 billion in FY 2018).
38. *See* Julian M. Alston et al., *The Economic Returns to U.S. Public Agricultural Research*, 93 AM. J. AGRIC. ECON. 1257, 1270 tbl.6 (2011).
39. *See* USDA ECONOMIC RESEARCH SERVICE, *Net Cash Income*, https://data.ers.usda.gov/reports.aspx?ID=17831 (last updated Dec. 2, 2020).

Chapter V. Transforming Farm Policy Toward Climate-Neutral Agriculture

be reduced if the research points to practices that make farms more resilient to climate change.

Congress should also increase funding for on-farm research and farmer-to-farmer information sharing regarding innovative carbon-farming practices. The Sustainable Agriculture Research and Education (SARE) program, which provides funding for on-farm research and efforts to increase knowledge about sustainable agricultural practices among farmers and agricultural professionals, provides an excellent model for future efforts.[40] Administered by NIFA, SARE is the only USDA competitive grants program that focuses exclusively on sustainable agriculture.[41] Its annual funding ranged from $27 to $37 million between 2016 and 2019.[42] Given SARE's important role in developing and disseminating sustainable practices—many of which are climate friendly—Congress should dramatically increase its annual budget, while also specifically appropriating funds for SARE to use to support the development of carbon farming. As discussed below, Congress should also create a program modeled after SARE for perennial crops. This program could be housed within SARE or another national body, or be administered by regional hubs, but there is a critical need for an on-farm competitive grants program that provides grant managers with specialized expertise on, and long-term funding specifically for, perennial systems.

The 2018 Farm Bill created a new funding source for on-farm research through the Conservation Innovation Grants program.[43] The new subprogram, called On-Farm Trials, provides $25 million for on-the-ground conservation activities, whose effectiveness will be evaluated by the Natural Resources Conservation Service (NRCS) and its research partners.[44] NRCS identified four high-priority funding categories for the program in 2020: irrigation management technologies; precision agriculture technologies and strategies; management technologies and strategies; and soil health demonstration trials.[45] The agency should create a new, permanent priority for climate mitigation and ensure that projects with mitigation components receive a plurality of available funding.

---

40. 7 U.S.C. §§5801-5832.
41. National Sustainable Agriculture Coalition (NSAC), *Sustainable Agriculture Research and Education Program*, https://sustainableagriculture.net/publications/grassrootsguide/sustainable-organic-research/sustainable-agriculture-research-and-education-program/ (last updated July 2019).
42. *Id.*
43. 16 U.S.C. §3839aa-8.
44. USDA NRCS, *Conservation Innovation Grants*, https://www.nrcs.usda.gov/wps/portal/nrcs/main/national/programs/financial/cig/ (last visited Nov. 12, 2020).
45. Grants.gov, *Notice of Funding Opportunity for NRCS's Conservation Innovation Grants (CIG) On-Farm Conservation Innovation Trials (On-Farm Trials) for Federal Fiscal Year (FY) 2020*, https://www.grants.gov/web/grants/view-opportunity.html?oppId=325306 (last updated Apr. 20, 2020).

2. Public-Private Research Programs

Agribusiness has increasingly sought federal funding for public-private research programs. The most significant example of this is a nonprofit nongovernmental organization called the Foundation for Food and Agriculture Research (FFAR), established by the 2014 Farm Bill to support "agricultural research focused on addressing key problems of national and international significance," including, among other focus areas, renewable resources, natural resources, and the environment.[46] Designed to spur public-private partnerships, Congress allocated FFAR $200 million to use as matching funds for nonfederal dollars.[47] FFAR has six "challenge areas" on which to focus its efforts, including "soil health," which is likely to benefit mitigation strategies due to the relationship between healthy soil and soil carbon sequestration.[48] FFAR announced its first healthy soils, thriving farms investment in March 2017, a $2.2 million grant that was matched with $4.4 million from a private foundation to create the National Cover Crop Initiative.[49]

Despite FFAR's potential to advance research into mitigation strategies, efforts to increase the private sector's involvement in public research should be evaluated carefully. Private-industry research funding is often driven by business interests. Not surprisingly, privately funded agricultural research is focused on technologies that are part of patentable products or services and it neglects research objectives—such as conservation of natural resources—that are unlikely to be profitable for private industry.[50] As a result, despite being advertised as win-win propositions for farmers and the environment, public-private research partnerships too often fail both. Unlike the public—or farmers—large corporations are able to advance their own priorities by directly contributing funds to public-private research efforts or by providing funding for private foundations and nonprofits to do so.

While FFAR shows some promise, it has yet to demonstrate its independence from industry interests. Rather than funding research that would benefit diversified, sustainable systems, it has largely funded projects friendly to agribusiness. For example, it awarded the National Pork Board almost $1 million for research aimed at reducing hog mortality in commercial concen-

---

46. 7 U.S.C. §5939.
47. *Id.*
48. FFAR, *What We Do*, https://foundationfar.org/what-we-do/ (last visited Jan. 13, 2020).
49. News Release, FFAR, New National Research Initiative Aims to Improve Cover Crop (Mar. 22, 2017), https://foundationfar.org/news/national-cover-crop-initiative/.
50. Clancy et al., *supra* note 22.

trated animal feeding operations (CAFOs),[51] and contributed $5 million to the industry-supported Irrigation Innovation Consortium.[52] Efforts such as these may sound laudable, but they effectively subsidize corporate interests and reinforce inherently unsustainable systems. Hog CAFOs will continue to cause significant environmental and public health harms regardless of hog mortality rates, and the type of projects generally funded by the Irrigation Innovation Consortium, such as precision irrigation,[53] have been found often to actually *increase* water usage rather than decrease it.[54] FFAR has also funded more promising projects, however. In 2020, the foundation awarded University of Minnesota a $1 million grant to help accelerate the development of a promising perennial wheatgrass, Kernza,[55] which researchers hope will be able to dramatically reduce net agricultural emissions.[56] Congress should ensure that FFAR funds primarily support research projects aimed at redesigning systems based on proven carbon farming principles, such as longer crop rotations, agroforestry, and integrated crop-livestock systems.[57]

3. State Research and Demonstration Programs

While federally funded research will be critical for the development of carbon farming, states and land-grant institutions can also play an important role in stimulating research into adaptation and mitigation strategies. State governments and land-grant institutions played a critical role in the growth of sustainable and organic agriculture before the federal government began providing consistent, if relatively meager, research funding in the 1990s.[58] The University of California, Iowa State University, the University of Maine, and others developed sustainable agriculture research and extension pro-

---

51. Jason Ross et al., An Integrated Approach to Improve Whole Herd Pig Survivability 3-4 (2019).
52. Press Release, Colorado State University, FFAR Awards $5 Million to Launch Public-Private Effort to Address Water Scarcity and Irrigation Innovation (Apr. 18, 2018) (on file with authors).
53. *See* Irrigation Innovation Consortium, *Funded Research Projects*, https://irrigationinnovation.org/funded-research-projects/ (last visited Nov. 12, 2020).
54. *See* Frank Ward & Manuel Pulido-Velazquez, *Water Conservation in Irrigation Can Increase Water Use*, 105 Proc. Nat'l Acad. Sci. U.S. Am. 18215 (2008); Lisa Pfeiffer & C.-Y. Cynthia Lin, *Does Efficient Irrigation Technology Lead to Reduced Groundwater Extraction? Empirical Evidence*, 67 J. Env't Econ. Mgmt. 189 (2014).
55. Foundation for Food & Agriculture Research, *FFAR Funds the Future of Sustainable Perennial Crops*, https://foundationfar.org/news/ffar-funds-the-future-of-sustainable-perennial-crops/ (last visited Jan. 23, 2021).
56. Daniel Cusick, *Grain May Take a Big Bite Out of Cropland Emissions*, E&E News (May 7, 2019), https://www.eenews.net/stories/1060290955/#:~:text=Scientists%20say%20Kernza%2C%20a%20trademarked,helping%20feed%20a%20hungry%20world.
57. *See* DeLonge et al., *supra* note 14, at 266-67 (discussing the capacity of agroecological systems to reduce environmental externalities, while providing ecosystem services and sufficient yields).
58. Niggli et al., *supra* note 36, at 55-56.

grams to help improve and spread sustainable practices. States are beginning to do the same for climate-friendly practices. As of mid-2021, 18 states have enacted laws to support "healthy soils," which largely align with climate-friendly practices, and many, such as California and Washington, encourage research as well as technical assistance. Similar bills are pending in 17 states.[59] Similarly, many states direct some of their clean water nonpoint source pollution abatement funds toward support for practices such as cover cropping, manure management, and riparian buffers that offer both water quality and climate benefits, while some states are also supporting practices to make farming more resilient to climate change. California has provided over $40 million in funding since 2016 for its Healthy Soils Program, an incentive and demonstration program for farmers and ranchers designed to increase soil carbon sequestration and reduce agricultural greenhouse gas emissions.[60] Other state legislatures, agencies, and land-grant institutions should expand on these efforts, giving programs designed to spread climate-friendly practices sufficient funding to develop robust research, education, and technical assistance arms.[61]

The Leopold Center for Sustainable Agriculture at Iowa State University provides an attractive example for future state efforts. While the Iowa Legislature eliminated the center's funding in 2017, substantially reducing its effectiveness,[62] the center's enabling legislation provides a compelling funding model for states with more favorable political environments. Established by the 1987 Iowa Groundwater Protection Act to conduct research designed to reduce the environmental harms of agriculture and to help promulgate sustainable practices,[63] the center received approximately $1.5 million annually until 2017 from a fund consisting of revenue from a small fee on nitrogen fertilizer sales and pesticide registrations. The fee on nitrogen fertilizer sales was set at 75 cents per ton of anhydrous ammonia—less than 0.2% of the average price paid by individual farmers.[64] While the center's $2 million annual

---

59. Soils 4 Climate, *US Healthy Soils Legislation Map and Table* (2020), https://policy.soil4climate.org/ (last visited Aug. 8, 2021).
60. California Department of Food and Agriculture, *Healthy Soils Program*, https://www.cdfa.ca.gov/oefi/healthysoils/ (last visited Sept. 4, 2021).
61. *An Act to Promote Healthy Soils: Hearing on H.3713 Before the Joint Comm. on Environment, Natural Resources, and Agriculture*, 2017 Leg., 190th Gen. Court (Mass. 2017) (statement of Peter Lehner, Director, Sustainable Food and Farming, Earthjustice) (identifying state legislative efforts to foster healthy soils).
62. Brianne Pfannenstiel & Jeff Charis-Carlson, *Branstad Defends Defunding of Leopold Center*, Des Moines Reg., May 15, 2017, https://www.desmoinesregister.com/story/news/politics/2017/05/15/branstad-defends-defunding-leopold-center/323382001/.
63. Iowa Code §266.39 (2017).
64. The retail price of anhydrous ammonia was $467 per ton in July 2017. Russ Quinn, *DTN Retail Fertilizer Trends: Anhydrous Breaks 8% Lower*, DTN/Progressive Farmer, July 7, 2017, https://www.

budget represented only a tiny portion of the amount spent nationally on agricultural research, it has an impressive record in fostering sustainable practices and has developed a national reputation for its groundbreaking research.

4. Data Collection and Analysis

Accurate and timely data are critical to the success of USDA's programs and to the understanding, expansion, and promotion of carbon farming. Such data allow the department, policymakers, and the public to monitor programs, ensure that USDA is using its resources effectively, and analyze trends in agriculture. The National Agricultural Statistics Service (NASS) and ERS collectively provide much of the data and analysis that policymakers, researchers, and agricultural firms use to inform their decisions. NASS conducts surveys and provides statistics on agriculture, while ERS provides social science information and analysis on a variety of issues relating to agriculture, the food system, and the rural United States. Both of these agencies will need to increase their focus on climate-friendly practices and carbon-farming systems in order to allow decisionmakers to effectively evaluate interventions, practices, and programs designed to reduce agricultural emissions.

In addition to improving its ability to collect and analyze survey and other research data, USDA also needs to collect—and make available—more data through its commodity, conservation, and crop insurance programs. The department provides billions of dollars to producers across the country each year through these programs, yet there is little data available that would allow policymakers and the public to meaningfully evaluate their impact. USDA should make a number of changes, including releasing timely data on conservation compliance spot checks, and collecting additional data on Federal Crop Insurance Program and conservation program participants, in order to ensure that these programs are effective and are aligned with national climate goals. Congress should also repeal §1619 of the 2008 Farm Bill, which prevents government agencies and the public from accessing critical information about businesses receiving federal funding through the department.[65] This guarantee of secrecy, which no other industry enjoys, will continue to hobble efforts to evaluate and improve federal farm programs until it is reversed.[66]

❏ *National Agricultural Statistics Service.* Every five years NASS conducts the Census of Agriculture (COA), which attempts to count every farm and ranch

---

dtnpf.com/agriculture/web/ag/news/article/2017/07/12/anhydrous-breaks-8-lower.
65. 7 U.S.C. §8791(b).
66. *See infra* notes 83-92.

in the United States, and provides a wealth of information about operator characteristics, finances, and practices.[67] It also conducts the Agricultural Resource Management Survey (ARMS), which provides an annual source of data on farm and ranch practices, resource use, and finances, and collects data on a number of other important issues, such as on-farm chemical use, expected yields, and farm labor.[68] These data sets have the potential to improve our understanding of how the agricultural sector is mitigating and adapting to climate change through the adoption of different practices and technologies and are critical for benchmarking progress towards conservation, climate, and environmental justice goals.

The COA and ARMS already include some questions about climate-friendly agricultural practices, yet these questions fail to meet the needs of farmers, policymakers, and researchers who need data about this quickly growing sector of the agricultural economy. This dearth of information makes it more challenging for policymakers and nonprofits to distribute much-needed resources, for researchers to investigate new developments and opportunities, and for farmers and industry groups to expand their markets or develop new ones. NASS should ensure that the agricultural census and ARMS provide a comprehensive data set on climate-friendly practices and the operations utilizing them. In addition, NASS should establish an advisory committee of researchers and practitioners who study or utilize climate-friendly practices in order to identify additional ways to better serve the sector. Congress should also allocate sufficient funding for a new Census of Carbon Farming to collect detailed data on the prevalence and use of climate-friendly practices such as alley cropping and range planting, as well as demographic, financial, and geographic data on the operations utilizing them. This new census, long overdue, will provide policymakers, the public, and farmers a reliable and detailed source of information about this critical sector of the industry.

❑ *Economic Research Service.* ERS received less than $85 million in funding in FY 2020,[69] a substantial reduction from previous levels of funding. In 1977, for example, ERS received in excess of $200 million after adjusting for inflation.[70] Its budget has slowly declined since the beginning of the Reagan

---

67. *See* NATIONAL AGRICULTURAL STATISTICS SERVICE, USDA, 2017 CENSUS OF AGRICULTURE, U.S. NATIONAL LEVEL DATA (2019).
68. *See* NASS, USDA, *Surveys: Agricultural Resource Management Survey (ARMS)*, https://www.nass.usda.gov/Surveys/Guide_to_NASS_Surveys/Ag_Resource_Management/ (last updated Oct. 19, 2020).
69. USDA, FY 2021 BUDGET SUMMARY 80 (2020).
70. ERS received $48.8 million in 1977 or $216.1 million after adjusting for inflation. Douglas E. Bowers, *The Economic Research Service, 1961-1977*, 64 AGRIC. HIST. 231, 242 (1990).

Administration, reducing the agency's capacity to produce original analysis in areas of critical concern, particularly those relating to poverty, pollution, and climate change. Indeed, the Trump Administration's decisions to move ERS staff from Washington, D.C., to Kansas City and to lower the agency's budget were reportedly designed to limit the agency's ability to analyze issues such as these.[71]

In FY 2017, ERS researchers devoted only 2% of their time to issues relating to climate adaptation and mitigation.[72] This share has likely decreased since then as the Trump Administration undercut ERS and climate-related research at USDA. The current administration should not only reverse the Trump Administration's restriction of climate-related research, but should also greatly increase the resources devoted to climate mitigation and resilience. Agricultural businesses lose billions of dollars of crops and livestock each year due to climate-related damages, and that amount will only increase as the climate crisis intensifies.[73] There is an urgent need for the industry to rapidly adopt climate-friendly practices to be able to withstand these growing challenges. ERS should quickly expand the number of researchers in its climate change program area, while ensuring that other program areas have researchers devoted to characterizing the specific impacts of climate change on their specialties.

Notably, ERS has historically lacked funding and support to expand its research beyond economics. Following the establishment of ERS in 1961, overtly white supremacist legislators blocked ERS efforts to study rural poverty and marginalized populations, particularly Black farmers, for decades.[74]

---

71. *See* Liz Crampton & Ryan McCrimmon, *Trump Administration to Move USDA Researchers to Kansas City Area*, POLITICO, June 13, 2019, https://www.politico.com/story/2019/06/13/usda-kansas-city-area-1529072; Phil McCausland, *Gutting of Two USDA Research Agencies Is Warning to All Federal Agencies, Ex-Employees Say*, NBC NEWS, Oct. 8, 2019, https://www.nbcnews.com/news/us-news/gutting-two-usda-research-agencies-warning-all-federal-agencies-ex-n1062726.
72. Calculated by the authors using Freedom of Information Act Response No. 2020-REE-04381-F From USDA to Authors (June 3, 2020) (on file with authors).
73. *See supra* Chapter III.A. Farms lost more than $1 billion in 2011 due to high temperatures alone. Jerry Hatfield et al., *Agriculture*, *in* U.S. GLOBAL CHANGE RESEARCH PROGRAM, CLIMATE CHANGE IMPACTS IN THE UNITED STATES: THE THIRD NATIONAL CLIMATE ASSESSMENT 150 (Jerry M. Melillo et al. eds., U.S. Government Printing Office 2014).
74. The chair of the of Agricultural Subcommittee of the House Appropriations Committee, Rep. Jamie Whitten (D-Miss.), known as the "permanent secretary of agriculture," threatened to cut ERS' funding when the agency was created if they produced any "hound dog studies" on rural Black residents. RICHARD A. LEVINS, WILLARD COCHRANE AND THE AMERICAN FAMILY FARM 52 (2000). The political scientist Don Hadwiger wrote in 1971 that Whitten and other southern legislators "have continued their role as research chiefs, overseeing . . . most of the available expertise on rural America," and have blocked research regarding human nutrition, farmworkers, and Black people, leaving "vast areas of the rural landscape . . . either off-limits or . . . covered on tiptoe." Don F. Hadwiger, *The Freeman Administration and the Poor*, 45 AGRIC. HIST. 21, 25-26 (1971). Whitten remained chair of agricultural appropriations until 1979 when he became chair of the House Committee on Appropriations,

While ERS was able to conduct important research in these areas at times, the congressional appropriations process constrained its ability to research issues outside of narrowly defined economic topics despite the agency's mission to provide social science analysis on a wide range of concerns. Decades of stagnant budgets have exacerbated this problem, leaving the agency with little flexibility to hire social scientist professionals outside of economics and unable to meet the research needs of rural communities. Ninety-eight percent of ERS' research scientists were agricultural economists in 2019.[75] ERS' research staff also lacks racial and ethnic diversity, further limiting its capacity to examine issues affecting people of color. Of the 83 research scientists on its staff in 2019, 62 were white (75%), 5 were Hispanic (6%), and 5 were Black (6%).[76] ERS will need to hire more social scientists with training outside of economics—while also recruiting more non-white ones—in order to provide the public with critical analysis on poverty, racial inequities, climate change mitigation, and other important topics affecting rural America.

❑ *Program and producer data.* As USDA works to advance climate-friendly practices, it will be critical for the department to collect additional data to understand how its commodity, crop insurance, and conservation programs impact farm finances, practices, and the climate. USDA's data on Federal Crop Insurance Program participants is especially limited due to a general prohibition against data disclosure that does not apply to other programs.[77] There is no compelling public policy justification for such a prohibition.[78] Congress should repeal it and require USDA to collect and report data on participating operations and, when possible, their practices. The department also does not collect data on how long operations maintain conservation activities after program enrollment ends. As a result, researchers, agencies, and the public cannot reliably estimate net emissions reductions from conservation programs. USDA should require producers who have received conservation funding to fill out short annual questionnaires for 5 to 10 years post-enrollment, while also conducting surveys to track longer term behavior.

---

a position he retained until 1990. David Binder, *Jamie Whitten Who Served 53 Years in House, Dies at 85*, N.Y. TIMES, Sept. 10, 1995, at 53. Congress renamed USDA's headquarters in honor of Whitten in 1994. Pub. L. No. 103-404, §2, 108 Stat. 4206 (1994).

75. Calculated by the authors using Freedom of Information Act Response No. 2020-REE-04647-F From USDA to Authors (June 19, 2020) (on file with authors).
76. Calculated by the authors using *id.*
77. 7 U.S.C. §1502(c)(1).
78. As discussed below in Section B.1, the Federal Crop Insurance Program is a subsidy and a substantial one at that: it covers more than 60% of the cost of a crop insurance premium on average. Information about this generous subsidy will be critical to reducing agricultural emissions.

As discussed below in Chapter V.B.6, USDA has consistently failed to enforce conservation compliance requirements, which prohibit producers enrolled in a number of federal farm programs from producing agricultural products on highly erodible land without a conservation plan or on unconverted wetlands under any circumstances.[79] The department's failure to enforce these basic conservation protections results in environmental degradation and, in some parts of the country, substantially higher agricultural emissions. There are a number of factors that have contributed to this failure, many of which we detail below, but a lack of data transparency plays an important role. A report by the Farm Bill Law Enterprise, a group of nonpartisan legal scholars and clinical programs, found that effective enforcement of conservation compliance requirements was hindered by insufficient data on compliance enforcement and program efficacy.[80] The Farm Bill Law Enterprise recommended that Congress require "compliance and enforcement data [to be made] available to the public and reported in a timely manner to Congress, with a granularity comparable to the Agricultural Census."[81] This would require the department to develop tracking and reporting procedures necessary to enforce conservation compliance, while ensuring that Congress and the public remain informed about these important conservation protections.[82]

What little data USDA does collect on producers and landowners is often kept in the dark due to a special-interest provision known as §1619, which was introduced in the 2008 Farm Bill.[83] Section 1619 prohibits USDA from disclosing information provided by producers or landowners about their operations, practices, or land in order to participate in USDA programs, as well as any geospatial information collected by the department about such agricultural land or operations. While already very broad, both courts[84] and USDA[85] have interpreted §1619 to provide an exemption to FOIA as well.[86]

---

79. 16 U.S.C. §§3811-3812 & 3821. Wetlands drained or filled before December 23, 1985, are not protected. *Id.* §3822(b)(1)(A).
80. Emily Broad Leib et al., Farm Bill Law Enterprise, Productivity and Risk Management 15-16 (2018).
81. *Id.* at 15-17.
82. *Id.* at 16.
83. 7 U.S.C. §8791.
84. *E.g.*, Zanoni v. USDA, 605 F. Supp. 2d 230 (D.D.C. 2009); Ctr. for Biological Diversity v. USDA, 626 F.3d 1113 (9th Cir. 2010); Cent. Platte Nat. Res. Dist. v. USDA, 643 F.3d 1142 (8th Cir. 2011).
85. Memorandum from Boyd K. Rutherford, Assistant Secretary, USDA, to Agency Freedom of Information Act Officers (July 30, 2008), *available at* https://www.sej.org/sites/default/files/USDA1619Memo073008.pdf.
86. 5 U.S.C. §552(b)(3). *See* Ann Havemann, The Center for Progressive Reform, *Going Dark on the Farm: Farm Bill Could Cloak Big Ag in Even More Secrecy*, Jan. 14, 2014, http://progressivereform.org/cpr-blog/going-dark-on-the-farm-farm-bill-could-cloak-big-ag-in-even-more-secrecy/ ("Congress has

The only exceptions allow the department to release information: (1) to a person or federal, state, local, or tribal agency working in cooperation with any USDA program when providing technical or financial assistance with respect to the agricultural operation, agricultural land, or farming or conservation practices; (2) when necessary for a response to a disease or pest threat to agriculture operations; (3) if the program information has been transformed into a statistical or aggregate form without naming any individual owner, operator, or producer; or (4) if the agricultural producer or owner consents to its release.

As a result of its broad language and narrow exceptions, §1619 has impeded USDA efforts to oversee its programs and conduct scientific research[87] and increased inefficiencies between federal and state conservation programs.[88] It has also reduced the effectiveness of state programs,[89] and state and regional partnerships.[90] Without access to USDA's data on Clean Water Act compliance at CAFOs, for example, EPA and state agencies are unable to effectively enforce the Clean Water Act.[91] Section 1619 also keeps civic and watchdog groups from tracking how USDA payments are distributed, reducing their ability to identify waste and fraud, or to evaluate whether programs are meeting their stated goals.

Agriculture is the only sector in which businesses receiving government payments for non-classified activities are shielded from public oversight. Congress should eliminate this loophole and repeal §1619. The section impedes interagency collaboration, the prevention of waste of

---

never extended FOIA's privacy exemption to coporations [prior to Section 1619]"); Rena Steinzor & Yee Huang, Going Dark Down on the Farm: How Legalized Secrecy Gives Agribusiness a Federally Funded Free Ride (2012).

87. Section 1619 "makes it impossible to assess the efficacy of the hundreds of millions of dollars that the U.S. taxpayer spends on conservation." (former USDA employee). Adena R. Rissman et al., *Public Access to Spatial Data on Private-Land Conservation*, Ecology & Soc'y 24 (June 2017); U.S. Gov't Accountability Office, Farm Bill: Issues for Consideration 14-18, GAO-12-338SP, at 16 ("almost half of USDA's field offices did not implement farm-bill conservation compliance provisions as required, in part because the offices reported that they were uncomfortable with their enforcement role. Some field office staff said it was difficult to cite farmers for noncompliance in the small communities where the staff and farmers both live and work. Furthermore, their noncompliance decisions were waived about 61 percent of the time, and the waiver decisions were not always adequately justified, providing further disincentive for issuing noncompliance decisions."). *See also* Laurie Ristino & Gabriela Steier, *Losing Ground; A Clarion Call for Farm Bill Reform to Ensure a Food Secure Future*, 42 Colum. J. Env't L. 79-106 (2016).
88. Jess R. Phelps, *Conservation, Regionality, and the Farm Bill*, 71 Me. L. Rev. 293, 339 (2019).
89. Adena R. Rissman et al., *Public Access to Spatial Data on Private-Land Conservation*, Ecology & Soc'y 24 (June 2017).
90. Jess R. Phelps, *Conservation, Regionality, and the Farm Bill*, 71 Me. L. Rev. 293, 338-39 (2019).
91. Rena L. Steinzor & Ling-Yee Huang, Agricultural Secrecy: Going Dark Down on the Farm: How Legalized Secrecy Gives Agribusiness a Federally Funded Free Ride 7-9 (Briefing Paper No. 1213) Center for Progressive Reform, 2012). *See* CAFO Reporting Rule, 76 Fed. Reg. 65, 445 (proposed Oct. 21, 2011).

public funds, and research into climate change, which poses a critical threat to agricultural production. Failing congressional action, USDA should expand the Conservation Cooperator program that allows confidential sharing of information, beyond the very few currently approved and tightly managed Cooperators.[92]

5. Extension Service

One of the most significant challenges facing carbon farming may be the difficulty inherent in learning, adopting, and disseminating new agricultural practices. Even large-scale farm operations may be loath to try new practices since they have previously invested significant sums in infrastructure and equipment designed for conventional practices.[93] And unlike in other industries where reducing emissions often entails the adoption of widely applicable practices or technology, each farm operation must contend with a range of unique variables, such as soil and climate conditions. Finally, some measures, like reducing nitrogen fertilizer overapplication, may seem too risky given all the other uncertainties of farming.

The Cooperative Extension System (CES) has often proven to be an effective mechanism for disseminating and perpetuating new agricultural practices.[94] For example, no-till farming has spread more deeply and more rapidly in states where extension services have advocated for its use.[95] Research also indicates that farmers are more receptive to learning new information and practices from extension programs than they are from other government bodies. An extensive 2012 survey of Corn Belt farmers found that 63% of the respondents believed that extension services should help farmers prepare for "increased weather variability"—despite the fact that only 41% of the surveyed farmers believed that climate change was caused by human activi-

---

92. Natural Resources Conservation Service, USDA, The Freedom of Information Act, The Privacy Act, Section 1619 of the Farm Bill (2011), *available at* http://www.nrcs.usda.gov/Internet/FSE_DOCUMENTS/stelprdb1166472.pptx.
93. Almost 60,000 family farms received more than $1 million in gross income in 2012; however, average production expenses for these large-scale farms exceeded $1 million as well. *See id.* at 1-2.
94. *See, e.g.*, Irwin Feller, *Technology Transfer, Public Policy, and the Cooperative Extension Service—OMB Imbroglio*, 6 J. Pol'y Analysis & Mgmt. 307, 307 (1987) ("The Cooperative Extension Service has come to represent the best of both an articulated but decentralized political arrangement and of a technology transfer system."); George McDowell, *Engaged Universities: Lessons From the Land Grant Universities and Extension*, 585 Annals Am. Acad. Pol. & Soc. Sci. 31, 35-36 (2003).
95. "We also struggle with the fact if a practice is not supported and sold by Oklahoma State University and Oklahoma State Extension, it's slow to be adopted." John Dobberstein, *No-Till Movement in U.S. Continues to Grow*, No-Till Farmer, Aug. 1, 2014, https://www.no-tillfarmer.com/articles/489-no-till-movement-in-us-continues-to-grow?v=preview.

ty.[96] In contrast, only 43% believed that state and federal agencies should help farmers to prepare for changing weather patterns.[97]

As with federal funding for agricultural research, funding for the extension system is historically low. The federal government spent approximately the same amount on the extension system in 2016 as it did in 1992 *without* accounting for inflation.[98] When the amount is adjusted for inflation, the extension service's budget was only about 60% of what it was in 1992. This drop in funding has accompanied a decline in the extension service's influence, and agribusiness and private consultants have filled the void.[99] Congress should at a minimum double the extension system's budget to $900 million, designating the additional funds for climate-related education, programming, and services. Distribution of these funds should favor states providing matching funds in order to reward states that invest in carbon farming and to help win local buy-in for the new extension program.

A long-term drop in funding has severely hindered extension's ability to disseminate information on new practices or to reach underserved populations. The number of county extension agents fell by 30% nationwide between 1980 and 2010,[100] and has likely continued to fall since then. Some states have also seen dramatic declines in extension specialists—research faculty that work with agents to provide programming.[101] As a result, farmers must increasingly turn to agribusiness dealers focused on making sales for information about crops, practices, and services.[102] Private-sector consultants generally either work for, or have close financial ties to, chemical and other conventional agribusiness firms and are poorly positioned—and often disincentivized—to help farmers implement cutting-edge climate-friendly

---

96. J. Gordon Arbuckle Jr., *Corn Belt Farmers Are Concerned, Support Adaptation Action in the Ag Community*, in Resilient Agriculture 22 (Lynn Laws ed., Sustainable Corn Project 2014).
97. Additionally, only 52% believed that farmer organizations should help farmers to prepare. *Id.*
98. *Compare* Consolidated Appropriations Act, 2016, Pub. L. No. 114-113, 129 Stat. 2250 (2015) (indicating that the federal government allocated $426 million to extension services in 2016), *with* National Research Council, Colleges of Agriculture at the Land Grant Universities: A Profile 68 (1995) (showing that the federal government allocated $401 million to extension services in 1992).
99. *See* Linda S. Prokopy et al., *Extension's Role in Disseminating Information About Climate Change to Agricultural Stakeholders in the United States*, 130 Climatic Change 261, 268 (2015); Mahdi M. Al-Kaisi et al., *Extension Agriculture and Natural Resources in the U.S. Midwest: A Review and Analysis of Challenges and Future Opportunities*, 44 Nat. Sci. Educ. 26, 27-28 (2015); Wendy Wintersteen et al., *Evaluation of Extension's Importance to Agribusinesses: A Case Study of Iowa*, 45 Am. Entomologist 6, 6 (1999).
100. Sun Ling Wang, *Cooperative Extension System: Trends and Economic Impacts on U.S. Agriculture*, 29 Choices 1, 2 (2014).
101. The number of extension specialists affiliated with the University of Illinois, for example, fell by 83% between 1986 and 2013. Al-Kaisi et al., *supra* note 99, at 29.
102. *Id.* at 28.

practices.[103] This is exacerbated by the fact that cash-strapped extension systems are increasingly turning to the private sector for funding, which further reduces their ability to devote resources to climate-friendly practices that do not directly benefit industry.[104]

NIFA, which received $484 million in 2018 to administer the extension system and help fund state extension services,[105] currently does little to support climate mitigation through the extension system. NIFA should immediately begin offering resources for carbon farming within the extension system, as it does for other issues, such as weed control and youth education. It should also work with states to ensure that all extension agents are knowledgeable about climate-friendly practices and should fund specialists who focus primarily on climate mitigation practices in order to ensure an in-house constituency and expertise.

Just as the extension service played an important role in disseminating modern agricultural practices in the 20th century, Congress should either expressly expand the mandate of existing extension services or fund a new climate extension service. This extension capacity can build on the base of the existing (as of 2021) Climate Hubs, 10 regional centers established by USDA in 2014 to provide much-needed support for climate mitigation and adaptation efforts by translating climate research into tools, materials, and methods for extension and outreach.[106] Although funding for the extension service largely comes from state and local sources, federal funding for climate-related extension will be critical—particularly in states where policymakers deny anthropogenic climate change.[107] Any climate-focused extension program will also need to have a clear climate-focused mission and retain institutional independence to ensure that its efforts are not compromised by local political dynamics.

---

103. *See, e.g., id.*
104. A 2014 survey of 212 extension professionals found that 81% of respondents had partnered with private industry on extension activities and that almost 80% had received private funding for at least some of their research. Rayda K. Krell et al., *A Proposal for Public and Private Partnership in Extension*, 7 J. Integrated Pest Mgmt. 1, 4 (2016).
105. Consolidated Appropriations Act, 2018, Pub. L. No. 115-141, 132 Stat. 356.
106. *See* USDA, USDA Regional Hubs for Risk Adaptation and Mitigation to Climate Change 1 (2015).
107. Federal funding currently accounts for about 10% of extension's funding—a historical low. Marsha Mercer, *Cooperative Extension Reinvents Itself for the 21st Century*, Pew Charitable Tr., Sept. 9, 2014, https://www.pewtrusts.org/en/research-and-analysis/blogs/stateline/2014/09/09/cooperative-extension-reinvents-itself-for-the-21st-century.

6. Technical Assistance

While the CES offers a number of educational services to farmers and other rural residents, NRCS' advising focuses exclusively on natural resource issues.[108] The agency has roughly 4,400 conservationists in field offices throughout the country who supply farmers with information about conservation practices and programs, conduct on-farm resource assessments and inventories, and provide assistance to farmers implementing conservation practices.[109] NRCS receives funding to provide technical assistance through a variety of programs. The single largest source is the Conservation Technical Assistance program (CTA), which received $726 million in FY 2018 to provide conservation planning and implementation assistance for any group or individual engaged in agriculture throughout the United States.[110] USDA's conservation programs, discussed below, also generally include funding for technical assistance to help farmers fulfill their program agreement or contract with the agency.[111] The three largest conservation programs, the Environmental Quality Incentives Program (EQIP), the Conservation Stewardship Program (CSP), and the Conservation Reserve Program (CRP), provided $768 million for technical assistance in FY 2019.[112] EQIP contributed the majority of these funds, devoting $428 million or 25% of its total funding to technical assistance.[113] CSP also included $267 million, representing 19% of CSP's budget, while only 4% of CRP funding ($73 million) was devoted to technical assistance.[114]

---

108. Sarah Wiener, *Ready, Willing, and Able? USDA Field Staff as Climate Advisors*, 75 J. Soil Water Conservation 62, 62 (2020).
109. Freedom of Information Act Response No. 2020-REE-04897-F From USDA to Authors (Aug. 5, 2020) (on file with authors) (including the number of NRCS conservationists employed each FY from 2013 to 2019). *Id*. at 63 (describing conservationists' activities). *See also* OBPA, USDA, FY 2021 USDA Explanatory Notes—Natural Resources Conservation Service 29-39 to 29-43 (2020) (describing NRCS' Conservational Technical Assistance Program).
110. Megan Stubbs, Congressional Research Service, FY2020 Appropriations for Agricultural Conservation 3 tbl.1 (2020).
111. *Id*. at 4.
112. Calculated by the authors using USDA NRCS, *NRCS Conservation Programs: Environmental Quality Incentives Program (EQIP)* [hereinafter NRCS EQIP], https://www.nrcs.usda.gov/Internet/NRCS_RCA/reports/fb08_cp_eqip.html (last visited Nov. 12, 2020); USDA NRCS, *NRCS Conservation Programs: Conservation Stewardship Program (CSP)* [hereinafter NRCS CSP], https://www.nrcs.usda.gov/Internet/NRCS_RCA/reports/fb08_cp_cstp.html (last visited Nov. 12, 2020); USDA NRCS, *NRCS Conservation Programs: Conservation Reserve Program (CRP)* [hereinafter NRCS CRP], https://www.nrcs.usda.gov/Internet/NRCS_RCA/reports/fb08_cp_crp.html (last visited Nov. 12, 2020).
113. *Compare* NRCS EQIP, *supra* note 112, *with* Congressional Budget Office, USDA's Mandatory Farm Programs—CBO's Baseline as of March 6, 2020, at 26 (2020).
114. *Compare* NRCS CSP & NRCS CRP, *supra* note 112, *with* Congressional Budget Office, *supra* note 113.

Technical assistance provided through Farm Bill conservation programs has increased steadily since FY 2009, when EQIP, CSP, and CRP collectively included only $364 million for technical assistance.[115] As conservation program funding for technical assistance has increased, however, CTA's budget has fallen substantially in real terms. CTA is funded through the Conservation Operations (which the Trump Administration proposed renaming "Private Lands Conservation Operations"), almost 90% of which goes to CTA. The 2018 Farm Bill appropriated $819 million annually to Conservation Operations in discretionary funding, making it USDA's largest discretionary conservation program.[116] Congressional appropriations for Conservation Operations has hovered around $800 million in nominal terms since 2002, which represents a substantial decline in real terms.[117] Inflation-adjusted funding for Conservation Operations reached more than $1.1 billion early in the George W. Bush Administration before reaching a low of $830 million in FY 2020.[118] As a result, the number of full-time staff positions funded by Conservation Operations has steadily declined from more than 8,000 in FY 2000 to a little under 5,000 in FY 2020.[119] Nonetheless, it remains a critically important program. Approximately one million farmers, ranchers, and foresters received technical assistance through CTA in FY 2019, affecting 111 million acres of agricultural land and forestland.[120]

While NRCS' national reach and focus on conservation practices make it an ideal agency to provide technical assistance to farmers using carbon-farming practices, it currently lacks the capacity to do so effectively. Its field staff generally focus on conventional conservation practices that have moderate climate effects, at best. Even more disconcerting, many of its conservationists are skeptical of anthropogenic climate change[121] and do not see increasing climate change mitigation or resilience as part of their job.[122] A 2017 nationwide survey of almost 2,000 NRCS field staff found that only 45% agreed that there will be an increased need for agency programs in their service area due to changing weather patterns and only 52% agreed that assisting farmers

---

115. Calculated by the authors using NRCS EQIP & NRCS CSP, *supra* note 112; CONGRESSIONAL BUDGET OFFICE, COMMODITY CREDIT CORPORATION ACCOUNT PLUS OTHER ACCOUNTS COMPARABLE TO THE USDA BASELINE, JANUARY 2010 CBO BASELINE 18 (2010).
116. STUBBS, *supra* note 110, at 2-3 tbl.1.
117. *Id.* at 6 fig.3.
118. *Id.* at 3 tbl.1.
119. *Id.* at 6 fig.3.
120. OBPA, *supra* note 109, at 29-58.
121. *See* Wiener, *supra* note 108, at 63.
122. *Id.* at 68.

with increased weather variability is part of their job.[123] Neither additional funding nor reframing climate change resilience as "soil health" will fully address these problems. Congress should dramatically increase funding for conservation technical assistance, but it will need to tie any new funds to the most environmentally beneficial conservation practices in order to ensure that the desired results are achieved.

7. Improving Coordination Among Research, Extension, and Technical Assistance Programs

In addition to expanding funding for research, extension, and technical assistance, USDA must improve its coordination of these services. USDA is a massive bureaucracy, with 35 agencies and offices, which often work at cross-purposes.[124] Yet in order to achieve climate neutrality in agriculture, USDA must address emissions in a systematic fashion, organizing its research, extension, and technical assistance arms around common goals and priorities. And it must work towards more ambitious national sequestration targets—set by Congress and updated at least every four years—to ensure that the sector achieves climate neutrality.

To accomplish these goals, USDA can build on its existing Climate Hubs, which translate new climate-related research into tools, materials, and methods for outreach and education, support applied research, and coordinate USDA's climate-related activities in each region.[125] Unfortunately, Climate Hubs do not have their own source of funding, but rather have received funding from other USDA agencies.[126] The Trump Administration also diverted resources away from the Climate Hubs, leaving them understaffed and underutilized.[127] The current administration and Congress should provide dedicated funding for Climate Hubs and make explicit their role in coordinating among research, extension, and technical assistance programs for the benefit of farmers and rural communities.

The department can also build upon its "Building Blocks" plan, released in 2015, to reduce or offset greenhouse gas emissions through agriculture and

---

123. *Id.*
124. USDA's support for agricultural commodities, for example, often undermines its dietary recommendations.
125. USDA, *supra* note 106.
126. Helena Bottemiller Evich, *"I'm Standing Right Here in the Middle of Climate Change": How USDA Is Failing Farmers*, Politico, Oct. 15, 2019, https://www.politico.com/news/2019/10/15/im-standing-here-in-the-middle-of-climate-change-how-usda-fails-farmers-043615 (noting that the Climate Hubs do not have their own line item).
127. *See id.* (explaining how USDA agencies have withdrawn resources from the Climate Hubs).

Chapter V. Transforming Farm Policy Toward Climate-Neutral Agriculture

forestry by 120 million metric tons of carbon dioxide equivalent (MMT CO_2 eq.) per year by 2025, by creating a department-wide agenda, drawing on a range of technologies and practices, and leveraging efforts by other government agencies and the private sector.[128] Though a promising idea, the plan has several weaknesses that USDA should work to remedy: its soil carbon sequestration goals are modest,[129] it favors practices preferred by agribusiness companies rather than those with demonstrated long-term climate benefits,[130] and it relies on voluntary incentives, which are often impermanent and ineffective at storing soil carbon.[131] And even if the planned emissions reductions materialized, the plan would not come close to achieving climate neutrality in agriculture due to its overwhelming reliance on nonagricultural sectors, such as forestry and housing energy, for greenhouse gas reductions.[132] USDA should release a new plan that lays out more ambitious goals and focuses on agricultural practices with the greatest potential to reduce net emissions.

B.   Public Subsidy and Conservation Programs

The federal government supports farms through five main avenues: crop insurance, commodity programs, conservation payments, credit, and trade. The first three categories—crop insurance, commodity, and conservation programs—are expected to provide farming operations with about $20 billion per year through 2023, making up 97% of Farm Bill appropriations outside of the Supplemental Nutrition Assistance Program (SNAP) (formerly "food stamps").[133] In 2018, the Trump Administration began providing farmers affected by retaliatory tariffs with direct payments through the new Market Facilitation Program (MFP).[134] The program was intended

---

128. *See* USDA, USDA Building Blocks for Climate Smart Agriculture and Forestry: Implementation Plan and Progress Report 2 (2016).
129. The plan calls for only 4-18 MMT CO_2 eq. of soil carbon to be sequestered each year through climate-friendly agricultural practices by 2025. *Id.* at 4.
130. Building Blocks prioritizes the synthetic fertilizer industry's best management practices, conventional no-till agriculture, and manure management systems for AFOs. *See id.* The climate benefits of these practices are much lower than other feasible options available to farm managers.
131. As discussed *infra* Section B.4, any increased carbon sequestration accomplished through voluntary programs such as CRP should be considered temporary. Conserving sensitive lands through CRP is one of the main elements of Building Blocks. *See* USDA, *supra* note 128, at 8.
132. Approximately 70% of the greenhouse gas reductions are in nonagricultural sectors. *See id.*
133. Congressional Research Service, The 2018 Farm Bill (P.L. 115-334): Summary and Side-by-Side 5 fig.1 (2010). The 2014 Farm Bill allocated a similar amount to the farm safety net between 2014-2018. Letter From Douglas Elmendorf, Director, Congressional Budget Office, to Frank Lucas, Chairman, House Committee on Agriculture (Jan. 28, 2014) (on file with Congressional Budget Office). Funding for these programs was set to expire with the 2014 Farm Bill at the end of FY 2018.
134. Congressional Research Service, Farm Policy: Comparison of 2018 and 2019 MFP Programs 1 (2019).

to serve as a temporary stopgap during trade negotiations between the United States and China, but due to its size—it distributed more than $14.5 billion for the 2019 program[135]—it has had important climate implications. The federal government also provides approximately $40 billion in subsidized financing to farm operations each year through the Farm Credit System (FCS) and Farm Service Agency (FSA).[136] Finally, USDA operates large foreign market development programs, which, while substantially smaller in size than USDA's farm safety net and lending programs, have helped spur the rapid growth of exports in carbon-intensive commodities. Each of these avenues is examined below, with an emphasis on how existing programs can be adapted to help decarbonize agriculture.

Ultimately, however, Congress should pass new legislation to optimize government support for carbon farming. We therefore include recommendations on how Congress could create a farm safety net that would more effectively meet the social and environmental needs of the nation. When crafting new agricultural legislation, regulations, or programs, it is important to recognize that the ability of farming operations to integrate new practices and absorb additional transactional costs varies considerably. While many climate-friendly techniques are cost effective regardless of a farm's scale, some requirements may nonetheless disadvantage small and midsized operations. The Food Safety Modernization Act attempted to account for this by exempting certain farms with gross sales below $500,000 from its requirements.[137] New regulations and requirements could also be similarly tiered so that farmers with small and midsized operations, or those who receive only a small portion of their household income from farming, face minimal new costs or paperwork.[138] Additionally, USDA and the extension service should

---

135. Press Release, USDA, USDA Issues Third Tranche of 2019 MFP Payments (Feb. 3, 2020) (on file with author).
136. Calculated by the authors. According to the Congressional Research Service, FSA provided 2.6% of all farm loans by volume in FY 2017 and FCS provided 41%. Since FSA direct loans totaled $2,328,150,069 in FY 2017, we can infer that FCS provided farmers with approximately $36.5 billion in loans that year. *See* JIM MONKE, CONGRESSIONAL RESEARCH SERVICE, AGRICULTURAL CREDIT: INSTITUTIONS AND ISSUES 1 (2018) (noting that FSA provides 2.6% of all farm credit through farm loans and FCS 41%); USDA, FSA, EXECUTIVE SUMMARY OF FARM LOAN PROGRAMS: FY 2017 AS OF SEPT. 30, 2017 4 (2017).
137. *See, e.g.*, FDA Food Safety Modernization Act, Pub. L. No. 111-353, §418(l)(1)-(2), 124 Stat. 3885, 3892 (2011).
138. Farms with gross earnings above $500,000 often have multiple employees, relationships with consultants, advisers, and extension staff, and are more likely to be able to afford accountants, attorneys, and other professionals to respond to new regulations and optimize their earnings. They also produce the majority of agricultural products in the United States and receive a disproportionate share of farm subsidies.

offer assistance and incentives to help small and midsized farms transition to climate-friendly practices.

1. Crop Insurance

Crop insurance is the largest component of the farm safety net.[139] Almost half of the estimated $20 billion in subsidies flowing to farms each year through farm bill programs now goes to crop insurance.[140] This is a relatively new development: between 1989 and 2014 the number of crop insurance policies doubled, the number of insured acres almost tripled, and federal spending on premium subsidies increased more than fifteen-fold.[141] This expansion was engineered by the agricultural industry, which created a new crop insurance system that is highly profitable for large-scale operations, politically more palatable to the general public than traditional subsidies,[142] and compliant with international trade agreements.

Proponents of the current crop insurance system often portray it as a safety net for farmers in the case of natural disaster,[143] but it operates more as an income subsidy. While some crop insurance policies only protect farmers from crop losses—routine or not—most provide revenue guarantees ensuring that covered crops remain lucrative. About 70% of federally subsidized crop insurance policies in 2019 were revenue-based; the remainder were mostly yield-based.[144] Despite large increases in funding in recent years, crop insurance continues to primarily serve large-scale producers of commodity crops. According to the Congressional Research Service, more than 70% of the acres covered by crop insurance are devoted to one of four crops—corn, cotton, soybeans, and wheat.[145] The 2014 Farm Bill opened crop insurance up to a wider range of products, and the USDA agency in charge of crop insurance programs, the Risk Management Agency (RMA), has taken important

---

139. *See* Congressional Research Service, *supra* note 133, at 6 tbl.3 (showing projected mandatory outlays by program area through FY 2023).
140. Ralph Chite et al., Congressional Research Service, The 2014 Farm Bill (P.L. 113-79): Summary and Side-by-Side 6, 17 (2014).
141. Nathan A. Rosenberg, The Farm Bill: A Negative History 16 (2019).
142. *See* North Star Opinion Research, National Survey of Registered Voters Regarding Crop Insurance (2016) (showing that voters support government-subsidized crop insurance by a four-to-one margin when told that claims are paid "only in the event of bad weather or low prices").
143. *See, e.g.*, Iowa Secretary of Agriculture Bill Northey on crop insurance: "Farmers rely on crop insurance as an important safety net and protection from devastating losses from natural disasters." Memorandum From Bill Northey, Iowa Secretary of Agriculture, to Iowa Reporters and Editors (Oct. 28, 2015) (on file with authors).
144. Stephanie Rosch, Congressional Research Service, Federal Crop Insurance: A Primary 17 (R46686) (2021).
145. *Id.* at 2.

steps to open up crop insurance to diversified and organic farms.[146] Nonetheless, many farms, particularly small- and medium-scale operations, continue to find it impractical or unavailable.[147] In addition to bolstering large-scale operations, crop insurance has also motivated farmers to bring more land under cultivation, particularly wetlands and other marginal lands, leading to increased emissions.[148]

There are four steps USDA should take to make its crop insurance programs more climate friendly. First, the RMA should ensure that its crop insurance policies do not interfere with cover cropping or other proven decarbonizing practices, or conversely encourage less beneficial practices. Farmers using innovative or sustainable methods often have difficulty receiving crop insurance, since the department requires producers to use "good farming practices" that are "generally recognized by agricultural experts" in their immediate geographic area.[149] This effectively disallows farmers from using many innovative climate-friendly practices, such as alley cropping or integrated crop-livestock systems, with which agricultural experts in their area are unlikely to be familiar.[150] Crop insurance requirements may push farmers to plant at suboptimal times and do not encourage wider rotations. In a 2014 report on climate change and federal insurance programs, the U.S. Government Accountability Office (GAO) noted that the department's good farming practices policies discourage climate adaptation and mitigation, while incentivizing practices that "increase vulnerability to climate change."[151]

---

146. *Have Access Improvements to the Federal Crop Insurance Program Gone Far Enough?*, NSAC, July 28, 2016, http://sustainableagriculture.net/blog/crop-insurance-access-data/.
147. *Id.*
148. Daniel Sumner & Carl Zulauf, Council on Food, Agricultural, and Resource Economics, Economic & Environmental Effects of Agricultural Insurance Programs 10-12 (2012).
149. *See* Chad G. Marzen & J. Grant Ballard, *Climate Change and Federal Crop Insurance*, 43 Env't Aff. 387, 398 (2016).
150. The most recent version of the RMA's *Good Farming Practices Handbook*, released in December 2015, included some important changes. For the first time, the RMA states that practices promoted by USDA's NRCS will generally be recognized by agricultural experts as "good farming practices." RMA, USDA, Good Farming Practice Determination Standards Handbook 33 (2016) [hereinafter Good Farming Practice Determination Standards Handbook]. This could make it much easier for farmers with crop insurance to adopt climate-friendly NRCS practices, since they are often deterred from doing so by the good farming practices requirement. However, the handbook considerably weakens the new provision by giving insurance companies the power to prohibit certain practices through the terms and conditions of their policies, and by indicating that both the RMA and insurance companies may prohibit practices that do not maximize yields. *Unified Support for Conservation as Good Farming Practice Needed at USDA*, NSAC, Dec. 16, 2016, http://sustainableagriculture.net/blog/gfp-updated-at-rma.
151. GAO, Climate Change: Better Management of Exposure to Potential Future Losses Is Needed for Federal Flood and Crop Insurance 24 (2014) (GAO-15-28).

In 2015, the RMA began allowing organic farmers to use opinions from organic agriculture experts outside of their immediate geographic area.[152] In part due to this change, the amount of organic acreage enrolled in crop insurance, while still small, increased by 34% during the first year of the new policy.[153] The RMA made this change as a result of the Agriculture Risk Protection Act of 2000, which provides that good farming practices include "scientifically sound sustainable and organic farming practices."[154] The RMA should likewise create a new standard for carbon farming—a scientifically sound sustainable farming system—that would allow farmers to use carbon-farming experts outside of their immediate area, while encouraging agricultural experts to take climate change into account when assessing "good farming practices." Ideally, basic conservation practices would be required to meet the "good farming practices" threshold on the basis that, over the long run, they will improve the financial soundness of the insurance system. The Senate version of the 2018 Farm Bill defined any conservation practice or enhancement promoted by NRCS as a good farming practice for crop insurance purposes;[155] however, the final 2018 Farm Bill was weakened, and only included language defining cover cropping as one.[156] Congress should pass legislation clarifying that all NRCS conservation practices and enhancements are good farming practices.

Current RMA guidelines also hinder beneficiaries from using cover crops and other conservation practices.[157] As a result, some farmers using cover cropping are unable to benefit from crop insurance, while others forgo cover cropping in order to receive crop insurance. While the agency has made it easier to adopt practices promoted by NRCS,[158] including cover cropping,[159]

---

152. 7 C.F.R. §457.8 (2015); GOOD FARMING PRACTICE DETERMINATION STANDARDS HANDBOOK, *supra* note 150, at 32.
153. Calculated by the authors using USDA data. *Compare* RMA, USDA, FEDERAL CROP INSURANCE SUMMARY OF BUSINESS FOR ORGANIC PRODUCTION 2 (2015) (showing 777,966 organic acres enrolled in federal crop insurance in 2014), *with* RMA, USDA, FEDERAL CROP INSURANCE SUMMARY OF BUSINESS FOR ORGANIC PRODUCTION 2 (2016) (showing 1,043,403 organic acres enrolled in federal crop insurance in 2015).
154. 7 U.S.C. §1508(a)(3)(iii).
155. S. 3042, 115th Cong. §11107 (2018).
156. Agriculture Improvement Act of 2018, Pub. L. No. 115-334, §11107, 132 Stat. 4490, 4921.
157. *See, e.g.*, Todd Neeley, *Grassley Asks Vilsack to Fix Crop Insurance, Cover Crops Glitches*, DTN/PROGRESSIVE FARMER, June 28, 2016, https://www.dtnpf.com/agriculture/web/ag/blogs/ag-policy-blog/blog-post/2016/06/28/grassley-asks-vilsack-fix-crop-cover. A 2015 survey found that the most commonly cited reason among farmers for not adopting cover cropping was the concern that doing so would interfere with crop insurance. John Dobberstein, *Crop Insurance Rules Still Hinder Cover Crop Adoption*, NO-TILL FARMER, Oct. 14, 2015.
158. *See supra* note 153.
159. USDA also established an interagency working group with NRCS, the RMA, and the Farm Security Administration to "develop consistent, simple, and flexible policy" on cover crop practices, making

it needs to further revise its guidelines to not only eliminate remaining barriers to conservation practices but actually to encourage them as risk-mitigating techniques.[160] The RMA should also conduct outreach encouraging conservation practices in order to dispel the widespread fear that they interfere with crop insurance coverage.

RMA's production history requirements further disadvantage many climate-friendly practices, particularly those used in diverse cropping systems. While the federal crop insurance program currently seeks a 10-year actual production history for each insured crop to provide coverage and set premium rates, this history requirement impedes adoption of diverse cropping systems because it could take decades to collect the necessary data. The RMA should therefore consider alternative methods to examine yield trends so as not to discourage these sustainable practices.

The second of the four broad measures USDA can take to make crop insurance more climate-friendly is to ensure that publicly funded crop insurance policies treat greenhouse gas-intensive practices as risk-enhancing and reduce or eliminate their premium subsidies accordingly.[161] In fact, the Federal Crop Insurance Corporation (FCIC) may be compelled to consider the climate impact of practices when establishing policies and premiums. Congress requires the FCIC to adopt rates and policies "that will improve the actuarial soundness" of its insurance operations.[162] Encouraging practices that both reduce climate change and make farms more resilient to it will clearly make the program more actuarially sound. Recent studies have demonstrated that climate change has already significantly increased crop insurance losses;[163] climate change-reducing practices are thus important risk mitigation measures. A 2014 GAO report on federal flood and crop insurance and climate change noted that crop insurance losses are expected to increase considerably by 2040 absent significant climate change mitigation.[164] FCIC regulations also require it to seek the RMA's assessment as to whether insurance policies and premiums are consistent with USDA's

---

it easier for operators to plant cover crops in accordance with federal rules. *See* RMA, USDA, NRCS Cover Crop Termination Guidelines I (2014).

160. *See* Jessica McKenzie, *Regenerative Agriculture Could Save Soil, Water, and the Climate. Here's How the U.S. Government Actively Discourages It*, Counter, Mar. 14, 2019, https://newfoodeconomy.org/regenerative-agriculture-cover-crops-no-till-usda/ (describing how the RMA continues to discourage cover cropping).

161. *See* Claire O'Connor, Natural Resources Defense Council, Soil Matters: How the Federal Crop Insurance Program Should Be Reformed to Encourage Low-Risk Farming Methods With High-Reward Environmental Outcomes 10 (2013).

162. 7 U.S.C. §1508(i)(1).

163. *See* Noah Diffenbaugh et al., *Historical Warming Has Increased U.S. Crop Insurance Losses*, 16 Env't Res. Letters 084025 (2021).

164. GAO, *supra* note 151, at 14.

Chapter V. Transforming Farm Policy Toward Climate-Neutral Agriculture   141

policy goals when reviewing them.[165] If the plan or premium under review is not consistent with USDA's policy goals, then the FCIC may reject it.[166] Climate change mitigation is an express policy goal of USDA and the FCIC should ensure crop insurance programs are furthering that goal.[167]

While operations using risk-enhancing practices should not receive government subsidies, risk-mitigating practices should be encouraged through premium discounts as an intermediary step. The Senate version of the 2018 Farm Bill included a provision that would have allowed the FCIC to provide performance-based crop insurance discounts for risk-reducing conservation practices, including "precision irrigation or fertilization, crop rotations, cover crops, and any other practices determined appropriate."[168] This provision was omitted from the final Farm Bill, but Congress should pass a strengthened version mandating that the RMA provide performance-based discounts to farmers using conservation practices that reduce risk to crops while mitigating climate change, including those practices listed in the Senate bill.

Third, USDA should collect and report data on the impact of various practices on risk. Without robust data on how conservation practices impact yield, soil health, water quality, and resilience, the RMA will not be able to accurately evaluate how practices affect risk. These data will also help farmers and ranchers better understand the degree to which different practices can make their operations more resilient. Also, as discussed above, Congress should eliminate data secrecy and agricultural data collection limitations.

Fourth, as discussed below in Chapter V.B.6, current conservation compliance requirements should be expanded to require key conservation practices for all operations receiving crop insurance.

2.   Commodity Programs

❏ *Agricultural Risk Coverage and Price Loss Coverage.* The commodities title of the 2014 Farm Bill replaced direct payments to farmers with two new programs: Agricultural Risk Coverage (ARC) and Price Loss Coverage (PLC).[169] Commodity programs are expected to distribute an average of $6 billion annually from 2019 to 2023—about 32% of average annual subsidies under

---

165. 7 C.F.R. §400.706(b)(4) (2016).
166. *Id.* §400.706(h)(5).
167. The agency's comprehensive expression of its policy goals, its 2014-2018 Strategic Plan, lists as a strategic goal: "Ensure our national forests and private working lands are conserved, restored, and made more resilient to climate change, while enhancing our water resources." It also states the agency's objective to "lead efforts to mitigate and adapt to climate change." USDA, STRATEGIC PLAN: FY 2014-2018, at 3, 15 (2014).
168. S. 3042, 115th Cong. §11109 (2018).
169. *See generally* Agricultural Act of 2014, Pub. L. No. 113-79, 128 Stat. 649.

the 2018 Farm Bill.[170] These programs supplement crop insurance for specified commodities, such as wheat, corn, sorghum, and rice, by enhancing price or revenue protection for producers.[171] Unlike crop insurance, ARC and PLC payments are generally made according to historical plantings, or "base acres," rather than planted acres.[172] This gives producers greater flexibility in their planting decisions because they can try new crops or use crop rotations while still receiving payments based on historic crop allocations. While PLC payments are triggered when a season average market price for a crop is below that crop's reference price (set by Congress), ARC compensates farmers when per-acre market revenue falls below the ARC's per-acre revenue guarantee.

In order to receive ARC or PLC payments, farm owners must agree to not grow crops on highly erodible land without a conservation plan or on unconverted wetlands under any circumstances due to statutory conservation compliance requirements discussed below. Under §9018 of the agriculture title, farmers must also control noxious weeds and "otherwise maintain the land in accordance with sound agricultural practices," which are determined at the discretion of the secretary of agriculture.[173] Congress gave USDA significant leeway to enforce this section, authorizing it to issue rules "the Secretary considers necessary to ensure producer compliance" with these requirements.[174] While there is no case law directly on this point,[175] §9018 gives USDA the authority to require farmers to implement mitigation strategies in order to receive commodity payments. This argument is strengthened by the law's explicit requirement that recipients of commodity payments must not degrade highly erodible land or wetlands, indicating that Congress intended to tie environmental protections to ARC and PLC. To help address the threat climate change poses to agricultural land, USDA should use its rulemaking authority to require farmers receiving commodity payments to adopt cost-effective climate-friendly practices. These requirements could be instituted slowly, ensuring that farmers have time to adapt.

---

170. Congressional Research Service, *supra* note 136, figs.1 & 2 and tbl.3, at 5-6.
171. *See id.* at 14-15 & CRS-34.
172. *See id.* at CRS-33.
173. 7 U.S.C. §9018(a)(1).
174. *Id.* §9018(a)(2). The agency has not yet used this authority.
175. Romany Webb & Steven Weissman, *University of California, Berkeley, School of Law, Addressing Climate Change Without Legislation*, 3 USDA 43 (2014).

3. The Commodity Credit Corporation

The Commodity Credit Corporation (CCC) is a federal corporation used to fund a variety of USDA activities, including conservation programs, export assistance, and international aid. The Commodity Credit Corporation Charter Act also gives USDA broad authority to use CCC to fund programs to support agricultural prices, help produce and market agricultural goods, and develop new markets for agricultural commodities.[176] The agricultural secretary has the authority to distribute up to $30 billion annually through CCC. While CCC is used to fund some mandatory programs, the secretary can also create new programs as long they support the production, distribution, or marketing of agricultural commodities.[177]

In 2018, the Trump Administration announced a major new subsidy program designed to help farmers affected by the United States' trade war with China.[178] The initiative, the MFP, provided direct payments to farm operations that produce a commodity affected by Chinese tariffs. The program was created through executive action, under §5 of the Commodity Credit Corporation Charter Act, which authorizes CCC to dispose of surplus agricultural commodities and aid in the development of new markets, marketing facilities, and uses of agricultural commodities.[179] The MFP quickly became the single largest source of subsidies for farmers, distributing $28 billion over the first two years.[180] In 2020, the MFP was replaced by the Coronavirus Food Assistance Program (CFAP), which was designed to assist producers who lost income due to the pandemic.[181] As of mid-2021, USDA has distributed over $24 billion through CFAP.[182]

While some tree nut producers were eligible for MFP and CFAP payments, most of the funds went to emissions-intensive operations produc-

---

176. 15 U.S.C. §714c.
177. *See id.*
178. Press Release, USDA, USDA Announces Details of Assistance for Farmers Impacted by Unjustified Retaliation (Aug. 27, 2018) (on file with authors).
179. 15 U.S.C. §714c.
180. CONGRESSIONAL RESEARCH SERVICE, *supra* note 134.
181. Coronavirus Aid, Relief, and Economic Stability Act, Pub. L. No. 116-136, §11002, 134 Stat. 281 (2020); Press Release, USDA, USDA Announces Coronavirus Food Assistance Program (Apr. 17, 2020) (on file with authors).
182. USDA, *CFAP 1.0 Dashboard: Program Payments by Category and Commodity*, https://www.farmers.gov/cfap1/data (last visited May 27, 2020); USDA, *CFAP 2.0 Dashboard: Program Payments by Category and Commodity*, https://www.farmers.gov/cfap2/data (last visited May 27, 2020); *e.g.*, H. Claire Brown, *Congress Is Negotiating the Next Round of Covid-19 Aid. Here's What That Means for the Food Industry*, COUNTER, July 21, 2020, https://thecounter.org/congress-negotiating-covid-19-aid-food-industry-businesses-ppp/.

ing dairy and livestock or commodity crops such as soybean, sorghum, and wheat.[183] Rather than respond to foreign tariffs or reductions in commodity prices by supporting the conventional production of annual crops, USDA should instead respond by developing new markets for products produced in carbon-farming systems, particularly those with the greatest climate benefits, such as alley cropping and silvopasture. It should also develop payment programs through CCC for farmers using climate-friendly practices, providing the largest benefits to those practices with the greatest sequestration potential. This would provide farms financial security during periods of low prices while reducing agriculture's net emissions.

Prior to his appointment as undersecretary for farm production and conservation in 2021, Robert Bonnie proposed using CCC to fund a carbon bank that would finance climate-friendly activities undertaken by farmers and foresters.[184] Specifically, the carbon bank would provide farmers with a guaranteed price for greenhouse gas emissions reductions through a reverse carbon credit auction.[185] Implementation of such a program will require credible, feasible, and accurate methods of determining the greenhouse gas impact of various practices; steps to address additionality concerns—whether the measures paid for would have been undertaken (or maintained for certain timeframes) even without the payment; and measures to ensure that financed reductions are either permanent (such as ongoing greenhouse gas emissions reductions) or for established long-term periods (such as through long-term contracts to continue certain practices). In addition, this approach should be designed not to disadvantage some of the most effective carbon-farming practices, such as alley cropping, which often require larger initial investments and could thus be less competitive in a reverse auction system. Likewise, a reverse auction system must be sure to address equity concerns and not favor larger-scale operations, which likely would require lower payments on a per acre basis than mid-sized and smaller-scale ones. If implemented well, such a carbon bank could play a major role in reducing net emissions.

---

183. USDA, Market Facilitation Program Data, https://www.farmers.gov/sites/default/files/documents/MFP%20Data%20-%20Feb%203%202020.pdf, (last updated Feb. 3, 2020) (showing that less than 2% of MFP payments by value went to specialty crops); USDA, *CFAP Dashboard: Program Payments by Category and Commodity*, https://www.farmers.gov/cfap/data (last visited Jan. 14, 2020) (showing that less than 17% of CFAP payments went to "sales commodities," a broad category of agricultural products that includes specialty crops, aquaculture, floriculture, and specialty livestock).
184. Robert Bonnie et al., Climate 21 Project, Transition Memo 9 (2020).
185. *Id.*

4. Conservation Payments

The 2018 Farm Bill allocated on average approximately $5.8 billion annually to USDA's conservation programs, with the majority of that funding going to CRP, EQIP, and CSP.[186] This section examines these three conservation programs and recommends both executive and legislative actions that would ensure they more effectively address climate change. In addition to programmatic changes, USDA should seek to reduce administrative barriers to signing up for conservation programs, including simplifying contracts, increasing administrative support for farmers and ranchers, and creating a comprehensive website to allow farmers to more easily access the wide range of incentives for the promotion of climate stewardship practices.

❑ *Conservation Reserve Program.* Under the 2018 Farm Bill, 34% of conservation spending went to CRP, which pays farmers an annual rental fee to take environmentally sensitive land out of agricultural production for 10-15 years, which generally improves the quality of that land, as well as the soil, air, wildlife habitat, and climate for at least the duration of the contract.[187] Fewer than 22 million acres (out of a possible 27 million acres, the cap set out in the Farm Bill) were enrolled in the program in May 2020.[188] USDA estimated that CRP sequestered more than 43 MMT CO_2 eq. in 2014, about 7% of agriculture's greenhouse gas emissions.[189] If accurate, this would translate into about 1.8 metric tons CO_2 sequestered per acre (or 4.5 tons per hectare).[190]

CRP's advertised climate benefits are often temporary, however. After their CRP contracts expire, many producers bring their CRP acres back into production, quickly releasing any carbon stored during the term of the contract.[191] Between 2006 and 2014, for example, an estimated 14 million acres

---

186. Congressional Research Service, *supra* note 133, fig.2 and tbl.3, at 5-6.
187. *Id.*
188. Farm Service Agency, USDA, Conservation Reserve Program: Monthly Summary May 2020, at 6 (2020).
189. USDA, Farm Service Agency Strategic Plan: Fiscal Year 2016-2018 Update 25, 28 (2016).
190. In contrast, a 2009 literature review of carbon sequestration rates on CRP acres estimated that they sequester slightly less than one metric ton CO_2 eq. per acre annually. Gervasio Piñeiro et al., *Set-Asides Can Be Better Climate Investment Than Corn Ethanol*, 19 Ecological Applications 277, 279 (2009).
191. Soren Rundquist & Craig Cox, Environmental Working Group, Fooling Ourselves: Executive Summary (2016) (finding that CRP water quality benefits were counteracted by losses from farmers exiting the program); Tyler Lark et al., *Cropland Expansion Outpaces Agricultural and Biofuel Policies in the United States*, 10 Env't Res. Letters 1, 9 (2015) (finding that up to 42% of all land converted to cropland came from land exiting CRP). Wetland acreage protected by CRP still have climate benefits, however, since their annual methane emissions while in the program are not lost if the land is converted back into production.

previously protected by CRP were returned to agricultural production.[192] Additionally, if any of the crop production on CRP land is displaced to other land, especially to newly converted land, the CRP-related climate benefits would be illusory. Congress should reform CRP to provide sustained climate benefits by offering farmers 30-year agreements or permanent easements to protect environmentally sensitive land.[193]

There have also been proposals to expand certain productive activities on CRP land in order to increase interest in the program;[194] Congress should consider this only for activities with proven climate benefits, such as agroforestry. In addition, legislators should reform CRP so that funding is directed to areas of higher ecological concern—land that consists of high-quality habitat, diverse native covers, and grassland at high risk of conversion—to maximize the conservation and climate benefits of the program.

Continuous CRP (CCRP), which pays farmers to plant conservation buffers or wildlife habitat on environmentally sensitive land, provides a model for Congress to expand on. The 2018 Farm Bill authorizes the FSA to offer CCRP enrollees a one-time incentive payment of up to 50% of the cost of establishing the practice, including seed costs.[195] However, FSA has offered only a 5% cost share during the past few sign-up periods.[196] USDA should instead offer the maximum permitted cost share under the Farm Bill for the next sign-up. Moreover, FSA has the authority to offer an additional rental rate incentive that provides landowners a supplemental payment of up to 20% of the annual rental payment for certain high-priority continuous practices.[197] USDA should use this flexibility to further incentivize the practices with the highest climate benefits.

The CLEAR30 pilot program—based on a more limited CCRP initiative started in 2016[198]—offers farmers and landowners an opportunity to enroll in a 30-year CRP contract. However, CLEAR30 is limited to expiring CRP water quality practice contracts, mostly for riparian, wildlife habitat, or wetland buffers.[199] USDA further limited it to contracts in the Great Lakes

---

192. Craig Cox et al., Environmental Working Group, Paradise Lost: Conservation Programs Falter as Agricultural Economy Booms 4 (2013).
193. *See id.* at 4-5; *USDA Freezes New Enrollments in Continuous Conservation Reserve Program*, NSAC, May 4, 2017, http://sustainableagriculture.net/blog/usda-freezes-ccrp-enrollment/.
194. *See, e.g.*, Press Release, Office of Sen. John Thune, Thune Farm Bill Proposals Would Improve Conservation Program Management (Apr. 10, 2017) (on file with authors).
195. Agriculture Improvement Act of 2018, Pub. L. No. 115-334, §2207, 132 Stat. 4490, 4547.
196. *E.g.*, FSA, USDA, CRP Continuous Signup Fact Sheet (2019).
197. 7 C.F.R. §1410.42(b).
198. Press Release, USDA, USDA Announces New Conservation Opportunities to Improve Water Quality and Restore Wildlife Habitat (Dec. 7, 2016) (on file with authors).
199. Press Release, USDA, USDA's New CRP Pilot Program Offers Longer-Term Conservation Benefits (Apr. 29, 2020) (on file with authors).

and Chesapeake Bay regions.[200] FSA should strengthen this pilot by allowing any eligible producer in the country to enroll in CLEAR30, while Congress should expand the program to all CRP contracts.

USDA should also prioritize the use of native vegetation on land enrolled where ecologically appropriate, which would likely increase and make more robust soil carbon sequestration, as well as improve native habitat for wildlife and pollinators. USDA should similarly prioritize applications for properties of higher ecological concern—land that consists of high-quality habitat, diverse native covers, and grassland at high risk of conversion.

In addition, USDA should increase incentive payments for participation. Under the Farm Bill, annual rental rates are capped at 85% of the average county soil rental rate for CRP enrollment and at 90% for CCRP enrollment.[201] Congress should remove the cap on rental rates to keep CRP attractive to landowners and, until it does so, USDA should expand its use of these discretionary incentives, including adjustments to the county soil rental rate for soil productivity, in order to boost payments. Similarly, while the Farm Bill currently limits CRP enrollment to no more than 25% of a county's cropland,[202] until Congress lifts that, USDA should use its waiver authority to the maximum extent possible to increase overall CRP enrollment.

Congress should also strengthen the Conservation Reserve Enhancement Program (CREP), which gives farmers higher CRP payments for participating in targeted conservation efforts organized by state and local officials.[203] Due to its higher annual payments, which are on average almost three times as high as general CRP payments,[204] CREP has remained popular with farmers even when increasing commodity prices have reduced acreage reenrollment in CRP overall.[205] Congress should increase funding for the program while prioritizing land with climate mitigation potential like forested land or herbaceous cover in order to enhance its climate benefits.

❏ *Environmental Quality Incentives Program.* Receiving about 38% of conservation spending, EQIP provides farmers with funding to plan and install structural, vegetative, and land management practices.[206] EQIP is managed

---

200. *Id.*
201. Agriculture Improvement Act of 2018, Pub. L. No. 115-334, §2207, 132 Stat. 4490, 4549.
202. 7 C.F.R. §1410.4.
203. *See Digging Deeper Into Continuous CRP Enrollments*, NSAC, Mar. 24, 2015, http://sustainableagriculture.net/blog/ccrp-enrollment-2015/; Cox et al., *supra* note 192.
204. States provide additional funding for CREP, bringing the average yearly CREP payments to $140 per acre. In contrast, general sign-up payments are $51 per acre. *Digging Deeper Into Continuous CRP Enrollments*, *supra* note 203.
205. Cox et al., *supra* note 192.
206. Congressional Research Service, *supra* note 133, at 17-18.

by USDA's NRCS, which was established by Congress in 1936 to reduce "the wastage of soil and moisture resources on farm, grazing, and forest lands."[207] Congress has authorized NRCS to pay producers up to 75% of the costs associated with the development and implementation of a new conservation practice and/or 100% of the income foregone by the producer as a result of a new practice.[208] Producers cannot receive funding for a preexisting practice through EQIP, regardless of how environmentally beneficial it may be. NRCS is required by statute to allocate at least 50% of its EQIP payments to livestock producers; Congress should reduce this percentage to advantage less carbon-intensive products.[209]

Funding decisions are made through a process that combines national, state, and local priorities. National priorities are determined by NRCS in accordance with the program's statutory guidelines, which require the agency to finance practices that promote one or more of the following:

- Soil health
- Water quality and quantity improvement
- Nutrient management
- Pest management
- Air quality improvement
- Wildlife habitat development
- Invasive species management[210]

State priorities are set by the head NRCS administrator of each state in consultation with stakeholders, while local priorities are set by "local working groups," composed of stakeholders such as producers and industry representatives. Given that farmers using sustainable management methods are generally in the minority in any given region, they are unlikely to be given equal representation—or even any representation—in local working groups or statewide stakeholder committees. As a result, the priority-setting process can disadvantage applications for truly innovative and sustainable practices.[211] According to an analysis of USDA data conducted by the Environmen-

---

207. 16 U.S.C. §590a. The agency was originally called the Soil Conservation Service but was renamed in 1994.
208. *Id.* §3839aa-2(d)(2).
209. *Id.* §3839aa-2(f)(1).
210. *Id.* §3839aa-2(d)(3).
211. For example, a 2008 report on EQIP in Iowa, Minnesota, and Missouri found that state-level priorities favored the worst polluting operations and disadvantaged applications from producers using

tal Working Group (EWG), only 14% of EQIP funding went to conservation practices that USDA has identified as producing the most environmental benefits for water quality, water quantity, soil health, air quality, and fish and wildlife.[212] Congress or NRCS should address this disparity by requiring greater representation of sustainable farming approaches on working groups and prioritizing practices demonstrated to have the greatest benefits.[213]

Environmentalists and small-scale farming advocates have also criticized NRCS for subsidizing large-scale, environmentally harmful operations through EQIP.[214] Since the program's inception in 1997, more than $1.6 billion has gone to support irrigation systems through EQIP, making them the most well-funded set of practices financed by the program.[215] Instead of conserving water, however, efficient irrigation systems often lead to land conversion and increased water usage as farmers use their savings to expand irrigated crop production, switch to more water-intensive crops, or both.[216] NRCS should prioritize making existing irrigation systems more efficient and ensure that the agency does not fund projects likely to lead to land conversion or increased water usage.

Similarly, waste storage facilities for concentrated animal facilities received a larger share of payments than any other single practice supported by the program; a survey of EQIP funding from 1997 to 2007 showed that these accounted for almost 15% of all EQIP payments.[217] NRCS' support for large-scale AFOs has continued since then: in FY 2016, about 12% of total EQIP funding—more than $110 million—went to such facilities.[218] While some waste management systems, such as anaerobic digesters, can reduce

---

sustainable management methods, such as crop rotation. ELANOR STARMER, CAMPAIGN FOR FAMILY FARMS AND THE ENVIRONMENT, INDUSTRIAL LIVESTOCK AT THE TAXPAYER TROUGH: HOW LARGE HOG AND DAIRY OPERATIONS ARE SUBSIDIZED BY THE ENVIRONMENTAL QUALITY INCENTIVES PROGRAM 14 (2008).

212. EWG, The EQIP Improvement Act (2018).
213. The EQIP Improvement Act, introduced in 2018, would prioritize funding at the state and county levels for the most effective conservation practices, while also reducing the maximum cost-sharing from 75% to 40% for less beneficial practices. EQIP Improvement Act, S. 2624, 115th Cong. (2018).
214. Andrew Martin, *In the Farm Bill, a Creature From the Black Lagoon?*, N.Y. TIMES, Jan. 13, 2008, http://www.nytimes.com/2008/01/13/business/13feed.html (suggesting that the program's name should be changed to the "Factory Farm Incentive Program"); Tom Laskaway, *Stop the Environmental Subsidy for Factory Farms*, GRIST, Apr. 17, 2009, http://grist.org/article/stop-the-environmental-subsidy-for-factory-farms/; *CAFOs and Cover Crops: A Closer Look at 2015 EQIP Dollars*, NSAC, Nov. 20, 2015, http://sustainableagriculture.net/blog/fy15-general-eqip-update/.
215. EWG, *Environmental Quality Incentives Program (EQIP) Practice Suite Payments in the United States, 1997-2015*, https://conservation.ewg.org/eqip_practice_suite.php (last visited Nov. 12, 2020).
216. Ward & Pulido-Velazquez, *supra* note 54; Pfeiffer & Lin, *supra* note 54.
217. Melissa Bailey & Kathleen Merrigan, *Rating Sustainability: An Opinion Survey of National Conservation Practices Funded Through the Environmental Quality Incentives Program*, 65 J. SOIL & WATER CONSERVATION 21A, 23A (2010).
218. *CAFOs and Cover Crops: A Closer Look at 2015 EQIP Dollars*, *supra* note 214.

emissions in feedlots using liquid manure management systems, sustainable agriculture and environmental justice groups have been highly critical of government efforts to finance them. Even if digesters reduce feedlot emissions, they argue, CAFOs are still bad for the environment, animal welfare, and rural communities. The National Sustainable Agriculture Coalition, for example, has come out against USDA's policy of supporting new or expanding CAFOs through EQIP, arguing that USDA is subsidizing fundamentally unsustainable practices by doing so.[219] This funding effectively subsidizes a carbon-intensive form of animal production and may be a factor in the continuing consolidation of the industry noted earlier.[220] A possible alternative, as noted above, are pasture-based or integrated crop-livestock systems, which have much lower rates of methane production.

With capital costs often exceeding $1 million, anaerobic digesters are beyond the price range of what most dairy or hog farmers in the United States can afford. According to EPA, digestion systems are generally not economically viable for operations with fewer than 500 cows, even with current cost-sharing programs.[221] This significantly limits their use—more than 90% of dairy farms in the United States have fewer than 500 cows, accounting for 40% of all dairy cows in the country.[222] Of these, many do not use liquid manure systems. However, the largest 10% of dairies—which account for 60% of the dairy cow population—could more feasibly be required to install digesters. Rather than subsidize concentrated animal facilities with EQIP funds, USDA or EPA should consider imposing regulatory methane emissions limits, which could drive most large-scale operations to install digesters or flaring systems or—a preferable long-term solution—eliminate liquid manure management altogether.[223]

---

219. *Id.*
220. *See* Jessica McKenzie, *The Misbegotten Promise of Anaerobic Digesters*, COUNTER, Dec. 3, 2019, https://thecounter.org/misbegotten-promise-anaerobic-digesters-cafo/.
221. EPA, *AgSTAR—Is Anaerobic Digestion Right for Your Farm?*, https://www.epa.gov/agstar/anaerobic-digestion-right-your-farm (last updated Aug. 17, 2020).
222. NASS, USDA, 2012 CENSUS OF AGRICULTURE, U.S. NATIONAL LEVEL DATA 21 tbl.17 (2014).
223. As discussed in Chapter VI, in 2021, a coalition of environmental groups petitioned the U.S. Environmental Protection Agency to impose methane emissions limits under the Clean Air Act on industrial dairy and hog operations—those with at least 500 cows or 1,000 hogs. The petitioners note that these facilities are "major sources of methane . . . and [account] for 33 percent of agricultural methane emissions, 13 percent of total U.S. methane emissions, and 1.3 percent of total U.S. greenhouse gas emissions." Petition to List Industrial Dairy and Hog Operations as Source Categories Under Section 111(b)(1)(A) of the Clean Air Act (Apr. 6, 2021), https://food.publicjustice.net/wp-content/uploads/sites/3/2021/04/2021.04.06-Industrial-Dairy-and-Hog-CAA-111-Petition-FINAL.pdf; Press Release, Public Justice, Climate, Environmental Justice Groups Call for Biden EPA to Hold Industrial Dairy and Hog Operations Accountable and to Reject Big Ag Technology, Apr. 6, 2021, https://food.publicjustice.net/methane-petition-press-release/.

Chapter V. Transforming Farm Policy Toward Climate-Neutral Agriculture

Rather than supporting environmentally harmful and carbon-intensive practices and production, Congress and USDA should instead redirect EQIP funds to support farms and ranches working to significantly reduce emissions or sequester carbon. Congress should lower the EQIP payment cap for all operations or, at a minimum, it should create a lower cap—or disallow any payments—for AFOs, as several rural, environmental, and family organizations have demanded.[224]

And while Congress should eliminate payments to environmentally harmful operations through legislative action, the agency can—and should—eliminate or reduce these payments before Congress acts. The agency itself has the ability to redirect funding to more climate-friendly practices. NRCS has significant leeway in determining which practices are prioritized and can set aside considerable funding for carbon-farming practices.[225] The agency's Organic Initiative provides an instructive example for how this might be accomplished. In the early 2000s, many organic producers were concerned that the program's reliance on local administrators and the high demand for EQIP funding from conventional producers disadvantaged applicants seeking funding for organic practices. In response, Congress included provisions in the 2008 Farm Bill requiring the agency to set aside EQIP funds specifically to assist organic producers or producers transitioning to organic production.[226] Producers applying for funds from the Organic Initiative are eligible for up to $140,000 over five years.[227] Farmers can still apply to the general funding pool for larger amounts, but the resulting Organic Initiative ensures that a pool of money is set aside for organic practices each year.

Given NRCS' broad authority to determine which practices to support, the agency should create a similar pool to support carbon farming, while also prioritizing applications with the greatest climate benefits. While there is no carbon-farming certification system equivalent to organic certification,

---

224. The EQIP Improvement Act would lower the payment cap to $150,000. EQIP Improvement Act, S. 2624, 115th Cong. (2018). In 2007, for example, a coalition of 26 organizations called on Congress to prohibit funding for AFOs with more than 1,000 animals. Letter From the Campaign for Family Farms and the Environment et al., to the Senate (May 8, 2007). Congress initially set the payment cap at $50,000, but then raised it ninefold in the 2002 Farm Bill to $450,000. Farm Security and Rural Investment Act of 2002, Pub. L. No. 107-171, §1240G, 116 Stat. 134, 257. The cap was lowered to $300,000 in the 2008 Farm Bill after Barack Obama promised to reduce it in his presidential campaign. Food, Conservation, and Energy Act of 2008, Pub. L. No. 110-234, §2508, 122 Stat. 923, 1063; see also OBAMA/BIDEN CAMPAIGN, REAL LEADERSHIP FOR RURAL AMERICA 2. However, it was ultimately raised back to $450,000 in the 2014 Farm Bill. Agricultural Act of 2014, Pub. L. No. 113-79, §2206, 128 Stat. 649, 730 (amending §1240G).
225. Each of the program's statutorily mandated objectives can be promoted through carbon sequestration and carbon farming.
226. 16 U.S.C. §3839aa-2(i).
227. Id.

EQIP covers a number of practices that are long-lasting and sequester significant amounts of carbon, such as alley cropping, tree buffer strips, and silvopasture, and the agency should establish simple guidelines for determining whether an applicant is transitioning to, or practicing, carbon farming. In FY 2018, EQIP paid out $1.4 billion in financial assistance.[228] The majority of EQIP funds should ultimately be used to support carbon farming, but even 10% of the total amount, $140 million, would significantly boost powerful sequestration methods, while advancing EQIP's statutory priorities.

Congress should also increase the co-share from 75% to 90% for the most effective sequestration practices, such as alley cropping and silvopasture. Socially disadvantaged, limited-resource, beginning, and veteran farmers are already eligible for a 90% cost-share rate.[229] By extending this rate to carbon farmers as well, legislators will not only convey that carbon-farming practices are a public policy priority, but they will also make the implementation of such practices more financially feasible for many farmers.

❑ *Conservation Stewardship Program.* NRCS also administers CSP, which pays farmers to improve, maintain, or adopt conservation practices on their farms. Unlike EQIP, which is focused on funding individual practices—some of which have dubious environmental benefits—CSP supports comprehensive, whole farm conservation in order to address high-priority sustainability concerns.[230] While CSP has the smallest budget of the three main conservation programs—it will receive about 17% of all conservation funding from the 2018 Farm Bill[231]—it is the largest USDA conservation program on an acreage basis.[232] It also has the biggest impact on a per dollar basis. According to a 2018 analysis, every dollar spent on CSP results in a return of $3.95 to farmers, rural communities, and the environment.[233] EQIP and CRP, in contrast, were estimated to have returns of $1.01 and $2.11, respectively.[234]

CSP pays farmers annually under a five-year contract with the option to renew for another five years if they agree to adopt additional conservation objectives.[235] NRCS revised CSP in fall 2016 by, among other things, offering farmers 67 new practices that are eligible for funding through the pro-

---

228. NRCS EQIP, *supra* note 112.
229. *CAFOs and Cover Crops: A Closer Look at 2015 EQIP Dollars*, *supra* note 214.
230. *See* Peter Lehner & Nathan Rosenberg, *A Farm Bill to Help Farmers Weather Climate Change*, 14 J. Food L. & Pol'y 8, 19-22 (2018).
231. Congressional Research Service, *supra* note 133, at 17 & fig.2.
232. NRCS CSP, *supra* note 112.
233. Union of Concerned Scientists, Farmers and Taxpayers Stand to Lose Billions With Elimination of the Conservation Stewardship Program: Appendix 2 (2018).
234. *Id.*
235. 7 C.F.R. §1470.26 (2016).

Chapter V. Transforming Farm Policy Toward Climate-Neutral Agriculture

gram, including "planting for high carbon sequestration rate."[236] As with EQIP, NRCS has the statutory authority to prioritize low-carbon practices and to create a funding pool within CSP for farmers transitioning to, or practicing, carbon farming.[237] NRCS should follow up on its revisions by doing both as quickly as possible, revising the CSP ranking tools so they prioritize applications that will have the greatest benefit on the climate and the environment. Congress should also expand funding for CSP in upcoming farm bills by raising the average payment rate per acre that is authorized for the program to ensure that higher-level conservation activities can be appropriately rewarded. In particular, climate-beneficial activities like resource-conserving crop rotations should be prioritized and receive a higher, supplemental payment to reflect the high-level environmental benefits of those practices.[238] Congress should also strengthen CSP sustainability standards for participation; increase the importance of environmental benefits in the application process; and create a funding pool for agroforestry and perennial operations.

❏ *Coordinating conservation efforts.* Despite spending billions of dollars each year on conservation programs, USDA does little to coordinate its conservation efforts or measure, evaluate, and report conservation outcomes. The 2018 Farm Bill requires USDA to coordinate procedures and processes for EQIP and CSP, such as applications, contracting, and conservation planning.[239] This will make it easier for farmers to move from EQIP to CSP. Congress should build on this step by requiring USDA to create an automatic graduation process from EQIP to CSP when eligibility requirements have been satisfied. This would increase the number of farmers "graduating" to CSP, which is a more effective conservation program than EQIP, and is a better tool for encouraging carbon-farming practices.

The federal government will need to direct funding to the most effective climate-friendly practices available—increased funding alone will not be sufficient to achieve net-zero emissions. In order to do this, however, USDA must closely monitor and evaluate its conservation programs and provide policymakers and the public with research on their effectiveness. The Healthy Fields and Farm Economies Act, a marker bill introduced in the

---

236. Press Release, USDA NRCS, USDA Announces Changes for Largest Conservation Program (Sept. 1, 2016); Marc Heller, *Revamps to Conservation Program Boost Options for Farmers*, GREENWIRE, Sept. 2, 2016.
237. 16 U.S.C. §3838g.
238. Lehner & Rosenberg, *supra* note 230, at 21-22.
239. *See* NSAC, *Conservation Stewardship Program*, https://sustainableagriculture.net/publications/grassrootsguide/conservation-environment/conservation-stewardship-program/ (last updated Apr. 2019) (discussing the benefits of coordinating between EQIP and CSP).

lead-up to the 2018 Farm Bill, included a number of policy changes designed to improve the ability of USDA, policymakers, and the public to evaluate conservation programs. The bill would have created a national technical committee to assist NRCS in monitoring and evaluating conservation programs, directed USDA to evaluate and quantify the environmental benefits of conservation practices, and required the department to submit a public report to Congress assessing each conservation program.[240] While these provisions were not included in the 2018 Farm Bill as sustainable agriculture advocates had hoped, they nonetheless lay the groundwork for future steps. Congress should pass legislation that will accomplish the goals set out in the Healthy Fields and Farm Economies Act, with additional provisions that provide funding for qualitative research aimed at improving farmer adoption, implementation, and retention of climate-friendly practices.

5. Conservation Easements

Conservation easements are legal agreements between a landowner and a third party—usually a land trust or a government agency—that are designed to permanently restrict the use of the land. The restrictions commonly protect natural areas or resources, such as wildlife habitats or water quality, but they are also increasingly being used to preserve farmland and prevent it from being converted to non-farm uses. As of 2017, farmland owners had protected more than 13 million acres from development through conservation easements.[241] Agricultural easements, which protect agricultural land from development, can also have important climate benefits since even conventional farms generally have much lower emissions than developed land. An analysis of emissions in California's Central Valley, for example, found that emission rates on urbanized land were 70 times higher than emissions on an equivalent area of irrigated cropland.[242]

Conservation easements that protect wetlands and other environmentally sensitive land from being converted to farmland offer substantial climate benefits. The Environmental Working Group, for instance, estimates that the conversion of wetlands to farmland between 2008 and 2012 resulted in greenhouse gas emissions totaling 25-74 MMT CO_2 eq. annually[243]—the

---

240. Healthy Fields and Farm Economies Act, H.R. 4751, 115th Cong. (2018).
241. NASS, *supra* note 67, at 58 tbl.47.
242. California Energy Commission Climate Change Center, University of California, Davis, Adaptation Strategies for Agricultural Sustainability in Yolo County, California 106 (2012).
243. EMILY CASSIDY, EWG, ETHANOL'S BROKEN PROMISE: USING LESS CORN ETHANOL REDUCES GREENHOUSE GAS EMISSIONS 4 (Nils Bruzelius ed., 2014).

Chapter V. Transforming Farm Policy Toward Climate-Neutral Agriculture 155

equivalent of adding five to 15 million cars to the road each year.[244] Others have studied the conversion of native grasslands to farmland, in large part to supply corn to ethanol plants, and similarly found significant soil carbon losses.[245] Expansion of programs to prevent such conversion can have substantial benefits and should be undertaken.

USDA has supported conservation easements on property owned by farm operators since 1990, preserving more than four million acres of farmland and environmentally sensitive lands in the process.[246] The 2014 Farm Bill consolidated USDA's three existing easement programs into the Agricultural Conservation Easement Program (ACEP).[247] Administered by NRCS, the program is designed to protect wetlands, grasslands, and productive farmland and has been fairly successful; it should be expanded.[248]

For permanent wetland easements, NRCS pays farm owners the lower of the fair market value of the land or an offer made by the farm owner.[249] NRCS can also set geographical caps limiting payments within specific regions.[250] Alternatively, farm owners can apply for "long-term" wetland easements, which typically run for 30 years, and provide 50%-75% of the compensation due to an equivalent permanent easement.[251] For agricultural land easements, which protect working agricultural land, NRCS generally pays farm owners up to 50% of the fair market value of the easement, although NRCS may contribute up to 75% of the fair market value of an easement protecting grasslands of "special environmental significance."[252]

ACEP receives substantially less funding than its predecessor programs. While the 2008 Farm Bill provided $691 million for easement programs each year from 2009-2013, the 2018 Farm Bill allocated $450 million annually to ACEP—almost $250 million less than its precursors enjoyed.[253] ACEP also receives significantly less funding than CRP, which the Congressional Budget Office projects will cost more than $2 billion annually on average

---

244. Calculated by the authors using EPA estimates for passenger vehicle emissions. *See* U.S. EPA, GREENHOUSE GAS EMISSIONS FROM A TYPICAL PASSSENGER VEHICLE (2018) (a typical passenger vehicle emits 4.6 metric tons of CO_2 annually).
245. Lark et al., *supra* note 191, at 5.
246. NSAC, *Agricultural Conservation Easement Program*, http://sustainableagriculture.net/publications/grassrootsguide/conservation-environment/agricultural-conservation-easement-program/ (last updated July 2019).
247. 16 U.S.C. §3865.
248. *Id.* §3865(b).
249. *Id.* §3865c(b)(6)(a)(i).
250. *Id.*
251. *Id.* §3865c(b)(6)(a)(ii).
252. *Id.* §3865b(2).
253. CONGRESSIONAL RESEARCH SERVICE, AGRICULTURAL CONSERVATION IN THE 2018 FARM BILL 9 (2019).

during the years addressed by the 2018 Farm Bill.[254] Unlike most ACEP easements, CRP protects environmentally sensitive land for a limited period, and therefore does not offer lasting climate benefits.

Congress should substantially expand ACEP and ensure that protecting environmentally sensitive lands that provide the greatest climate benefits is among the program's priorities. Wetlands, for example, are estimated to emit between 405 and 1,215 metric tons CO_2 eq. per acre when converted to agricultural land.[255] Similarly, it should eliminate ACEP's long-term (but not permanent) easement option since its long-term climate benefits are dubious. The program should also be expanded to allow for easements on additional types of environmentally sensitive land, allowing USDA to protect terrestrial carbon pools in a wider variety of ecosystems.[256]

Finally, given the amount of forestland on farms, Congress should increase funding for the Healthy Forests Reserve Program, an NRCS program that helps landowners restore and protect forests through easements and financial assistance such as 10-year restoration agreements or 30-year or permanent easements for specific conservation actions.[257] As with other programs, this should be administered so as to prioritize climate benefits.

6. Conservation Compliance Requirements

Producers enrolled in a number of federal farm programs are prohibited from producing agricultural products on highly erodible land without a conservation plan[258] or on unconverted wetlands under any circumstances.[259] The requirement protecting highly erodible land is commonly referred to as the "Sodbuster" provision, while the one protecting wetlands is known as the "Swampbuster" provision. Compliance oversight is split between FSA and NRCS; FSA is responsible for determining compliance

---

254. Calculated by the authors using Congressional Research Service, Budget Issues That Shaped the 2018 Farm Bill CRS-6 (2019).
255. Richard Plevin et al., *Greenhouse Gas Emissions From Biofuels' Indirect Land Use Change Are Uncertain but May Be Much Greater Than Previously Estimated*, 44 Env't Sci. & Tech. 8018 (2010). The EWG estimated that 25 to 74 MMT CO_2 eq. were emitted each year between 2008 and 2012 due to the conversion of wetlands to farmland. Cassidy, *supra* note 243.
256. *See* Todd Neeley, *Conservation Controversy*, Progressive Farmer, Winter 2014, *available at* http://dtnpf-digital.com/article/Conservation_Controversy/1888630/239474/article.html.
257. USDA NRCS, *Healthy Forests Reserve Program*, https://www.nrcs.usda.gov/wps/portal/nrcs/main/national/programs/easements/forests/ (last visited Nov. 12, 2020).
258. 16 U.S.C. §§3811-3812.
259. *Id.* §3821. Wetlands drained or filled before December 23, 1985, are not protected. *Id.* §3822(b)(1)(A).

and exemptions, while NRCS is responsible for making wetland and highly erodible land determinations.[260]

The 2008 Farm Bill allowed states to implement a third type of compliance requirement, called "Sodsaver," to protect native sod in the Prairie Pothole National Priority Area.[261] The 2014 Farm Bill eliminated the opt-in requirement and amended the provision, which now reduces crop insurance premium subsidies by 50% for production on native sod for the first four years of planting in six states—Minnesota, Iowa, North Dakota, South Dakota, Montana, and Nebraska.[262]

While the Sodsaver provision only applies to the crop insurance program, the Sodbuster and Swampbuster requirements also apply to each of the conservation programs, as well as many of the smaller programs administered by FSA and NRCS. All three provisions offer potentially important climate benefits since conventional farming on native sod, highly erodible land, and wetlands results in significant greenhouse gas emissions.[263] Despite the clear environmental benefits of these programs, however, the agency has failed to consistently enforce them.[264] A 2016 USDA Office of Inspector General (OIG) report, for example, found that the agency's auditing process had completely bypassed at least 10 states in 2015, apparently in error.[265] At a minimum, USDA should vigorously enforce the Farm Bill's current conservation compliance provisions, withholding benefits from farmers that fail to meet their requirements.[266]

FSA should work with NRCS to develop, publish, and implement a plan to address the current weaknesses in the enforcement of conservation compliance provisions. Such a plan should address FSA's failure to collect and report reliable data, which has crippled enforcement efforts. Seasonal wetlands are often missed by FSA due to the agency's reliance on decades-old wetland determinations, aerial imagery from the hottest time of the year—

---

260. *See* Laurie Ristino & Gabriela Steier, *Losing Ground: A Clarion Call for Food System Reform to Ensure a Food Secure Future*, 42 Colum. J. Env't L. 59, 96-102 (2016) (providing an in-depth discussion of compliance oversight).
261. Food, Conservation, and Energy Act of 2008, Pub. L. No. 110-234, §12020, 122 Stat. 923, 1381.
262. Agricultural Act of 2014, Pub. L. No. 113-79, §11014, 128 Stat. 649, 961.
263. As mentioned above, wetlands are estimated to emit between 405 and 1,215 metric tons of CO_2 eq. per acre when converted to agricultural land. Plevin et al., *supra* note 255.
264. Ristino & Steier, *supra* note 260.
265. OIG, USDA, Audit Report No. 50601-0005-31, USDA Monitoring of Highly Erodible Land and Wetland Conservation Violations—Interim Report 3 (2016); *see* Ristino & Steier, *supra* note 260, at 97.
266. In addition to failing to hold operators accountable, the agency has also failed to sufficiently monitor operators subject to the requirement. It must do both. *See* Joshua Ulan Galperin, *Trust Me I'm a Pragmatist: A Partially Pragmatic Critique of Pragmatic Activism*, 42 Colum. J. Env't L. 426, 487-89 (2017).

when many seasonal wetlands have dried out—and precipitation data from historically dry periods.[267] In addition, FSA should study and implement best practices for remote sensing technologies that can allow it to detect violations and identify issues for on-site inspections. Since compliance is often relatively easy to determine visually, including by satellite,[268] USDA should be able to increase inspections at little additional cost. And, finally, FSA should publicly report on conservation compliance violations (without releasing private information) and resume reporting on conversion of land to cropland, also known as "new breakings," which has not occurred since 2013. If USDA fails to do so, Congress should consider shifting enforcement responsibility to EPA, which has a stronger record of enforcing regulations, while also enabling states, localities, and citizens to enforce the requirements, as is possible under most federal environmental statutes.[269] As discussed in Chapter V.A.4, Congress should also require USDA to provide the public with regular reports on compliance and enforcement data to ensure that FSA and NRCS develop effective tracking and reporting procedures, while also keeping the public and policymakers informed about these critical conservation protections.

Congress should also extend the conservation compliance requirement to farm programs that are not currently covered by the requirement, ensuring that all producers who receive federal subsidies are not causing significant environmental harm. In addition, Congress should expand the required practices to include those that protect soil carbon and water. For example, requiring buffer zones around streams or, where appropriate, cover crops (perhaps with the initial switch partially funded through EQIP), would have significant climate benefits and co-benefits such as improved soil health, nutrient cycling, pest regulation, and crop productivity.[270] The EWG has proposed a "new and stronger conservation compact" that would require farmers to plan and apply an approved conservation system on all annually tilled cropland

---

267. *See, e.g.,* Letter from Dan Ash, Director, Fish and Wildlife Service, Department of the Interior, to Jason Weller, Chief, Natural Resources Conservation Service, USDA (Feb. 20, 2015) (on file with authors) (detailing numerous problems with USDA's wetland determination process, including that "wetland acreages could be substantially reduced" because FSA's imagery is captured during the "natural dry-down during the summer months.").
268. *See, e.g.*, Regrow Agriculture, https://www.regrow.ag/ (last visited Aug. 7, 2021), one of several companies providing agronomic and environmental analysis through analysis of satellite imagery.
269. Most federal environmental statutes empower citizens to enforce compliance through citizen suit provisions, which have proven to be among the most effective methods available for holding regulatory subjects and government agencies accountable. Galperin, *supra* note 266, at 487.
270. Meagan Schipanski et al., *A Framework for Evaluating Ecosystem Services Provided by Cover Crops in Agroecosystems*, 125 Agric. Sys. 12, 13 (2014).

in order to remain eligible for farm programs.[271] Each operation's conservation system would be designed to (1) achieve a rate of soil erosion no greater than the soil loss tolerance level on all annually planted cropland, (2) prevent ephemeral gully erosion, and (3) establish and maintain a minimum of 50 feet between annually tilled cropland and waterways.[272] Congress should adopt the EWG's outcome-based proposal, which offers a cost-effective and quick way to catalyze widespread change.

Despite Sodsaver, agricultural policy still incentivizes the conversion of natural grassland. This must be addressed if the sector is to achieve net-zero emissions. Natural grasslands provide a number of important ecosystem services, especially in the prairie states, including water supply and flow regulation, erosion control, pollination, and wildlife habitat.[273] They also mitigate climate change and sequester carbon.[274] A 2019 study found that almost 90% of emissions from cropland expansion in the United States between 2008 and 2012 came from grassland conversion.[275] In early 2018, almost 200 conservation, agriculture, and wildlife organizations urged Congress to extend Sodsaver to the entire country, including states that are experiencing high rates of conversion of native grassland, such as Texas and Kansas.[276] Congress should make Sodsaver a national program and require USDA to track and report on grassland loss by county, allowing policymakers to effectively evaluate the program's efficacy.

USDA does not maintain a comprehensive inventory of intact native prairies, hampering efforts to conserve native prairies. USDA should immediately begin to distinguish between native and non-native prairies and grasses, while coordinating monitoring and reporting frameworks, which include the NASS' Cropland Data Layer, NRCS' National Resources Inventory, the U.S. Geological Survey's National Land Cover Database, and the U.S. Fish and Wildlife Service's National Wetlands Inventory. This will allow policymakers and USDA to prioritize native sod, increasing Sodsaver's effectiveness as it expands.

---

271. EWG, Less Farm Pollution, More Clean Water: An Agenda for Conservation in the 2018 Farm Bill 3-4 (2017).
272. *Id.*
273. Jan Bengtsson et al., *Grasslands—More Important for Ecosystem Services Than You Might Think*, 10 Ecosphere 1 (2019).
274. *Id.* at 10.
275. Seth A. Spawn, *Carbon Emissions From Grassland Expansion in the United States*, 14 Env't Res. Letters 5 (2019).
276. Letter From Organizations in Support of the American Prairie Conservation Act, to Chairman Pat Roberts and Ranking Member Debbie Stabenow, Senate Committee on Agriculture, Nutrition, and Forestry, and Chairman K. Michael Conaway and Ranking Member Collin Peterson, House Committee on Agriculture (Jan. 18, 2018) (on file with authors).

7. Lending Programs

The federal government has traditionally subsidized financing to farm operations through the Farm Credit System, a privately owned, federally chartered network of lending institutions that focus on agricultural loans, and USDA's lending agency, FSA. Together, FCS and FSA provide almost 45% of all farm credit or about $40 billion annually.[277] The Small Business Administration (SBA) has become an increasingly important lender for some types of farm operations in recent years due to its relatively high lending caps, which we discuss in Chapter VI.C. We also discuss private financing in Chapter VII.B. This section provides an overview of the Farm Credit System and FSA and proposes policy changes designed to make credit more accessible to carbon farmers and other operators using climate-friendly practices.

Created by Congress in 1916 to provide a reliable source of credit for agricultural producers,[278] the Farm Credit System now holds nearly 41% of the farm sector's total debt—a larger share than that held by commercial banks.[279] The Farm Credit System benefits from a range of publicly funded guarantees, subsidies, and exemptions. It is exempt from all taxes on profits earned from real estate transactions, gets funding capital from the Federal Farm Credit Banks Funding Corporation, and enjoys USDA guarantees on many of its loans.[280]

USDA also manages FSA, which, among other things, acts as a lender of last resort for farmers and ranchers. In addition to offering direct loans to farmers, the agency also issues guarantees on loans made by commercial lenders for farmers that would not otherwise qualify. While FSA's overall impact on the agricultural credit market is relatively small—it holds about 2.6% of all farm debt through direct loans and guarantees another 4%-5% of loans—it has come to play an important role in supporting beginning and female farmers,[281] and, to a much lesser degree, farmers of color.[282] The Agricultural Credit Improvement Act of 1992 required FSA to reserve funds

---

277. *See supra* note 136.
278. *See* Farm Credit Act of 1933, Pub. L. No. 73-75, 48 Stat. 257.
279. Jim Monke, Congressional Research Service, Agricultural Credit: Institutions and Issues 1 (2018).
280. Total farm debt was approximately $374 billion at the end of 2016. *Id.*
281. *Id.*
282. *See* Nathan Rosenberg & Bryce Wilson Stucki, *How USDA Distorted Data to Conceal Decades of Discrimination Against Black Farmers*, Counter (June 26, 2019), https://thecounter.org/usda-black-farmers-discrimination-tom-vilsack-reparations-civil-rights/ (showing racial disparities in USDA lending).

Chapter V. Transforming Farm Policy Toward Climate-Neutral Agriculture 161

for beginning farmers and to target disadvantaged farmers,[283] a population that it had long discriminated against.[284]

In 2013, FSA created a microloan program aimed at smaller farmers through its existing direct operating loan program. In the first three years of the program, the agency distributed more than $350 million through almost 17,000 microloans.[285] The program caps loans at $50,000 instead of $300,000 and has a streamlined application and approval process, making it easier for small and diversified farmers to participate.[286] FSA's microloans are designed to support "non-traditional" farms, such as urban, organic, and direct-to-consumer operations.[287] The current $50,000 cap is too low to benefit most farmers engaged in commercial production, but a streamlined loan program with a ceiling of $100,000 could help carbon farmers get off the ground or expand their operations, since many climate-friendly practices do not require large capital expenditures. Congress should also require FSA to create a funding pool for agroforesters and other farmers with perennial crops and produce a report identifying how the agency could better serve them. With sufficient outreach to farmers and farm groups, and training for its loan officers, FSA could become the "lender of first opportunity" for carbon farmers.

Congress should require FSA and the Farm Credit System lending institutions to offer programs providing favorable credit to farmers and ranchers using climate-friendly practices recognized by NRCS and to require minimum climate-friendly practices relating to all loans. Both FSA and the Farm Credit System are already required to offer services to young, beginning, and small farmers and ranchers.[288] Congress should extend this mandate

---

283. *See generally* Agricultural Credit Improvement Act of 1992, Pub. L. No. 102-554, 106 Stat. 4142. FSA must dedicate 75% of its funding for direct farm ownership loans and 50% of its funding for direct operating loans to beginning farmers and ranchers during the first 11 months of the FY. 7 U.S.C. §1994(b)(2)(A). FSA is also required to reserve 40% of its funding for guaranteed ownership and operating loans to beginning farmers during the first half of the FY. *Id.* "Low-income, limited resource" farmers must receive at least 25% of FSA's guaranteed ownership and operating loans. *Id.* §1994(d).
284. A USDA task force in 1997 found low participation rates in FSA programs among minorities as well as evidence of long-running discrimination. Civil Rights Action Team, Civil Rights at the United States Department of Agriculture 21-27 (1997); *see also* Stephen Carpenter, *The USDA Discrimination Cases:* Pigford, *In re* Black Farmers, Keepseagle, Garcia, *and* Love, 17 Drake J. Agric. L. 1 (2012) (discussing credit discrimination claims against USDA).
285. *Expanded USDA Microloans Program Increases Opportunity for Small and Beginning Farmers*, NSAC, Jan. 25, 2016, http://sustainableagriculture.net/blog/expanded_usda_microloans/.
286. USDA FSA, *Microloan Programs*, https://www.fsa.usda.gov/programs-and-services/farm-loan-programs/microloans/index (last visited Nov. 12, 2020).
287. *Id.*
288. 12 U.S.C. §2207; 7 U.S.C. §1994(b)(2)(A). *See generally* Farm Credit Administration, 2015 Annual Report on the Farm Credit System 26-32 (2016) (describing the Farm Credit System's efforts to serve young, beginning, and small farmers).

to carbon farmers, giving FSA and the Farm Credit System reasonable, but escalating goals.

8. Trade Policy

Exports have played an increasingly important role in the domestic production of agricultural goods in recent years, accounting for roughly 20% of U.S. agricultural production by volume. Commodities such as cotton, rice, soybeans, and wheat generally have much higher export rates, often relying on foreign markets for the majority of their sales.[289] The United States' marketing of these products abroad not only has a significant impact on the nation's emissions, but it has also had negative economic and health consequences for many of its trade partners.[290]

Title III of the 2018 Farm Bill, which covers trade, authorizes and funds a number of export programs. It includes a large market development program that is designed to assist industry efforts to expand market demand for U.S. agricultural products abroad, and two export credit guarantee programs that guarantee loans made by U.S. private financial institutions to buyers of U.S. agricultural products in emerging markets.[291] The federal government will spend an average of $362 million on these programs annually from 2019 to 2023,[292] although this is likely to increase in future years as agribusiness groups have increasingly focused on expanding export markets. Industry groups successfully lobbied for a provision in the 2014 Farm Bill, for example, requiring the creation of an undersecretary of trade and foreign agricultural affairs position at USDA,[293] which the Trump Administration subsequently established in 2017.[294]

U.S. trade policy should be aligned with the need to curb climate change and other environmental challenges. To continue to expand demand abroad for carbon-intensive products such as grain-fed meat, while simultaneously encouraging U.S. farmers to produce climate-friendly products through other policies, would send contradictory signals to farmers, industry groups, and citizens about U.S. agricultural policy, and would directly undermine

---

289. USDA ERS, *Exports Expand the Market for U.S. Agricultural Products*, https://www.ers.usda.gov/data-products/chart-gallery/gallery/chart-detail/?chartId=58396 (last updated Apr. 11, 2016).
290. *See, e.g., Exporting Obesity: U.S. Farm and Trade Policy and the Transformation of the Mexican Consumer Food Environment*, 18 INT'L J. OCCUPATIONAL & ENV'T HEALTH 53 (2012).
291. CONGRESSIONAL RESEARCH SERVICE, 2018 FARM BILL PRIMER: AGRICULTURAL TRADE AND FOOD ASSISTANCE (2019).
292. *Id.*
293. 7 U.S.C. §6935.
294. Press Release, USDA, Secretary Perdue Announces Creation of Undersecretary of Trade and USDA Reorganization (May 11, 2014) (on file with authors).

efforts to achieve climate neutrality in agriculture. Congress should integrate climate concerns into agricultural trade policy, mandating that USDA and other government agencies focus on developing markets for climate-friendly products and discontinue support for carbon-intensive commodities.

9. Transforming the Farm Safety Net Through Legislative Action

The federal government radically transformed the farm sector in the 1930s through a series of laws that created a robust system of subsidies for commodity crop production, and provided for an ambitious set of new research and loan programs.[295] This flurry of legislation saved countless farms from bankruptcy during the Great Depression, but it also led to the rapid expansion of large-scale, capital-intensive farms and feedlots,[296] with scant concern for agriculture's environmental and social impacts. This policy shift was accompanied by significant technological change and mechanization as well.[297] These laws have since been modified, but their basic framework persists today—as does their emphasis on the large-scale production of commodity crops and meat.[298]

Agricultural law is long overdue for another transformation for a number of reasons, including the need to incorporate climate stability and resilience as major goals. The new framework must recognize that the agricultural sector is now vastly different than it was when the laws were first shaped. It has evolved from a diversified and labor-intensive enterprise to a capital-intensive, specialized, and heavily mechanized operation, typically conducted on a massive scale.[299] The pastoral "family farm"—which has always been

---

295. By 1935, USDA's budget had expanded twelvefold from pre-Depression levels, making it the single largest agency in the United States. ERNEST C. PASOUR JR., AGRICULTURE AND THE STATE: MARKET PROCESSES AND BUREAUCRACY 235 (1990). In contrast, a Congressional Research Service report from 2006 found that only 3.4% of federal outlays went to USDA between 2001 and 2005, making it the fifth largest federal agency in terms of spending. The majority of USDA funding went to SNAP, however, rather than to agricultural programs. When spending was analyzed by budget function, agriculture was found to be 12th, immediately following international affairs. PHILLIP D. WINTERS, CONGRESSIONAL RESEARCH SERVICE, FEDERAL SPENDING BY AGENCY AND BUDGET FUNCTION, FY2001-FY2005, at 10 (2006).
296. Nathan A. Rosenberg & Bryce Wilson Stucki, *The Butz Stops Here: Why the Food Movement Needs to Rethink Agricultural History*, 13 J. FOOD L. & POL'Y 12, 13-14 (2017).
297. The benefits of technological changes and mechanization were disproportionately distributed to large-scale landowners as the result of highly favorable federal programs. *Id.* at 20-21.
298. As historian Paul K. Conkin writes, the details of agricultural policy changed over the years, but "aspects of every policy option undertaken in the 1930s have endured until the present." PAUL K. CONKIN, A REVOLUTION DOWN ON THE FARM: THE TRANSFORMATION OF AMERICAN AGRICULTURE SINCE 1929, at 51 (2008).
299. CAROLYN DIMITRI ET AL., USDA, THE 20TH CENTURY TRANSFORMATION OF U.S. AGRICULTURE AND FARM POLICY (2005) (EIB-3).

more myth than reality—is of little relevance to today's agricultural industry: more than 80% of agricultural products are produced by only 7% of farms and only 42% of farms earn a gross income of $10,000 or more.[300] Indeed, the majority of low-sales "farms" are hobby, retirement, or paper farms that produce few to no agricultural goods. Environmental laws typically exempt (or have been interpreted to exempt) most aspects of agricultural production from pollution limits and other safeguards. These exemptions are sometimes presented as protecting small and midsized farms, but they often instead externalize the costs of large-scale, capital-intensive operations. The new framework should further recognize that industrial agriculture is now the largest source of water quality impairments, a major source of air pollution resulting in 17,900 annual air quality related deaths,[301] and a driver for much degradation of natural resources. This pollution often threatens human health, as do the predominant crops grown and subsidized;[302] about 60% of federal farm subsidies support corn, soy, and wheat, which are often processed into less healthy foods.[303]

As climate change intensifies, the need for programs designed around a different set of goals will become even more pressing. Instead of serving to expand the capital-intensive production of commodities, the farm safety net should directly compensate farmers for protecting the environment, mitigating climate change, growing healthy food, and strengthening rural communities.[304] As demonstrated above, USDA has significant leeway under current law to revise programs and move agriculture toward, and even to, climate neutrality. However, a system providing for robust payments for ecosystem

---

300. Calculated by the authors using USDA data. *See* NASS, *supra* note 67, at 9 tbl.1. *See generally* Chapter II.
301. Nina Domingo et al., *Air Quality-Related Health Damages of Food*, 118 PROC. NAT'L ACAD. SCI. 20 (May 2021), https://www.pnas.org/content/118/20/e2013637118. Industrial livestock facilities, often called CAFOs, produce the majority of ammonia emissions in the United States, in addition to large amounts of hydrogen sulfide, silica dust, and noxious odors. *See* D. BRUCE HARRIS ET AL., EPA, AMMONIA EMISSION FACTORS FROM SWINE FINISHING OPERATIONS 1 (2001) (noting that livestock facilities are responsible for 73% of ammonia emissions). *See generally* Dick Heederik et al., *Health Effects of Airborne Exposures From Concentrated Animal Feeding Operations*, 115 ENV'T HEALTH PERSP. 298 (2007) (summarizing research on toxic gases, vapors, and particles emitted from CAFOs).
302. Tamar Haspel, *Farm Bill: Why Don't Taxpayers Subsidize the Foods That Are Better for Us?*, WASH. POST, Feb. 18, 2014, https://www.washingtonpost.com/lifestyle/food/farm-bill-why-dont-taxpayers-subsidize-the-foods-that-are-better-for-us/2014/02/14/d7642a3c-9434-11e3-84e1-27626c5ef5fb_story.html.
303. Diet-related diseases are responsible for more than a million deaths and hundreds of billions of dollars in medical costs in the United States each year. INSTITUTE OF MEDICINE, A FRAMEWORK FOR ASSESSING EFFECTS OF THE FOOD SYSTEM 3-6 (2015); *see also* Centers for Disease Control and Prevention, *Leading Causes of Death*, https://www.cdc.gov/nchs/fastats/leading-causes-of-death.htm (last reviewed Oct. 30, 2020).
304. *See* Alison Power, *Ecosystem Services and Agriculture: Tradeoffs and Synergies*, 365 PHIL. TRANSACTIONS ROYAL SOC'Y B 2959, 2966-67 (2010) (noting that farm management can considerably enhance the ecosystem services provided by agriculture).

services (PES) could help realize this goal more quickly and efficiently than the current farm safety net, even with the changes recommended above.

Ecosystem services are benefits that humans derive from ecological resources such as farms, including food, carbon sequestration, wildlife habitat, and recreational enjoyment, among others.[305] A PES program is one that provides incentives to farmers or other landowners for provisioning such services. A 2014 study examining the societal value of soil carbon determined that farmers should be compensated at a rate of $16 an acre for implementing best management practices.[306] It would cost less than $15 billion annually to implement a PES program at this rate for all 914 million acres of farmland in the United States—billions less than we currently spend on crop insurance, commodity, and conservation programs each year. If larger payments could accelerate even faster adoption of climate-friendly practices, they would still provide the public with greater returns than the current system.

Carbon farming will require new infrastructure and equipment, both off and on the farm. Paying farmers for implementing climate-friendly practices—while regulating polluting practices—will facilitate this transition, helping to offset decades of experience and sunk costs in conventional agricultural practices. Reducing the waste that runs through the entire agriculture and food system would provide ample land and resources for a PES system.[307] Replacing a portion of the current farm safety net with a PES program would reduce or eliminate payments for crops with high climate impacts, especially those grown for animal feed, while increasing payments for crops with lower climate impacts, thus helping to make healthy food more affordable. Adopting a progressive payment system could also help midsized farms, thus increasing the economic well-being of rural communities, and reduce costs. Limiting payments to the first 1,000 acres of a farm, for example, would reduce the number of eligible acres by more than half.

While such a large-scale shift of the farm subsidy system may seem like a radical proposal, progressive groups and legislators have increasingly proposed major changes to agricultural policy in order to mitigate climate change. Representative Chellie Pingree (D-Me.) released a major bill in February 2020, the Agriculture Resilience Act, designed to reduce agricultural emissions to net zero within 20 years. Its introduction highlights how quickly climate mitigation has become a major objective of agriculture

---

305. J.B. Ruhl, *Agriculture and Payments for Ecosystem Services in the Era of Climate Change*, in Research Handbook on Climate Change and Agricultural Law 315-16 (Mary Jane Angelo & Anél Du Plessis eds., Edward Elgar 2017).
306. Rattan Lal, *Societal Value of Carbon*, 69 J. Soil & Water Conservation 186A, 190A (2014).
307. *See* Peter Lehner, *Feed More With Less*, 34 Env't F. 42 (2017).

policy for progressive groups and legislators.[308] The bill was also designed to serve as a model for the House Select Committee on the Climate Crisis' recommendations, which were released in June 2020. The Agriculture Resilience Act calls for a number of ambitious policy changes, including the quadrupling of funding for agricultural research and extension, significantly greater federal support for agroforestry, the integration of soil health into the federal crop insurance program, and a massive reduction in the use of liquid manure management systems.[309]

Paying farmers directly for positive ecological services needed by society would also avoid problems associated with the "submerged state"—the collection of indirect subsidies and incentives granted to private parties by the government.[310] Crop insurance, for example, appears to be a private-sector form of insurance, but instead acts as a generous subsidy to large-scale farmers.[311] Such indirect funding can undermine public support for robust government action, making policy reform more difficult.[312] A PES approach has the advantages of fostering transparency, while increasing efficiency. Congress should reform the farm safety net as soon as possible to shift to greater reliance on payments based on what the country now needs most—climate stabilization and a healthier environment. In so doing, Congress would also be supporting a substantially more transparent, equitable, and sustainable agricultural system.

C. Grazing Practices on Government Land

Overgrazing by livestock increases soil erosion, water pollution, and the loss of soil carbon.[313] While grazing occurs on hundreds of millions of acres of private land, more than 40% of all grazing lands in the United States—approximately 330 million acres—are on federal public lands,[314] managed

---

308. H.R. 5861, 116th Cong. (2020). The bill achieved broad support among environmental and progressive rural advocacy organizations. *See* Office of Rep. Chellie Pingree, *Statements of Support for the Agriculture Resilience Act*, https://pingree.house.gov/netzeroagriculture/ara-statements-of-support.htm (last visited Nov. 12, 2020).
309. H.R. 5861 §101 (research), §308 (agroforestry), §301 (crop insurance), & tit. V (pasture-based livestock systems).
310. Suzanne Mettler, The Submerged State: How Indivisible Government Policies Undermine American Democracy 4 (2011).
311. *See* David Dayen, *The Farm Bill Still Gives Wads of Cash to Agribusiness. It's Just Sneakier About It.*, New Republic, Feb. 4, 2014, https://newrepublic.com/article/116470/farm-bill-2014-its-even-worse-old-farm-bill.
312. Mettler, *supra* note 310.
313. Richard T. Conant & Keith Paustian, *Potential Soil Carbon Sequestration in Overgrazed Grassland Ecosystems*, 16 Global Biogeochemical Cycles 90-1, 90-1 (2002).
314. *See* USDA USFS, *About Rangeland Management*, https://www.fs.fed.us/rangeland-management/aboutus/index.shtml (last visited Nov. 12, 2020).

by BLM and USFS. While BLM is an agency within the U.S. Department of Interior, we discuss its land management policies in this chapter along with USFS, an agency within USDA, because our recommendations for both agencies overlap considerably.

A reduction in grazing intensity would help restore lost carbon.[315] Although Congress has repeatedly pushed the agencies to reduce overgrazing,[316] both agencies now do little to reduce overgrazing and indeed have policies that affirmatively thwart independent efforts by ranchers to reduce grazing to more sustainable rates. It would make economic and environmental sense to increase federal grazing fees to reflect fair market value.[317]

BLM and USFS lease land to ranchers on the condition that they will uphold conservation values,[318] including soil health. However, public interest groups allege that BLM and USFS have done little to enforce these lease provisions.[319] These agencies should not only enforce these provisions, but should also add new ones designed to reduce the climate impacts of grazing systems. Even small improvements in practices could have a significant impact due to the immense size of federal grazing lands. Just as on private lands, intensive rotational or carefully managed grazing can have numerous ecological and climate benefits, so BLM and USFS should, through pricing or other preferences, seek to incentivize such practices.

Congress has directed that leasing consider not only production but "ecological, environmental, air and atmospheric," and other values.[320] A key term in any lease or grazing permit is the "grazing intensity"—how many animals can graze a certain allotment in a certain period. Grazing intensities should be based upon accurate assessments of forage consumption by livestock and forage availability.[321] It appears that the grazing intensity established in many leases is now outdated, in part because beef cattle live weights (and so forage consumption) have increased by about 30% over the past 30 years,[322] and decades of overgrazing and now climate change reduce forage availability in

---

315. *Id.*
316. *See* Taylor Grazing Act, 43 U.S.C. §315; Federal Lands Policy and Management Act, *id.* §1751. *See also* Public Rangelands Improvement Act, *id.* §1901; National Forest Management Act, 16 U.S.C. §1600; Multiple-Use Sustained-Yield Act, *id.* §538.
317. In 2013, federal grazing fees were less than 7% of the fees charged for equivalent grazing lands on private property. CHRISTINE GLASER ET AL., CENTER FOR BIOLOGICAL DIVERSITY, COSTS AND CONSEQUENCES: THE REAL PRICE OF LIVESTOCK GRAZING ON AMERICA'S PUBLIC LANDS 1 (2015).
318. 43 C.F.R. §4180.2 (2016).
319. *About the BLM Grazing Data*, PUB. EMPLOYEES FOR ENV'T RESP., Sept. 22, 2014, http://www.peer.org/campaigns/public-lands/public-lands-grazing-reform/blm-grazing-data.html.
320. 43 U.S.C. §1701(a)(8).
321. NRCS, USDA, ESTIMATING INITIAL STOCKING RATES 7-8 (2009).
322. Bryan McMurry, *Cow Size Is Growing*, BEEF MAG., Feb. 1, 2009, http://www.beefmagazine.com/genetics/0201-increased-beef-cows.

many regions.[323] This leads to overgrazing and also has important economic consequences for ranchers, who must purchase supplemental feed to sustain cattle that the land itself cannot support.[324] Both BLM and USFS should undertake a process to update the grazing intensity limit in leases to reflect current conditions.

Even if they do not update leases, the agencies should give ranchers the flexibility to graze fewer than the allotted number of animals in order to preserve the range over the longer term and increase their profitability. However, BLM regulations provide for canceling permits of ranchers who fail to make "substantial use" of allotted forage for two consecutive years.[325] The term "substantial use" is undefined and this ambiguity has prompted many ranchers to maximize their use of allotted forage to ensure compliance with BLM requirements.[326] Similarly, USFS generally requires ranchers to graze at least 90% of allotted forage or risk revocation of their leases.[327] BLM and USFS should revise their policies to allow ranchers to graze at intensities they believe are optimal, allowing them to restore the range and increase soil carbon.

Finally, courts have held that, under the existing law governing grazing on land that is "chiefly valuable for grazing and raising forage crops,"[328] permits and leases cannot be used solely for conservation.[329] This has prevented even those who have paid fair market value for leases to retire the allotments from grazing. Congress should clarify that the purchaser of a lease or permit can graze as few animals as desired in order to preserve ecological values such as soil carbon.

D. Perennial Agriculture

Perennial agriculture uses crops that do not need to replanted each year, which results in a number of environmental and climate benefits.[330] As discussed above, current perennial crop production in the United States typi-

---

323. Daniel W. McCollum et al., *Climate Change Effects on Rangelands and Rangeland Management: Affirming the Need for Monitoring*, 3 Ecosystem Health & Sustainability 1, 7 (2017).
324. M. Rebecca Shaw et al., *The Impact of Climate Change on California's Ecosystem Services*, 109 Climatic Change 465, 478 (2011).
325. 43 C.F.R. §4170.1-2 (2016).
326. Steven C. Forrest, *Creating New Opportunities for Ecosystem Restoration on Public Lands: An Analysis of the Potential for Bureau of Land Management Lands*, 23 Pub. Land & Resources L. Rev. 21, 39 (2002).
327. USFS, USDA, Range Management ch. 2230, at 18 (2005).
328. Taylor Grazing Act, 43 U.S.C. §315.
329. Public Lands Council v. Babbitt, 529 U.S. 728 (2000).
330. *See supra* Chapter IV, for more on the climate and ecological benefits of perennial systems.

Chapter V. Transforming Farm Policy Toward Climate-Neutral Agriculture 169

cally focuses on fruit- or nut-bearing trees for specialty markets or perennial forages for grazing livestock. Nonetheless, there is also a wide variety of grains, forages, vegetables, and other types of perennial crops either available or in development for both specialty and staple crop scales.[331] Perennial agriculture merits its own set of recommendations for several reasons, including its unrivalled potential to naturally sequester carbon (as shown in Figure 2 on the next page, the top six sequestering practices are all perennial); the federal government's current lack of serious support for perennial research, financing, and outreach;[332] and perennial agriculture's unique characteristics, which make many current programs for annual crops—the dominant mode of production in contemporary agriculture—ill-suited for perennial practices.[333] In order to achieve net-zero emissions in agriculture, policymakers will need to create new programs and policies designed specifically to boost perennial practices, while revising the priorities of existing programs to ensure they no longer disadvantage perennial farming. This section describes why perennial production requires different services than annual crop production, outlines disparities between funding for annual and perennial crop production, and concludes with policy proposals aimed at addressing current disparities in federal funding.

Federal agricultural research, extension, financing, and safety net programs rarely meet the needs of farmers with perennial crops. Agricultural research, even within public universities, is increasingly focused on the priorities of private-sector corporations,[334] which sell inputs that are used less intensively—or not at all—in perennial systems. Federal funding for agricultural research is also generally focused on short-term projects.[335] This presents a major barrier to research regarding perennial crops, which requires longer funding periods due to the longer life-span of perennials.[336] While agricultural extension agents and specialists are often available to advise farmers with annual crops on how to troubleshoot problems or optimize production, few have the requisite training to provide advice regarding perennial ones.[337]

---

331. *Id.*
332. Lingxi Chenyang et al., *Farming With Trees: Reforming U.S. Farm Policy to Expand Agroforestry and Mitigate Climate Change*, 48 Ecology L.Q. (forthcoming 2021).
333. *Id.*
334. *See, e.g.*, Food & Water Watch, Public Research, Private Gain: Corporate Influence Over University Agricultural Research (2012).
335. Chenyang et al., *supra* note 332.
336. *Id.*
337. A 2011 survey of extension professionals found that about half did not have programs focused on agroforestry—the most common perennial system in the United States outside of perennial forages—in their state extension system. Michael Jacobson & Shiba Kar, *Extent of Agroforestry Extension Programs in the United States*, 51 J. Extension (2013). Of the extension professionals with at least

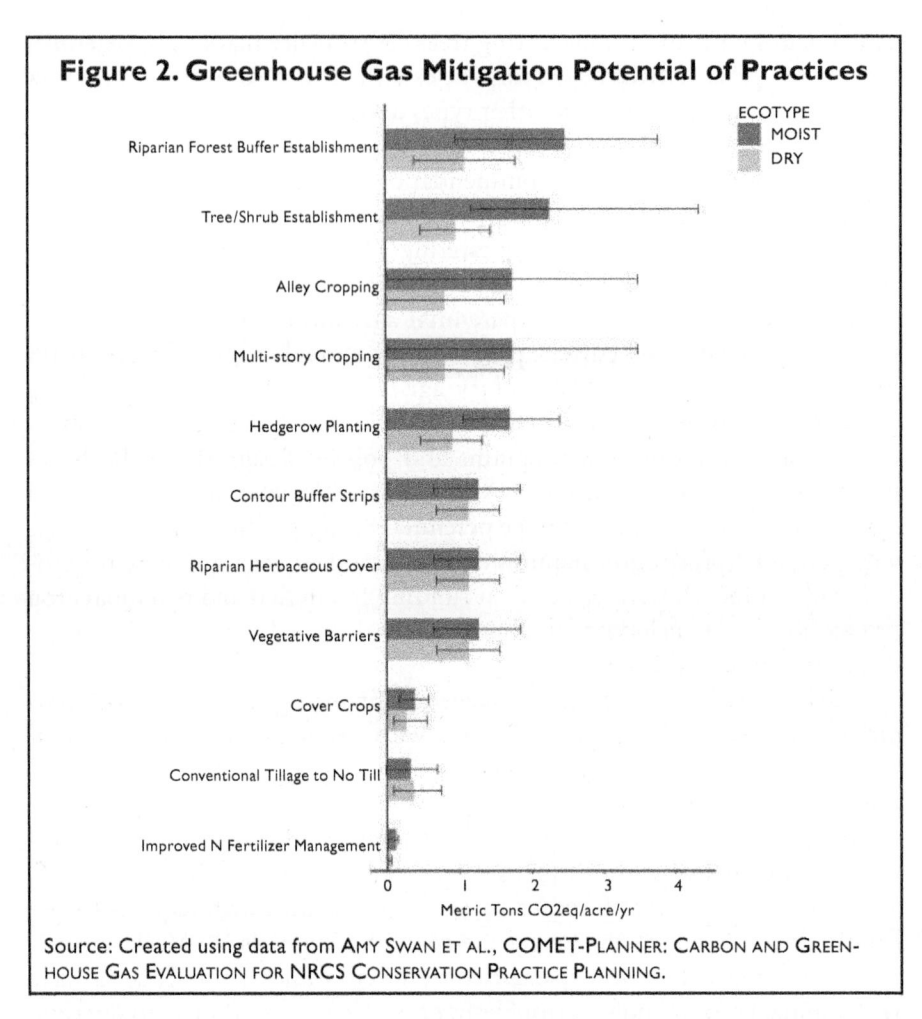

Figure 2. Greenhouse Gas Mitigation Potential of Practices

[Bar chart showing mitigation potential in Metric Tons CO2eq/acre/yr for Moist and Dry ecotypes across practices: Riparian Forest Buffer Establishment, Tree/Shrub Establishment, Alley Cropping, Multi-story Cropping, Hedgerow Planting, Contour Buffer Strips, Riparian Herbaceous Cover, Vegetative Barriers, Cover Crops, Conventional Tillage to No Till, Improved N Fertilizer Management.]

Source: Created using data from AMY SWAN ET AL., COMET-PLANNER: CARBON AND GREENHOUSE GAS EVALUATION FOR NRCS CONSERVATION PRACTICE PLANNING.

Perennial producers also face heavy hurdles when trying to process, market, or distribute their products, while producers of common annual crops, like corn, wheat, and soy, benefit from decades of government support for such infrastructure. FSA and Farm Credit System loans are not designed for perennial farmers, who often have higher upfront costs—even if their average annual costs are the same or even lower than similarly positioned farmers with annual crops. Federal subsidy programs likewise fail to take into account the higher upfront costs of perennial production, or, in some cases, simply exclude perennial crops.

---

one agroforestry program in their state, only one-quarter said that their state's program(s) were very or fairly successful. *Id.*

While USDA does not track how much support perennial practices or systems receive, available data indicate that they receive an incredibly small share of federal funding for agriculture. An analysis of USDA's 2014 research budget found that less than 0.1% of USDA's research, education, and economics budget went to agroforestry research.[338] Similarly, only a tiny amount of USDA's conservation funding goes to perennial production practices. EQIP provided almost $1.4 billion in financial assistance to farm operations for conservation practices in FY 2018,[339] but less than .05% of that amount went to the only two eligible perennial production practices, silvopasture and alley cropping.[340] This low level of funding undercuts climate-neutrality goals since these two practices have the greatest potential for carbon sequestration among contemporary agricultural practices.

Federal farm subsidies are generally distributed through commodity programs, crop insurance, and conservation programs. Most subsidy programs only support a limited range of perennial monocultures, such as almond trees, if they support any at all. They are designed to support production of a single crop on a field at any given time, limiting their utility for farms using alley cropping or silvopasture, which integrate multiple agricultural products on a single field. A Missouri producer that grows winter wheat in a monoculture, for example, could choose from a number of programs to subsidize their operation in 2019, including both of the main commodity programs,[341] the MFP,[342] and a range of crop insurance options.[343] However if that same producer wanted to intercrop Chinese chestnuts with their winter wheat, their chestnuts would not be eligible for either of the two main commodity programs or the MFP, and it would be much more difficult for the producer to receive crop insurance.

Given the lack of funding that perennial agriculture receives, and the challenges that perennial farmers face, it is notable that farmers have expressed a growing interest in perennial practices in recent years. For example, regional agroforestry groups have been organized throughout the country,[344] while many farmers have enthusiastically embraced the peren-

---

338. DeLonge et al., *supra* note 14, at 266.
339. NRCS EQIP, *supra* note 112.
340. Calculated by the authors using *id.* Approximately $667,000 went to support silvopasture and $19,000 to alley cropping. *Id.*
341. PLC has been more advantageous for Missouri wheat producers in recent years. *See* David Widmar, *ARC-CO vs. PLC: Which Won in the 2014 Farm Bill?*, Agric. Econ. Insights, Sept. 16, 2019, https://aei.ag/2019/09/16/arc-co-vs-plc-which-won-in-the-2014-farm-bill/.
342. USDA, Market Facilitation Program: 2019 County Per Acre Payment Rate 1, 12 (2019).
343. FCIC, Commodity Year Statistics for 2018, at 20 (2020).
344. USDA National Agroforestry Center, *Agroforestry Working Groups*, https://www.fs.usda.gov/nac/resources/working-groups.shtml (last visited Nov. 12, 2020).

nial grain Kernza, despite the fact that its breeders at the Land Institute note that it is not yet "economical for farmers to produce at large scale."[345] Farmers are clearly looking for perennial options. However, agricultural policy will need to offer perennial agriculture the same type of support it currently provides annual farming in order for it to be viable for most farm operations. This will require specialized programs ensuring long-term land tenure for perennial farms; funding for research, development, and extension; financial support for individual operations; and institutions capable of coordinating these efforts. The remainder of this section addresses each of these needs in turn.

1. Land Tenure

Almost 40% of farm acreage in the United States is leased and 70% of those leases are annual rather than multi-year.[346] This provides a substantial barrier to perennial production since perennial crops require farmers to invest over longer periods of time. Alley cropping, for example, generally takes five to seven years to turn a profit.[347] Congress should address this barrier by providing zero-interest farm ownership loans with low payments during the initial years of the loan and by funding a land bank that would purchase land suitable for perennial production and lease it to farmers utilizing perennial practices for up to 99 years.[348]

2. Research, Development, and Extension

Government investment in breeding and agronomic research will be critical to increasing perennial agriculture's commercial viability.[349] At the same time, farmers will need to be taught how to use new crops and practices. Authorized in the 1990 Farm Bill, the National Agroforestry Center (NAC) provides a base from which to provide these services.[350] A partnership between USFS and NRCS, NAC conducts research, develops new technologies and tools, trains natural resource advisors, and provides information

---

345. Land Institute, *Kernza® Grain*, https://landinstitute.org/our-work/perennial-crops/kernza/ (last visited Nov. 12, 2020).
346. DANIEL BIGELOW ET AL., USDA, U.S. FARMLAND OWNERSHIP, TENURE, AND TRANSFER 25 (2014) (EIB-161).
347. Chenyang et al., *supra* note 331.
348. *See id.*
349. Kevin Wolz et al., *Frontiers in Alley Cropping: Transformative Solutions for Temperate Agriculture*, 24 GLOBAL CHANGE BIOLOGY 6 (2018).
350. Food, Agriculture, Conservation, and Trade Act of 1990, Pub. L No. 101-624, §1243, 104 Stat. 3359, 3546.

about agroforestry to natural resource professionals and agroforesters.[351] Despite a renewed interest in agroforestry, however, NAC's funding has remained low, leaving it unable to match the growing need for agroforestry research, extension, and technical assistance.[352] NAC's initial 1990 budget authorization of $5 million has not increased in the intervening 30 years, meaning that its funding authorization has effectively declined by more than 50% after adjusting for inflation.[353] Further, NAC is a discretionary program whose actual funding has consistently been below its budget authorization. Since its creation in 1992, NAC's budget has stayed between $1-1.9 million; its 2020 funding was below $1.4 million, less than half of its 1994 budget after adjusting for inflation.[354] Congress should increase NAC's funding to $10 million and make it mandatory, ensuring that farmers and ranchers have access to the tools and expertise they need to adopt agroforestry practices.

Congress should also fund regional perennial agriculture centers throughout the country to train extension agents, provide technical assistance to producers growing perennial crops, and conduct research. Developing regional expertise in perennial crops and practices will be critical to expanding perennial systems because farmers are often constrained by the ecoregions and markets within which they operate. It will be important for regional perennial centers to collaborate with, and, when possible, be located within existing institutions that support perennial research and outreach to ensure that they strengthen established programs.

Research funding for perennial systems should be expanded across all major USDA research programs. In addition, Congress should create two competitive grant research programs focused exclusively on perennial agriculture. One program should be housed within the department's primary competitive research grant program, the Agriculture and Food Research Initiative (AFRI). This program should distribute at least $10 million in mandatory funding to researchers doing both basic and applied research on perennial crops and practices.[355] Congress should also create a Perennial Agriculture Research and Education (PARE) program modeled after the highly successful SARE program to provide funding for on-farm research and efforts to increase knowledge about perennial practices among producers and extension professionals. SARE's current funding process, which utilizes

---
351. USDA, AGROFORESTRY STRATEGIC FRAMEWORK: FISCAL YEARS 2019-2024, at 20 (2019).
352. Chenyang et al., *supra* note 332.
353. Food, Agriculture, Conservation, and Trade Act of 1990, Pub. L No.. 101–624, §1243, 104 Stat. 3359, 3546; 16 U.S.C. §1642 note.
354. Freedom of Information Act Response No. 2021-FS-WO-01912-F from USDA to Authors (Feb. 26, 2021) (on file with authors).
355. *See* Chenyang et al., *supra* note 332.

short-term grants, is not well-suited for perennial agriculture.[356] Whether housed within SARE or created as an independent program, PARE would offer longer-term grants, while building in-house expertise on perennial agriculture. PARE and the perennial regional centers would work in conjunction to support groundbreaking new research, while ensuring that farmers and extension officials are able to access that research.

3. Input, Distribution, and Marketing Infrastructure

Conventional farmers that grow annual crops are able to easily procure, distribute, and market their products through systems that have benefited from government support for decades, if not longer. Perennial farmers, meanwhile, generally have limited access to suppliers, distributors, and marketing opportunities, limiting their ability to expand production. Congress should address this bottleneck by providing states with annual block grants to enhance the production, distribution, and marketing of perennial crops. Modeled after the Specialty Crop Block Grant Program (but unlike the SCBGP, it would not be restricted to specialty crops), this new initiative would help develop regional markets, while ensuring that perennial production proliferates throughout the country. To spur the development of new products and markets, Congress should also create a funding pool for perennial operations within the Value-Added Producer Grant (VAPG) program, which helps producers create or expand value-added farm businesses.[357] Finally, as discussed above, USDA should utilize its funding authority under CCC to directly support perennial crops.[358]

4. Farm Finance and Support

More than a third of net farm income in 2019 came from government payments and programs,[359] yet farms using perennial practices receive almost no public support. Congress should address this in part through a new federally administered crop insurance program for agroforestry and other perennial operations. Diversified perennial operations can receive crop insurance

---

356. *Id.*
357. A USDA analysis found that businesses receiving VAPG funds were more likely to survive and increase employment. Anil Rupasingha et al., USDA, USDA's Value-Added Producer Grant Program and Its Effect on Business Survival and Growth 19 (Economic Research Report No. 248, 2018).
358. *Supra* Chapter V.B.3.
359. Adam Belz, *More Than a Third of U.S. Farm Income in 2019 Will Come From the Government*, Minn. Star Trib., Nov. 5, 2019, http://www.startribune.com/more-than-a-third-of-u-s-farm-income-in-2019-will-come-from-the-government/564525932/.

through the Whole-Farm Revenue Protection program, but private crop insurance providers do not like to administer such policies due to the additional time and expertise required.[360] A federal program would reduce overhead costs, while creating a national agency with the expertise necessary to effectively serve perennial operations. Congress should also create a funding pool for perennial practices within EQIP, modeled after the Organic Initiative, to ensure that any eligible operation applying for funds to transition to, or expand, perennial production is able to receive funding through the program.

Federal policies designed to expand perennial systems should not ignore field edges. While practices that introduce perennial plants to field edges, such as windbreaks and riparian buffers, do not have the same sequestration potential in the aggregate as those that transform core productive activities in the field, they nonetheless offer a number of important ecological and climate benefits.[361] In addition, programs designed to expand windbreaks and riparian buffers could introduce many farmers to perennial plants, and if designed well, encourage the adoption of perennial crops elsewhere on the operation.

During the New Deal, the Great Plains Forestry Project planted 220 million trees to, among other reasons, reduce soil erosion.[362] Nonetheless, Aldo Leopold in 1949 observed that federally funded conservation practices such as windbreaks had been "widely forgotten" by farmers after their five-year contract periods had expired.[363] By 2017, a reported 57% of the New Deal-era windbreaks in Nebraska had been removed and other Great Plains states likely have similar rates.[364] We need a similarly ambitious program today to expand windbreaks and riparian buffers, yet we must ensure that these new windbreaks and buffers are maintained over the long term. Whether funded through conservation programs or a new payment-for-ecosystem-services program, farmers receiving funding should sign long-term contracts—and receive funding sufficient to incentivize a long-term commitment.

---

360. *See* Mary Beth Miller & D. Lee Miller, *Insuring a Future for Small Farms*, 14 J. Food L. & Pol'y 56, 68-69 (2018).
361. *Supra* Chapter IV.A.1
362. We are grateful to Lingxi Chenyang for bringing Aldo Leopold's observations regarding the Great Plains Forestry Project to our attention and for emphasizing its relevance to contemporary policy discussions. We summarize her analysis here for our readers but it can be found in full in Chenyang et al. *supra* note 332.
363. Aldo Leopold, *The Land Ethic*, *in* A Sand County Almanac and Sketches Here and There (Oxford Univ. Press, 1949).
364. Carson Vaughan, *Uprooting FDR's "Great Wall of Trees"* (2017), https://features.weather.com/us-climate-change/nebraska/.

5.   Coordinating Efforts

Tracking current funding and initiatives will be critical to coordinating government efforts to support perennial agriculture. Congress should provide USDA's NAC with sufficient funding to document, coordinate, and evaluate government efforts to expand perennial agriculture. In addition, policymakers should require USDA to publish a detailed annual report documenting the federal government's support for perennial practices, with total funding organized, when applicable, by specific practices, geographic regions, and programs. The report should also include an in-depth assessment of the federal government's efforts to support perennial agriculture, a side-by-side comparison of funding for annual and perennial crops, and an overview of USDA's activities to coordinate efforts between federal, state, and local agencies. Finally, Congress should require USDA to release a long-term strategy for expanding perennial practices every five years, while detailing how it can improve coordination between agencies at all levels of government.

E.   *Toward Climate-Neutral Agriculture*

Farm policy, with its myriad programs, offers unparalleled opportunities to accelerate the shift toward climate-neutral agriculture. We offer above many, but surely not all, the ways that USDA farm policies could be revised to achieve this goal. In early 2021, President Biden issued an executive order directing federal agencies to develop a government-wide approach to combating the climate crisis and tasked the secretary of agriculture to make recommendations for a "climate-smart agriculture and forestry strategy." USDA issued its 90-day progress report in May 2021.[365] The report identifies many of the tools discussed here, including increased research and data collection, reforms to existing programs, and strengthening education, training, and technical assistance. As USDA continues its efforts to develop and implement a more detailed strategy, we hope the information and recommendations here will be helpful.

Unlike in other sectors of the economy, there are already extensive government agencies, research and outreach organizations, and financial institutions that are knowledgeable about farm practices. The federal government also spends tens of billions of dollars through a wide range of programs to directly support agricultural production. The farm economy is profoundly

---

365. USDA, Climate-Smart Agriculture and Forestry Strategy: 90-Day Progress Report (May 2021), https://www.usda.gov/sites/default/files/documents/climate-smart-ag-forestry-strategy-90-day-progress-report.pdf.

shaped by these forces. This has led to massive public health and environmental problems, but it also provides Congress and USDA with a clear opportunity to modify these programs for the benefit of the public—and achieve climate-neutrality in agriculture. In doing so, they will help make the United States a leader not only in carbon farming, but also in the fight against catastrophic climate change.

> **Key Recommendations**
>
> - USDA must address emissions in a systematic fashion, organizing its research, extension, and technical assistance arms around common goals and priorities. It must further work toward more ambitious national sequestration targets—set by Congress and updated at least every four years—to ensure that the sector achieves climate neutrality.
>
> - Agricultural research in several critical areas, including agriculture's impact on climate change and the environment, relies almost entirely on public spending, making publicly funded research on carbon farming practices and systems especially critical. The private sector is unlikely to focus research on ways to reduce agriculture's environmental and climate change impacts.
>
> - Congress should increase funding for climate-related agricultural research and for climate-related outreach, education, and technical assistance.
>
> - USDA should improve its ability to collect, report, and analyze data related to climate change to allow policymakers and the public to effectively evaluate the climate impact of farm practices and programs.
>
> - Congress should repeal §1619 of the 2008 Farm Bill, which prevents government agencies and the public from accessing critical information. Current law also allows USDA to provide data covered under §1619 to other government agencies in a variety of circumstances, which the department should take full advantage of until the exclusion's repeal.

- Federal crop insurance policies should treat greenhouse gas-intensive practices as risk-enhancing and reduce or eliminate their premium subsidies accordingly, while ensuring that climate-friendly practices are not discouraged by federal guidelines and requirements.

- USDA should use its rulemaking authority to require farmers receiving commodity payments to adopt cost-effective climate-friendly practices.

- USDA should use its broad authority to support the production, distribution, or marketing of agricultural commodities to develop payment programs for farmers using climate-friendly practices, providing the largest benefits to those practices with the greatest sequestration potential.

- USDA should transition to longer term or permanent land conservation programs while prioritizing land with the greatest climate mitigation potential.

- Congress or USDA should eliminate or reduce conservation payments to concentrated animal feeding operations (CAFOs), which are a fundamentally carbon-intensive form of animal production.

- Congress should expand funding for comprehensive, whole farm conservation, which research shows has the biggest environmental and economic returns on a per dollar basis.

- USDA should reduce administrative barriers to signing up for conservation programs, including simplifying contracts, increasing administrative support for farmers and ranchers, and creating a comprehensive website to allow farmers to more easily access the wide range of incentives for the promotion of climate stewardship practices.

- Agricultural operations that do not follow basic conservation practices such as cover crops, riparian buffers, and managed grazing, for example, should not be eligible to receive funds through USDA, whether through commodity, crop insurance, or conservation programs.

- Congress should require agricultural lending institutions receiving federal subsidies to offer programs providing favorable credit to farmers and ranchers using climate-friendly practices.

- Congress should integrate climate concerns into agricultural trade policy and mandate that USDA focus on developing markets for climate-friendly products rather than carbon-intensive ones.
- Agricultural law is long overdue for another transformation for a number of reasons, including the need to incorporate climate stability and resilience as a major goal.
- Congress should create, or USDA should develop, a system that pays for ecosystem services in place of much or all of the current farm safety net programs (especially the commodity and crop insurance programs) to more tightly link these subsidies to sound ecological and climate outcomes.
- Federal grazing fees should reflect fair market value to ensure better stewardship of grazing land to reduce greenhouse gas emissions.
- Policies regarding grazing on federal lands should encourage managed rotational grazing. Current policies that require ranchers and grazers to stock at close to the allocated level regardless of range and weather conditions should be revised to allow them to stock at reduced levels when better for soil, water, and climate change outcomes.
- Policymakers should create new programs and policies designed specifically to boost perennial practices and revise the priorities of existing programs to ensure they no longer disadvantage perennial farming.

Chapter VI.
Public Policy Pathways Beyond USDA for Advancing Climate-Neutral Agriculture

A variety of federal, state, and local agencies outside of the U.S. Department of Agriculture (USDA) support or regulate agricultural production. We examine the most important programs and policies among these agencies in this chapter, offering a wide range of recommendations designed to reduce emissions and expand carbon farming. We begin with a summary of the federal government's regulatory options. While the U.S. Environmental Protection Agency (EPA) has the authority to regulate methane and nitrous oxide emissions from many agricultural operations, regulating only the relatively few largest facilities could address the vast majority of the pollution without affecting the vast majority of producers.[1] Congress can also reform the rewnewable fuel standard and pass legislation regulating fertilizer manufacturers through efficiency standards or other measures. In addition, state and local governments can also stop the most harmful agricultural practices using similar approaches to those available to the federal government.

The public sector also provides significant benefits to farms through a number of non-regulatory avenues outside of traditional farm programs. We identify policy pathways to reform tax policy, the Small Business Administration (SBA) subsidized lending programs, and biogas subsidies. We conclude the chapter by examining the government's role in greenhouse gas pricing (usually focused solely on carbon dioxide from the power and industrial sectors) and other market approaches.

A. Regulatory Options

Methane and nitrous oxide are the two main greenhouse gases emitted by agricultural sources. EPA has several direct regulatory tools available to

---

1. EPA could use an emissions threshold, as is commonly done in other EPA regulatory programs, to target the largest facilities.

reduce emissions of these greenhouse gases, including recognizing the harm or "endangerment" caused by these pollutants and promulgating regulatory programs to require or support their reduction. These regulatory programs could include direct limits, prohibitions on certain activities or practices known to emit significant amounts, or increased support for known practices that reduce emissions. EPA should also reform the renewable fuel standard (RFS) to stop the conversion of nonagricultural land and to only support biofuels with substantial climate benefits. In addition to EPA's current regulatory tools, we also assess the potential for new legislation to reduce emissions from nitrogen fertilizer, and discuss how state and local governments can improve upon federal regulatory schemes.

As we discuss in Chapters II and III, agriculture is now highly consolidated in the United States. The largest 0.4% of farms in the United States produce more than a third of all agricultural products in the country, while the top 7% are responsible for more than 80%.[2] The top 7% of producers also owns 60% of the harvested cropland,[3] receives almost half of all government farm payments,[4] and takes in almost 90% of all net farm income.[5] Policymakers should be attentive to the genuine challenges farming operations face when transitioning to climate-friendly practices, but most of these large commercial farms, which often earn millions each year, can afford to adopt basic conservation practices. As we discuss in Chapter V, Congress and USDA should require large-scale operations to curb their most environmentally damaging practices in exchange for support from government programs, while maintaining a robust regulatory approach focused on the largest farms. Small and midsized farms should also be required to adopt basic conservation practices in order to receive government support, but additional funds should be made available to provide them with the financial means to adopt climate-friendly practices.

❑ *CAFO air pollution regulation.* The Clean Air Act (CAA) is the principal federal statute regulating air pollution. It is administered by EPA and the states through delegations from EPA with the states having authority to be more, but not less, protective than EPA.[6]

---

2. Calculated by the authors using data from NATIONAL AGRICULTURAL STATISTICS SERVICE, USDA, 2017 CENSUS OF AGRICULTURE, U.S. NATIONAL LEVEL DATA 9 tbl.2 (2019).
3. *Id.* at 124 tbl.72.
4. They receive 45% of farm subsidies and 55% of crop insurance payments. *Id.* at 94, 124 tbl.72.
5. *Id.* at 124 tbl.72.
6. 42 U.S.C. §§7401 et seq.; §7402.

Section 111 of the CAA requires the EPA Administrator to set and revise "a list of categories of stationary sources" that "cause, or contribute significantly to, air pollution which may reasonably be anticipated to endanger public health or welfare."[7] After the EPA Administrator lists a category, the CAA requires that EPA set standards of performance for new and modified (e.g., expanded) sources in that category.[8] These standards must reflect "the best system of emission reduction" that has been "adequately demonstrated" taking into account "the cost of achieving such reduction and any non-air quality health and environmental impact and energy requirements."[9] Further, when a category is listed, the CAA then requires that states develop plans with "standards of performance" to be achieved by existing sources in that category.[10]

In 2009, the Humane Society and other groups petitioned to set emissions standards for CAFOs on the grounds that they are stationary sources that emit significant quantities of a number of air pollutants, including the greenhouse gases methane and nitrous oxide.[11] Although EPA had already recognized that these greenhouse gases endanger public health and welfare and that enteric fermentation and livestock manure are responsible for a large portion of overall methane and nitrous oxide emissions,[12] EPA denied the petition in 2017 citing "the EPA's current comprehensive strategy to address CAFO emissions and the agency's limited resources and the need to prioritize its regulatory activity and to use its resources efficiently."[13] In 2021, Public Justice and other environmental organizations revived this issue and petitioned EPA to "list industrial dairy and hog operations as source categories" that cause or contribute significantly to dangerous pollution," noting

---

7. *Id.* §7411(b)(1)(A).
8. *Id.* §7411(b)(1)(B).
9. *Id.* §7411(a)(1); *see, e.g.,* Lignite Energy Council v. U.S. E.P.A., 198 F.3d 930, 932 (D.C. Cir. 1999).
10. *Id.* §7411(d); The requirement applies to pollutants that are not covered with National Ambient Air Quality Standards or Hazardous Air Pollutants program. *See also* Am. Elec. Power Co., Inc. v. Connecticut, 564 US 410, n.7 (2011).
11. Humane Soc'y of the United States et al., Petition to List Concentrated Animal Feeding Operations Under Clean Air Act Section 111(b)(1)(A) of the Clean Air Act, and to Promulgate Standards for Performance Under Clean Air Act Sections 111(b)(1)(B) and 111(d), at 3 (Sept. 21, 2009), http://www.humane society.org/assets/pdfs/litigation/hsus-et-al-v-epa-cafo-caa-petition.pdf.
12. U.S. ENVIRONMENTAL PROTECTION AGENCY, INVENTORY OF U.S. GREENHOUSE GAS EMISSIONS AND SINKS, 1990-2016 (Apr. 2018), https://www.epa.gov/sites/production/files/2018-01/documents/2018_executive_summary.pdf. Endangerment and Cause or Contribute Findings for Greenhouse Gases Under Section 202(a) of the Clean Air Act, 74 Fed. Reg. 66496, at 66497, 66516-22 (Dec. 15, 2009).
13. Letter from E. Scott Pruitt, Adm'r, U.S. Envtl. Prot. Agency, to Tom Frantz, President, Association of Irritated Residents (Dec. 15, 2017), https://www.regulations.gov/document?D=EPA-HQ-OAR-2017-0638-0003.

that they are major sources of methane . . . and [account] for 33 percent of agricultural methane emissions."[14]

In setting emissions standards, EPA could assess achievable limits by looking at a number of actions new CAFOs can take to reduce methane emissions, including methane-reducing feed additives,[15] grazing animals,[16] cover and flare manure management systems (a cost-effective method of reducing GHG emissions while reducing odor and protecting water quality),[17] or using anaerobic digestors to convert the methane into energy.[18] The Public Justice petition, for example, urges EPA to "base subsequent regulations on the emission reductions achievable with widespread application of sustainable, pasture-based practices [that] not only significantly reduce methane, [but] also remove carbon dioxide from the atmosphere through healthy soils [and] reduce nitrous oxide emissions from feed crops and manure disposal."[19]

If a standard of methane emissions for new CAFOs proves to be impractical (for instance if emissions monitoring difficulties precluded adequate enforcement), the CAA also allows the EPA Administrator to promulgate a "design, equipment, work practice, or operational standard, or combination thereof" to limit emissions.[20] For example, EPA implemented work practice standards to control particulate matter emissions from open coal piles at preparation and processing plants, rather than numerical emissions limitations, on the basis that it would be "difficult and prohibitively expensive to measure actual emissions from individual open storage piles or roadways."[21] This approach could be used to require, for example, dry manure handling practices, which generate less methane.

---

14. Public Justice et al., Petition to List Industrial Dairy and Hog Operations as Source Categories Under Section 111(b)(1) (A) of the Clean Air Act (May 2021), https://food.publicjustice.net/wp-content/uploads/sites/3/2021/04/2021.04.06-Industrial-Dairy-and-Hog-CAA-111-Petition-FINAL.pdf; https://food.publicjustice.net/methane-petition-press-release/.
15. J. Merint, *How Eating Seaweed Can Help Cows to Belch Less Methane*, Yale Env't 360, (July 2, 2018), https://e360.yale.edu/features/how-eating-seaweed-can-help-cows-to-belch-less-methane.
16. K. Tomas, *Manure Management for Climate Change Mitigation: Regulating CAFO Greenhouse Gas Emissions Under the Clean Air Act*, 73 U. Miami L. Rev. 2, 531, https://repository.law.miami.edu/cgi/viewcontent.cgi?article=4568&context=umlr.
17. J. Wightman & P. Woodbury, Cornell University, New York Agriculture and Climate Change: Key Opportunities for Mitigation, Resilience, and Adaptation, Final Report on Carbon Farming Project for the New York State Department of Agriculture and Markets (May 1, 2020), https://cpb-us-e1.wpmucdn.com/blogs.cornell.edu/dist/2/7553/files/2020/07/CarbonFarming_NYSAGM_FINAL_May2020.pdf.
18. EPA, *How Does Anaerobic Digestion Work*, https://www.epa.gov/agstar/how-does-anaerobic-digestion-work.
19. *See supra* Petition note 14, at 4.
20. 42 U.S.C. §7411(h)(1).
21. Standards of Performance for Coal Preparation and Processing Plants, 74 Fed. Reg. 51950, 51950, 51954 (Oct. 8, 2009) (to be codified at 40 C.F.R. pt. 60).

While EPA's regulatory oversight responsibilities would be increased by the addition of a new category of sources, EPA could minimize this burden by prioritizing regulation of the largest CAFOs. The largest 10% of CAFOs emit around 80% of CAFO greenhouse gas pollution. Section 111 allows for EPA to "distinguish among classes, types, and sizes within categories of new sources for the purpose of establishing … standards."[22] EPA could distinguish sizes based on animal units, as EPA does under the Clean Water Act,[23] or amounts of pollution emitted as it does under §114 of the CAA (see below).

Additionally, EPA could regulate ammonia emissions, which could also lead to decreased nitrous oxide emissions, depending on the control technology. One option is for EPA to designate and regulate ammonia under CAA §§108 and 109. Section 108 requires EPA to designate pollutants generated by "numerous or diverse mobile or stationary sources" that "may reasonably be anticipated to endanger public health or welfare" as "criteria" pollutants.[24] Under §109, this designation further requires EPA to establish a national ambient air quality standard (NAAQS) for the pollutant.[25] Following this, states must provide a plan to implement, maintain and enforce the NAAQS.[26] Ammonia is emitted in large quantities by CAFOs and poses significant health risks.[27] As a result, Environmental Integrity Project and other organizations petitioned EPA to list ammonia as a criteria pollutant in 2011.[28] However EPA failed to respond and litigation to compel a response was dismissed on procedural grounds.[29]

22. 42 U.S.C. §7411(b)(1)(B)(2). *See also* J. England, *Saving Preemption in the Clean Air Act: Climate Change, State Common Law, and Plaintiffs Without a Remedy*, 43 Env't L. 701, 719 (2013) (noting that EPA has limited the reach of NSPS regulation by setting facility size thresholds for approximately half of categories).
23. *See* 40 C.F.R. §122.23 (defining a subset of Animal Feeding Operations (AFOs) as Concentrated Animal Feeding Operations (CAFOs) based on species-dependent thresholds).
24. 42 U.S.C. §7408(a). At present, EPA has designated six criteria air pollutants: ozone, particulate matter (PM), nitrogen oxides (NO$_x$), sulfur dioxide, carbon monoxide, and lead. 40 C.F.R. §50.4-12.
25. *Id.* at §7408(a)(2).
26. 42 U.S.C. §7410(a)(1).
27. Environmental Integrity Project, Hazardous Pollution From Factory Farms: An Analysis of EPA's National Air Emissions Monitoring Study Data 2 (2011), https://www.ciwf.org.uk/media/7436155/hazardouspollutionfromfactoryfarms.pdf. At present, EPA regulates airborne ammonia under CERCLA as a hazardous substance, and under EPCRA as an extremely hazardous substance. *See* 40 C.F.R. §§302.4 and 355 App. A.
28. Environmental Integrity Project et al., Petition for the Regulation of Ammonia as a Criteria Pollutant Under Clean Air Act Sections 108 and 109 (2011), https://web.archive.org/web/20120917005115/http://environmentalintegrity.org/documents/PetitiontoListAmmoniaasaCleanAirActCriteriaPollutant.pdf.
29. Environmental Integrity Project et al. v. United States Environmental Protection Agency et al., No. 15-0139 (ABJ) (D.D.C., Dec. 1, 2015), http://blogs2.law.columbia.edu/climate-change-litigation/wp-content/uploads/sites/16/case-documents/2015/20151201_docket-15-cv-139_memorandum-opinion.pdf.

A second option is regulation under §112, the Hazardous Air Pollutants (HAP) program. Under §112, EPA can undertake rulemaking to add pollutants that "present, or may present, through inhalation or other routes of exposure, a threat of adverse human health effects...or adverse environmental effects . . ." to the HAP list.[30] Once a pollutant is listed, EPA must set maximum achievable control technology (MACT) standards for "major sources" of the HAP (emitting 10 tons per year or more of any single HAP or 25 tons per year of any combination of HAPs).[31] For new major sources, MACT standards for a category must be equally or more stringent than "the emission control that is achieved in practice by the best controlled similar source."[32] For existing major sources, MACT standards for a given category must be at least as strict as the average emissions limitation achieved by a best performing subset of that category.[33] In addition to major sources, EPA can also regulate "area sources" (groups of sources whose aggregate emissions present a threat to health) by imposing a less strict requirement that the sources use generally available control technologies to limit their emissions.[34]

Such limitations on ammonia emissions could also limit nitrous oxide emissions depending on the method of reduction. For example, reductions in quantities of manure would likely result in decreased emissions of both pollutants. On the other hand, certain methods of reducing ammonia may result in an increase in nitrous oxide, so the precise regulatory approach must be carefully tailored.

To effectively regulate CAFO emissions, EPA will need to establish methods by which CAFOs can measure and monitor their emissions. Accordingly, Congress should direct EPA to complete its development of emission-estimating methodologies (EEMs) for CAFOs, as the agency committed to do by 2009 under a 2005 consent agreement.[35] In addition, EPA should adopt a process-based modeling approach to developing the EEMs, since, as its Science Advisory Board has stated, such a model is better able to "represent the chemical, biological and physical processes and constraints associated with emissions."[36] These steps will allow the public and policymakers to more

---

30. 42 U.S.C. §7412(b)(2). Only pollutants not designated under CAA §108 may be regulated under the HAP program. *Id.*
31. *Id.* §7412(a)(1).
32. *Id.* §7412(d)(3).
33. *Id.* §7412(d)(3)(A), (B).
34. *Id.* §7412(a)(2), §7412(d)(5).
35. *See* OIG, EPA, Report No. 17-P-0396, Eleven Years After Agreement, EPA Has Not Developed Reliable Emission Estimation Methods to Determine Whether Animal Feeding Operations Comply With Clean Air Act and Other Statutes 5 (2017).
36. *Id.* at 14.

accurately evaluate the role of CAFOs in the climate crisis, while giving regulators the tools they need to reduce their emissions.

❏ *Row crop air pollution regulation.* Row crop agriculture is a major source of nitrous oxide emissions (as well as particulate matter pollution which can be regulated by states to achieve the particulate matter NAAQS). One approach to address nitrous oxide would be under §615 of the CAA. Section 615 states that:

> If, in the Administrator's judgment, any substance, practice, process, or activity may reasonably be anticipated to affect the stratosphere, especially ozone in the stratosphere, and such effect may reasonably be anticipated to endanger public health or welfare, the Administrator shall promptly promulgate regulations respecting the control of such substance, practice, process or activity, and shall submit notice of the proposal and promulgation of such regulation to the Congress.[37]

Currently, nitrous oxide is considered "the most important ozone-depleting substance emission"[38] and is projected to be the most significant contributor to stratospheric ozone pollution during the twenty-first century.[39] In addition, stratospheric ozone-depletion increases human exposure to UV radiation.[40] This results in health impacts including an increased prevalence of skin cancers,[41] cataracts and other eye diseases,[42] and immune system suppression.[43] Agriculture is responsible for more than 80% of U.S. nitrous oxide emissions.[44] The vast majority of these agricultural nitrous oxide emissions come from soil management, including fertilization, tillage,

---

37. 42 U.S.C. §7671n.
38. United Nations Environment Programme, Drawing Down N_2O to Protect Climate and the Ozone Layer: A UNEP Synthesis Report 7, (2013), https://wedocs.unep.org/bitstream/handle/20.500.11822/8489/-Drawing%20down%20N2O%20to%20protect%20climate%20and%20the%20ozone%20layer_%20a%20UNEP%20synthesis%20report-2013UNEPN2Oreport.pdf.
39. A.R. Ravishankara et al., *Nitrous Oxide (N_2O): The Dominant Ozone-Depleting Substance Emitted in the 21st Century*, 326 Science 123, 123-25 (2009).
40. A. McMichael at al., *Stratospheric Ozone Depletion: Ultraviolet Radiation and Health*, in Climate Change and Health (World Health Organization 2003), https://www.who.int/globalchange/publications/climchange.pdf.
41. U.S. Department of Health and Human Services, The Surgeon General's Call to Action to Prevent Skin Cancer 1-2 (2014), https://www.surgeongeneral.gov/library/calls/prevent-skin-cancer/call-to-action-prevent-skin-cancer.pdf.
42. United Nations Environment Programme, Environmental Effects of Ozone Depletion and Its Interactions With Climate Change: 2014 Assessment 57-60 (2014), https://ozone.unep.org/sites/default/files/2019-05/eeap_report_2014.pdf.
43. *Id.* at 62.
44. EPA, *Overview of Greenhouse Gases: Nitrous Oxide Emissions*, https://www.epa.gov/ghgemissions/overview-greenhouse-gases#nitrous-oxide.

drainage, irrigation, and fallowing of land.[45] EPA could prescribe regulations for these activities to reduce the emissions of nitrous oxide. In this context, "regulations" could require work practices rather than any numerical emission limit, which would likely be infeasible.[46] Some work practices with demonstrated efficacy in reducing nitrous oxide emissions include matching fertilizer and manure application rates to crop nitrogen requirements[47] and planting cover crops.[48]

❏ *Greenhouse gas regulation under water pollution programs.* Federal and state governments can also reduce greenhouse gas emissions as incidental to their regulation of water or other pollution. Programs to reduce nitrate runoff from fields into rivers would (depending on the precise practices incentivized) likely reduce nitrous oxide emissions; programs to reduce erosion and sediment pollution from grazing could increase soil carbon; and programs to change manure management could reduce methane emissions.

The Clean Water Act (CWA) establishes a national pollutant discharge elimination system (NPDES) to regulate operations that discharge pollutants directly into waters. While most field operations and irrigation water return flows are exempted from direct regulation,[49] other agricultural operations including CAFOs that do, or are likely to, discharge are covered.[50] The law requires point source dischargers to obtain an NPDES permit from EPA or authorized state authorities in order to operate.[51] States that have been authorized to act as a permitting authority may impose more stringent requirements than the federal government.[52] In addition, the CWA requires states to develop programs to address nonpoint source (runoff) pollution, including agricultural sources.[53]

EPA should strengthen its nationwide regulations in ways that would reduce greenhouse gas emissions as well as water pollution. Moreover, since states can be more stringent than the federal government, states with NPDES permitting authority should strengthen their programs in similar ways. For

---

45. EPA, Inventory of U.S. Greenhouse Gas Emissions and Sinks: 1990- 2016, at 5-21, 5-22 (2018) (EPA 430-R18-003).
46. 42 U.S.C. §7411(h)(1).
47. U. Sehy et al., *Nitrous Oxide Fluxes From Maize Fields: Relationship to Yield, Site-Specific Fertilization, and Soil Conditions*, Agriculture, 99 Ecosystems & Env't 97 (2003).
48. G. Robertson & P. Vitousek, *Nitrogen in Agriculture: Balancing the Cost of an Essential Resource*, 34 Ann. Rev. Env't & Resources 111 (2009).
49. 33 U.S.C. §1362(14).
50. *Id.*
51. *Id.* §1342.
52. 40 C.F.R. §123.25(a) (2020).
53. 33 U.S.C. §1329; 40 C.F.R. §130.6.

example, NPDES programs should clearly prohibit CAFOs from spreading manure on frozen or saturated lands, insist on vegetated buffer zones along water courses, limit application rates, or require dry manure management, which can also reduce methane emissions. Similarly, management of crop production should require or incentivize buffer zones to reduce nitrate emissions, and thus also nitrous oxide emissions.

❏ *Greenhouse gas regulation under waste programs.* Other statutes also give EPA regulatory options for reducing agricultural greenhouse gas emissions. The most common waste management systems at industrial livestock facilities produce massive quantities of toxic fumes of ammonia and hydrogen sulfide in addition to the greenhouse gases methane and nitrous oxide. EPA estimates that livestock facilities are responsible for 73% of the country's ammonia air emissions.[54] Many of the practices that would reduce these hazardous air emissions would also reduce methane and nitrous oxide emissions, and EPA should thus use its regulatory tools to achieve such reductions.

The Comprehensive Environmental Response, Compensation, and Liability Act of 1980 (CERCLA) and the Emergency Planning and Community Right-to-Know Act of 1986 (EPCRA) require all facilities that release hazardous substances to report these emissions to federal, state, and local governments and emergency responders.[55] In 2008, EPA exempted livestock facilities from this reporting requirement.[56] In 2017, the D.C. Circuit struck down EPA's loophole as illegal.[57] Responding to pressure from the animal production industry, Congress passed a rider to the March 2018 budget bill excluding livestock facilities from CERCLA reporting requirements.[58] The following month, EPA asserted that, as a result of the CERCLA exemption, the facilities were also exempt from EPCRA reporting, an action that has been challenged by community organizations.[59] Congress should pass new legislation eliminating both exemptions, ensuring that an estimated 33,000 facilities are covered, or, at a minimum, require reporting by medium and

---

54. D. BRUCE HARRIS ET AL., EPA, AMMONIA EMISSION FACTORS FROM SWINE FINISHING OPERATIONS 1 (2001).
55. 42 U.S.C. §§9603(a), 11004.
56. 40 C.F.R. §§302.6(e)(3); 355.31(g), (h) (2016).
57. Waterkeeper Alliance v. Environmental Prot. Agency, 853 F.3d 527 (D.C. Cir. 2017).
58. *See* Fair Agricultural Reporting Method (FARM) Act, 42 U.S.C. §9603.
59. EPA, *CERCLA and EPCRA Reporting Requirements for Releases of Hazardous Substances From Animal Waste at Farms,* https://www.epa.gov/epcra/cercla-and-epcra-reporting-requirements-air-releases-hazardous-substances-animal-waste-farms (last updated June 13, 2019). *But see* Memorandum From the Congressional Research Service to Senate Committee on Environment and Public Works (Mar. 13, 2018) (on file with authors) (explaining that the CERCLA exemption does not affect EPCRA reporting requirements). Citizens groups challenged this rule in *Rural Empowerment Ass'n for Community Help v. United States Environmental Protection Agency,* No. 1:18-cv-02260-TJK (D.D.C.).

large CAFOs, which would impose reporting only on the largest facilities that produce the vast majority of the waste.[60] Of course, reporting alone does not reduce emissions, but it can raise awareness of the issue, leading to reductions of waste.

Similarly, the Resource Conservation and Recovery Act (RCRA)[61] has been successfully used by neighbors of a large animal facility to require the better management of stored and spread manure to limit groundwater contamination.[62] As a result of this case, industry is urging Congress to amend RCRA to exempt animal manure, and EPA and Congress must resist this pressure as this law provides an important opportunity to reduce water and air pollution. As noted, manure management changes instigated by concerns for groundwater, including more significant changes such as switching to dry manure handling or installation of digesters, can also reduce greenhouse gases.

❑ *Renewable fuel standard.* In 2017, close to 30 million acres of corn were grown in the United States as feedstock for ethanol.[63] As noted above, the purported climate change benefits of corn ethanol are widely disputed and modest at best. Ideally, Congress should reform the renewable fuel standard (RFS) to support only those biofuels with significant climate benefits. Short of congressional reform of the RFS program, however, EPA should revise its "aggregate compliance" mechanism to ensure that nonagricultural land is not converted to growing corn as ethanol feedstock.

Conversion of native ecosystems for cultivation releases vast amounts of CO_2. A 2008 study found that converting forest, grassland, or peatland for biofuel production can release 17-420 times more CO_2 than the annual greenhouse gas reductions these biofuels would provide by replacing fossil fuels.[64] To prevent this conversion of natural ecosystems, Congress in 2007

---

60. EPA estimated that 33,000 facilities were exempted by its rule from CERCLA reporting. *See* EPA, Final Economic Analysis for CERCLA/EPCRA Administrative Reporting Exemption for Air Releases of Hazardous Substances From Animal Waste at Farms (EPA-HQ-SFUND-2007-0469-1361) (Dec. 18, 2008), https://www.regulations.gov/document?D=EPA-HQ-SFUND-2007-0469-1361.
61. 42 U.S.C. §6901.
62. Community Ass'n for Restoration of the Env't, Inc. v. Cow Palace LLC, 80 F. Supp. 3d 1180 (E.D. Wash. 2015). *See* Caroline Simson, *Wash. Dairy Settles Enviros' Manure Contamination Suit*, Law360, May 12, 2015, https://www.law360.com/articles/654586.
63. This figure was estimated for marketing year 2015/2016 as the proportion of 88 million acres planted to corn equal to the proportion of corn production used for ethanol for fuel. In that year, 43% of the corn supply was used for ethanol for fuel, and 88% of the corn supply was produced in the same year (88 million × 0.43 × 0.88 = 33 million). All data were obtained from USDA ERS, *Feed Grains: Yearbook Tables*, https://www.ers.usda.gov/data-products/feed-grains-database/feed-grains-yearbook-tables (last updated Oct. 15, 2020).
64. Joseph Fargione et al., *Land Clearing and the Biofuel Carbon Debt*, 319 Science 1235, 1235 (2008).

revised the 2005 RFS to exclude crops "harvested from land cleared or cultivated" after December 19, 2007, from its definition of "renewable biomass."[65]

EPA regulations implementing this provision, however, have rendered it meaningless. Though EPA's proposed rule required crop producers to comply with recordkeeping requirements to verify that feedstocks met Congress' definition, the agency then worked with USDA to write a final rule that differed significantly from that proposal. In the final rule, the agency adopted an "aggregate compliance" approach that instead deems all producers compliant with the standard as long as the net land area used for agriculture in the United States does not exceed its 2007 level of 402 million acres.[66] This approach has demonstrably failed to prevent significant land conversion. A 2016 analysis of satellite data estimated that 4.2 million acres of land have been converted to agriculture for biofuel production since the adoption of the standard, and EPA itself found in a 2018 report that there has been "an increase in actively managed cropland by roughly 4–7.8 million acres" since 2007, some amount of which is attributable to biofuel production.[67] An earlier study estimated that between 2008 and 2012, 1.6 million acres of long-term grasslands (that is, grasslands that were uncultivated since 1992 and likely earlier) were converted.[68] Separately, the World Wildlife Fund (WWF) estimated that between 2009 and 2015, 53 million acres of uncultivated grassland in the United States were converted.[69] (The WWF estimate likely exceeds the other estimates because it did not account for grassland that, while uncultivated in 2008, had been cultivated in earlier years.)

This "aggregate compliance" approach is also facially ineffective as millions of acres of agricultural land are taken out of production each year for many reasons, such as urban development, roads, or energy production. Thus, a static acreage cap on cropland cannot prevent conversion. Given that this approach violates both Congress' stated intent and clear language by eliminating the requirements on what land may be used to produce renewable biomass, EPA should repeal the "aggregate compliance" standard and replace it with a mandate to demonstrate feedstock was produced on land cleared before December 7, 2007, replace the static acreage cap with one reflecting current active cropland acreage, or otherwise reform the program.

---

65. 42 U.S.C. §7545.
66. 40 C.F.R. §80.1454(g) (2016).
67. Christopher K. Wright, *Recent Grassland Losses Are Concentrated Around U.S. Ethanol Refineries*, 12 ENV'T RES. LETTERS 1 (2017); U.S. EPA, BIOFUELS AND THE ENVIRONMENT: SECOND TRIENNIAL REPORT TO CONGRESS 37 (EPA-HQ-OAR-2018-0167-1334) (June 2018).
68. Tyler Lark et al., *Cropland Expansion Outpaces Agricultural and Biofuel Policies in the United States*, 10 ENV'T RES. LETTERS 1, 5 (2015).
69. WWF, 2016 PLOWPRINT REPORT 2 (2016).

❏ *Nitrogen fertilizer standards.* Conventional nitrogen fertilizer practices are quite inefficient—more than half of the nitrogen applied does not contribute to plant or animal growth.[70] In addition to regulating nitrogen runoff and other on-farm behavior contributing to nitrous oxide pollution, policymakers can lower emissions by increasing the market share of so-called enhanced-efficiency fertilizers, which are designed to reduce emissions relative to conventional fertilizers. Enhanced-efficiency fertilizers take one of two forms: inhibitors, which can help retain nitrogen in soil for longer periods; and slow- and controlled-release fertilizers, which delay the release of nitrogen into the soil.[71] Despite their potential to reduce emissions, these products have been understudied and underutilized, and researchers do not expect their use to increase through voluntary initiatives alone.[72] As a result, a pair of researchers have proposed applying new requirements to gradually increase the proportion of enhanced-efficiency fertilizers sold over time, while simultaneously providing incentives for companies to improve the climate profile of their products.[73] They propose modeling the program on the Corporate Average Fuel Economy (CAFE) standards, which set field consumption standards for vehicles, and have proven to be a durable and effective tool for increasing fuel economy.[74]

The researchers propose two possible mechanisms for enhanced-efficiency fertilizer standards. One approach would be to require that manufacturers increase the share of enhanced-efficiency fertilizer sales over time.[75] Alternatively, regulators could set a national average efficiency level, which would take into account both the share of enhanced-efficiency fertilizer sold as well as how effective those enhanced-efficiency fertilizers are at reducing emissions.[76] As discussed in Chapter IV.A.1, there are concerns regarding the environmental impacts of nitrification inhibitors, which should be studied further.[77] In addition, as the researchers warn, such standards must be coupled with other actions to reduce nitrous oxide emissions. Nonetheless, the CAFE standards provide an intriguing model for reducing nitrous oxide emissions.

---

70. David S. Kanter & Timothy D. Searchinger, *A Technology-Forcing Approach to Reduce Nitrogen Pollution*, 1 Nature Sustainability 544, 544 (2018).
71. *See supra* Chapter IV.A.1; *see also id.*
72. *Id.*
73. Kanter & Searchinger, *supra* note 70, at 548.
74. *Id.* at 548-49.
75. *Id.* at 548.
76. *Id.* at 549.
77. *See* Ch. IV.A.1.

❑ *State and local regulatory tools.* Finally, state and local governments should improve on current federal regulations by passing their own legislation designed to reduce emissions from agricultural operations. The California State Legislature, for example, passed a law in 2014 directing the California Air Resources Board (CARB) to develop a comprehensive strategy to reduce short-lived climate pollutants, including methane.[78] Subsequent legislation required CARB to begin implementing the plan by 2018.[79] CARB's strategy calls for significant decreases in emissions from dairy manure management with reductions of at least 20% in 2020, 50% in 2025, and 75% in 2030.[80] In 2015, Minnesota passed a pioneering law requiring permanent vegetative buffers on farmland abutting lakes and streams.[81] The law was designed to reduce runoff, but will also increase soil carbon sequestration on the new strips, thereby reducing greenhouse gas emissions within the state. There are a variety of practices that state legislatures and environmental agencies and local governments should require, such as riparian buffers, or prohibit, such as spreading manure on frozen land, in order to further reduce the environmental harms of modern industrial agriculture. This would provide models for future federal initiatives, while also producing immediate climate and environmental benefits.

B. Tax Policy

While many aspects of tax policy may influence farming or ranching decisions, most are too complicated, indirect, or uncertain to allow generalizations as to how they would effectuate climate-friendly practices. However, there are a few direct taxing approaches that would be effective in enhancing climate-friendly practices.[82]

The majority of agricultural emissions are from nitrous oxide produced in soils, much of which is caused by the application of nitrogen fertilizer. In section A, we discussed regulatory options to reduce emissions from nitrogen fertilizer; however, state and federal policymakers should also use their taxing power to address fertilizer emissions. Since, as noted in Chapter IV.A.1,

---

78. CAL. HEALTH & SAFETY CODE §39730 (West 2017).
79. *Id.* §39730.5.
80. CARB, CALIFORNIA ENVIRONMENTAL PROTECTION AGENCY, PROPOSED SHORT-LIVED CLIMATE POLLUTANT REDUCTION STRATEGY 7 (2016).
81. *See generally* 2016 Minn. Sess. Law ch. 85, S.F. No. 2503 (to be codified at scattered sections of MINN. STAT. ANN. chs. 103A-114B).
82. Tax incentives for climate-friendly practices should be considered with caution since tax expenditures often erode support for direct government action on the issues they are designed to address. SUZANNE METTLER, THE SUBMERGED STATE: HOW INDIVISIBLE GOVERNMENT POLICIES UNDERMINE AMERICAN DEMOCRACY (2011).

most producers routinely apply excess fertilizer, federal or state legislators should consider adopting a fertilizer fee that could both encourage more judicious use of fertilizer and help fund training on climate-friendly agricultural practices.[83] Economists long considered demand for nitrogen fertilizers to be relatively inelastic, meaning that farmers generally continued to buy about the same amount of nitrogen fertilizer barring drastic price changes.[84] However, more recent evidence indicates that rising fertilizer prices have made farmers examine fertilizer use more carefully.[85] A 2011 study in the United States estimated that for every 1% increase in price for synthetic fertilizers, demand for the product would drop 1.87%.[86] At this rate, a 10% tax on nitrogen fertilizers would reduce application rates by 2.4 million tons annually,[87] and result in hundreds of millions of dollars of revenue, while having an insignificant effect on overall costs and prices.[88]

Just as a fertilizer fee could reduce emissions from commodity crops, a small fee on conventional grain feed for cattle could reduce livestock emissions while funding programs designed to support climate-friendly management methods, such as perennial pasture-based systems, methane-reducing vaccines, or feed with methane-reducing additives. As discussed in Chapter IV.A.3, the longterm safety and efficacy of methane-reducing vaccines and additives still need to be demonstrated.[89] Nonetheless, a tax on grain cattle feed could be used to fund public research into such methods, and, if proven effective, to incentivize their use. It could also be structured to be lower on any feed with proven additives that reduce enteric or manure methane

---

83. A 2012 report commissioned by the California Water Resources Control Board examining nitrate in California's drinking water found that a fee on fertilizer equal to the state's sales tax rate of 7.2% would raise $28 million in revenue annually and reduce nitrogen application by 1.6%. Fertilizer sales are currently exempt from California's sales tax. THOMAS HARTER ET AL., CALIFORNIA STATE WATER RESOURCES CONTROL BOARD, ADDRESSING NITRATE IN CALIFORNIA'S DRINKING WATER 33 (2012).
84. Sweden's tax on synthetic fertilizer, which lasted from 1984 to 2010, is estimated to have reduced the application of synthetic nitrogen fertilizers by only 2%. The estimated price elasticity of the average nitrogen application rate varied by crop, but it was estimated to have ranged from -0.3 to -0.5, meaning that for every 1% increase in the price of synthetic fertilizers, the application rate only dropped 0.3%-0.5%. ANNE PRESTVIK ET AL., NORDEN, AGRICULTURE AND THE ENVIRONMENT IN THE NORDIC COUNTRIES: POLICIES FOR SUSTAINABILITY AND GREEN GROWTH 72 (2013).
85. After fertilizer prices rose in 2006, 32% of surveyed farmers in the United States reported reducing their fertilizer use. JAYSON BECKMAN ET AL., USDA, AGRICULTURE'S SUPPLY AND DEMAND FOR ENERGY AND ENERGY PRODUCTS 17 (2013) (EIB-112).
86. James Williamson, *The Role of Information and Prices in the Nitrogen Fertilizer Management Decision: New Evidence From the Agricultural Resource Management Survey*, 36 J. AGRIC. & RESOURCE ECON. 552, 568 (2011).
87. A total of 12,840,000 tons of nitrogen fertilizer were applied in the United States in 2011. ERS, USDA, U.S. CONSUMPTION OF NITROGEN, PHOSPHATE, AND POTASH, 1960-2011, at 1 tbl.1 (2013).
88. Nitrogen fertilizer prices have ranged from $351 to $847 per ton in recent years. *Id.*
89. *See, e.g.,* Ula Chrobak, *The Inconvenient Truth About Burger King's "Reduced Methane" Whopper,* POPULAR SCI., July 20, 2020, https://www.popsci.com/story/environment/burger-king-reduced-methane-whopper-debunk/.

generation. While the majority of cattle ranches are owned by small-scale hobbyists or retirees, they purchase relatively little grain feed.[90] Instead such a tax would target large-scale Animal-Feeding Operations (AFOs), which consume most of the country's cattle feed.[91] Cattle AFOs are also highly lucrative, earning an average net income of $377,000 in 2017, making a small fee on conventional grain feed both financially feasible for individual feedlots and politically feasible for policymakers.[92]

States and local governments can also discourage carbon-intensive practices through taxation (although local governments usually need state authority to levy fees and taxes). Many states and local governments currently provide significant property tax reductions for farm owners, regardless of how large or profitable their farm operations are.[93] In Utah, for instance, property taxes can be reduced by more than 99% for farms and ranches.[94] These tax benefits can keep farms viable in areas where encroaching development might otherwise make property taxes unaffordable or inordinately burdensome. While protecting farmland from development can have climate benefits, governments should also take farm practices into account when assessing farmland values. Highly profitable, highly polluting hog CAFOs are often eligible to receive agricultural use exemptions, for example. States and local governments should condition tax reductions for agriculture on the adoption of more climate-friendly practices, perhaps targeting more stringent requirements on larger farms or those with a larger than average (perhaps analyzed by size range) carbon impact.[95] States and localities can also explore ways to expand tax incentives for carbon-friendly practices. For example, in New York, former Governor Cuomo proposed to amend the Real Property Tax Law to allow certain forestland owners, who now can get a property tax

---

90. *See, e.g.,* Aerin Einstein-Curtis, *USDA: As Livestock Producers Expand, They Buy Rather Than Grow Feed*, FEEDNAVIGATOR, Mar. 28, 2018, https://www.feednavigator.com/Article/2018/03/28/USDA-As-livestock-producers-expand-they-buy-rather-than-grow-feed.
91. *See* NATIONAL AGRICULTURAL STATISTICS SERVICE, USDA, 2017 CENSUS OF AGRICULTURE, U.S. NATIONAL LEVEL DATA 202 tbl.75 (2019).
92. *Id.* at 207 tbl.75.
93. *See supra* Chapter II; *see also, e.g.,* N.M. STAT. ANN. §7-36-20 (2016). For a complete list, see Lincoln Institute of Land Policy & George Washington Institute of Public Policy, *Significant Features of the Property Tax—Tax Treatment of Agricultural Property*, https://www.lincolninst.edu/pt-br/research-data/data-toolkits/significant-features-property-tax/access-property-tax-database/tax-treatment-agricultural-property (last visited Nov. 12, 2020).
94. CLARK ISRAELSEN ET AL., UTAH STATE UNIVERSITY COOPERATIVE EXTENSION, UTAH FARMLAND ASSESSMENT ACT (2009).
95. Many states have similar tax reduction programs for lands held for forestry. *See* JANE MALME, PREFERENTIAL PROPERTY TAX TREATMENT OF LAND 9-11 (Lincoln Institute of Land Policy, Working Paper Product Code No. WP93JM1, 1993). Originally designed to encourage forest products industries, these programs should also be redesigned to prioritize carbon-friendly forestry programs and to require carbon-friendly core practices.

reduction if they have a plan to harvest timber, to get an equivalent tax reduction if they manage their forest to improve carbon sequestration and water quality.[96]

A number of federal, state, and local tax expenditures also support conservation easements. In 2015, Congress permanently extended an enhanced tax deduction for landowners donating a conservation easement to a land trust or government agency.[97] Among other benefits, the enhanced deduction allows farmers and ranchers to deduct up to 100% of their income.[98] Conservation easement donations also reduce state and local taxes by reducing the assessed value of the land, and, in some cases, through tax deductions and credits.[99] Thirty states allow tax deductions for conservation easement donations,[100] while 16 states grant tax credits, including New York and California.[101]

In order to be eligible for the federal enhanced tax deduction, a conservation easement must be created exclusively for "conservation purposes," as defined in the Internal Revenue Code.[102] The definition is broad, however, and includes the "preservation of open space," including farmland.[103] Maintaining the rural character of an area, for example, can be a sufficient conservation purpose.[104] State and local governments generally have similar requirements for tax deductions or credits. Federal, state, and local governments should all consider requiring farm owners to comply with basic climate-friendly practices, such as installing buffer strips next to streams, in order to receive tax benefits for agricultural easements or, at a minimum, discount the value of any agricultural easement that does not ensure the implementation of such practices.

---

96. In New York, Real Property Tax Law §480-a provides the existing tax reduction. In the 2017 State of the State report, Governor Cuomo proposed enactment of a §480-b to allow for expanded eligibility for tax reductions. GOVERNOR ANDREW M. CUOMO, 2017 STATE OF THE STATE 240 (2017). The amendment did not pass in 2017 or 2018. This idea was recently supported by an advisory panel convened to make recommendations on how New York can achieve its climate goals. *See* Agriculture and Forestry Advisory Panel, New York Climate Action Council, *Emissions Reduction and Carbon Sequestration Recommendations* (Apr. 2021), https://climate.ny.gov/Climate-Action-Council/Meetings-and-Materials.
97. I.R.C. §170(b)(1)(E) (2016).
98. *Id.* §170(b)(1)(E)(iv).
99. Gerald Korngold, *Government Conservation Easements: A Means to Advance Efficiency, Freedom From Coercion, Flexibility, and Democracy*, 78 BROOK. L. REV. 467, 471 (2013).
100. JEFFREY O. SUNDBERG, STATE INCOME TAX CREDITS FOR CONSERVATION EASEMENTS: DO ADDITIONAL CREDITS CREATE ADDITIONAL VALUE? 3 (Lincoln Land Institute Working Paper No. WP11JSS1, 2011).
101. Land Trust Alliance, *Income Tax Incentives for Land Conservation*, https://www.landtrustalliance.org/topics/taxes/income-tax-incentives-land-conservation (last visited Nov. 12, 2020).
102. I.R.C. §170(h)(4)(A) (2016).
103. *Id.*
104. INTERNAL REVENUE SERVICE, CONSERVATION EASEMENT AUDIT TECHNIQUES GUIDE 18 (2016).

C. Small Business Administration Lending Programs

SBA is a cabinet-level agency that serves small businesses by providing federal contracts, counseling, and credit.[105] SBA's main credit program is the 7(a) Loan Program, which guarantees loan amounts of up to $5 million to entities that meet its small business size standards.[106] The requirements differ depending on the type of business: cattle feedlots with less than $8 million in annual receipts are eligible for 7(a) loans, for example, while eligible poultry producers cannot have more than $1 million in annual receipts.[107] However, a 2018 investigation by the SBA Office of Inspector General (OIG) found that poultry producers regularly received SBA guaranteed loans even when they did not meet SBA size standards or other requirements for eligibility.[108] In part due to these lax standards, the agency guaranteed more than 1,500 7(a) loans for poultry producers between FY 2012 and FY 2016, providing them with approximately $1.8 billion in financing.[109] In addition to poultry operations, SBA also guarantees large loans to other highly polluting agricultural operations, such as hog and dairy CAFOs.[110] The value of SBA-guaranteed loans to CAFOs doubled overall during President Obama's second term, increasing from $224 million in FY 2012 to more than $652 million in FY 2015.[111]

CAFOs are highly polluting operations—often with high greenhouse gas emissions—that depress land values, while only providing a small number of low-wage jobs.[112] Subsidizing their production is contrary to SBA's mission to invest in communities and create jobs. The agency should immediately ensure that it is enforcing its own regulations when providing loans or loan guarantees to CAFOs. Congress should also make CAFOs ineligible for the 7(a) Loan Program and other SBA assistance programs or otherwise

---

105. SBA, FY 2020 Congressional Justification 2-3 (2020).
106. *Id.* at 27.
107. 13 C.F.R. §121.201.
108. OIG, SBA, Evaluation of SBA 7(A) Loans Made to Poultry Farmers 7-9 (2018).
109. *Id.* at 2.
110. In one such case, Arkansas paid an SBA-financed hog CAFO $6.2 million to close after it was found to pose a threat to the Buffalo River watershed. Emily Walkenhorst, *C&H Hog Farms Takes State Buyout; $6.2M Deal Cut to Preserve Buffalo River*, Ark. Democrat Gazette, June 14, 2019, https://www.arkansasonline.com/news/2019/jun/14/c-h-hog-farms-takes-state-buyout-201906/. Local conservation and civic groups had strongly opposed the SBA loan guarantee for the CAFO, arguing that the agency's environmental assessment of the loan ignored critical information. Letter From the Buffalo River Watershed Alliance et al., to Val Dolcini, Administrator, FSA, and Maria Contreras-Sweet, Administrator, SBA (Jan. 29, 2016) (on file with authors).
111. Catherine Boudreau, *Feds Hit Brakes on Loans to Big Farms*, Politico, Oct. 24, 2016, https://www.politico.com/story/2016/10/slow-loans-over-green-woes-put-cafos-in-limbo-230234.
112. *See supra* Chapter II.

impose strong environmental conditions for eligibility (such as lower methane emissions).

D. Biogas Subsidies

Anaerobic digestion uses microorganisms to break down organic waste and produce biogas, an energy source largely made up of CO_2 and methane. This discussion focuses on biogas produced using animal manure, which has become a major focus of industry efforts to reduce greenhouse gas emissions in agriculture.[113] On-farm digesters are generally found on dairy and swine operations, which account for 90% of methane emissions from manure management and at least 9% of *all* direct agricultural emissions according to EPA.[114]

Agribusiness advertises on-farm anaerobic digestion as a win-win-win for family farmers, rural communities, and the climate, claiming that it produces a reliable source of income for farmers, helps control pollution, and reduces greenhouse gas emissions.[115] These claims, however, rely on unrealistic models and misleading data. In reality, subsidies for manure-derived biogas hurt small-scale farmers, incentivize polluting practices, allow polluters to externalize the cost of their polluting practices, and result in greenhouse gas emissions at levels far above more practical and sustainable alternatives.

When Dominion Energy, owner of one the nation's largest natural gas storage systems,[116] announced a $200 million dairy biogas partnership, it claimed it would be "providing a new source of long-term revenue for family farmers across the country."[117] However, a 2018 study found that anaerobic digesters are not economically viable on dairy herds with fewer than 3,000 cows.[118] Only about 1% of dairy farms in the country have herds larger than

---

113. The U.S. Department of Energy's National Renewable Energy Laboratory estimates that animal manure generates 24% of the methane suitable for biogas in the United States. NATIONAL RENEWABLE ENERGY LABORATORY, U.S. DEPARTMENT OF ENERGY, ENERGY ANALYSIS: BIOGAS POTENTIAL IN THE UNITED STATES 1 (2013).
114. *See* EPA, INVENTORY OF U.S. GREENHOUSE GAS EMISSIONS AND SINKS: 1990-2017, at 5-11 tbl.5-7 (2019) (EPA 430-P-19-001). This share is likely even higher since EPA calculates the global warming potential of methane using an outdated formula, which understates methane's potency. *Supra* Chapter III.
115. *See, e.g.*, Press Release, Dominion Energy, Dominion Energy and Vanguard Renewables Form Strategic Partnership to Develop First Nationwide Network of Dairy Waste-to Energy Projects (Dec. 11, 2019) (on file with author).
116. Dominion Energy, *Natural Gas Storage Systems*, https://www.dominionenergy.com/our-company/moving-energy/natural-gas-storage-systems (last visited Jan. 14, 2020).
117. Press Release, Dominion Energy, *supra* note 115.
118. Mark Lauer et al., *Making Money From Waste: The Economic Viability of Producing Biogas and Biomethane in the Idaho Dairy Industry*, 222 APPLIED ENERGY 621, 632 (2018).

2,500 cows, holding about one-third of the country's dairy cows.[119] Another 2018 study found that anaerobic digestion is not profitable on swine operations, regardless of size, without government subsidies.[120] Nor are digesters likely to become financially attractive without major public subsidies: biogas produced with manure is much more expensive than other truly renewable sources of energy.[121] And while solar, wind, and other renewable sources of energy are becoming less expensive due to economies of scale, the opposite is happening with digesters, which will need to be built in increasingly inefficient locations in order to expand production.[122]

CAFOs—large-scale industrial facilities—are major sources of air and water pollution, which cause higher rates of respiratory and other diseases in the communities in which they are located. They also lower property values in surrounding communities: one study in Missouri found that each CAFO built in the state lowered property values in surrounding communities by an average of $2.68 million.[123] As a result, they are often built within marginalized communities that suffer public health harms and noxious odors emanating from these facilities, neither of which is remedied by digesters. Pixley, California, is a prime example: it has more anaerobic digesters than any other community in the state,[124] but its 3,300 residents, almost 40% of whom live in poverty[125] and 90% of whom are Latinx, continue to suffer from the resulting pollution and benefit little from the presence of the digesters.[126] This dynamic is replicated throughout the state's anaerobic digestion hub, the San Joaquin Valley, where livestock operations are the top source of

119. *See* NASS, *supra* note 2, at 23 tbl.18.
120. Cortney Cowley & B. Wade Brorsen, *Anaerobic Digester Production and Cost Functions*, 152 ECOLOGICAL ECON. 347, 355-56 (2018).
121. *See, e.g.*, David Kesmodel, *Energy Prices Steer Farmers Away From Power Generators*, WALL ST. J., Feb. 18, 2016 (discussing the high costs of manure-to-energy digester), https://www.wsj.com/articles/energy-prices-steer-farmers-away-from-power-generators-1455814921. A 2011 study found that electricity costs for anaerobic digesters using manure were between $0.128 per kilowatt hour (/kWh) and $0.204/kWh. David P.M. Zaks et al., *The Contribution of Anaerobic Digesters to Emissions Mitigation and Electricity Generation Under U.S. Climate Policy*, 45 ENV'T SCI. & TECH. 6735, 6738 (2011). The U.S. Energy Information Administration estimates that other forms of renewable energy have costs ranging from $0.038/kWh to $0.121/kWh. U.S. ENERGY INFORMATION ADMINISTRATION, U.S. DEPARTMENT OF ENERGY, LEVELIZED COST AND LEVELIZED AVOIDED COST OF NEW GENERATION RESOURCES IN THE ANNUAL ENERGY OUTLOOK 8 tbl.1b (2020).
122. David Roberts, *The False Promise of "Renewable Natural Gas,"* Vox, Feb. 20, 2020, https://www.vox.com/energy-and-environment/2020/2/14/21131109/california-natural-gas-renewable-socalgas.
123. MUBARAK HAMED ET AL., UNIVERSITY OF MISSOURI—COLUMBIA, REPORT NO. R-99-02, THE IMPACTS OF ANIMAL FEEDING OPERATIONS ON RURAL LAND VALUES 8 (1999).
124. *See* EPA, *AgSTAR—Livestock Anaerobic Digester Database*, https://www.epa.gov/agstar/livestock-anaerobic-digester-database (last updated Aug. 16, 2020).
125. U.S. Census Bureau, *Pixley CDP*, https://data.census.gov/cedsci/profile?g=1600000US0657512 (last visited Mar. 15, 2021).
126. U.S. Census Bureau, *American Community Survey: Data Profiles*, https://www.census.gov/acs/www/data/data-tables-and-tools/data-profiles/ (last visited Nov. 12, 2020).

ozone-causing pollutants and leading sources of nitrate, ammonia, and fine particulate pollution.[127] CAFOs are not only strategically placed within poor communities, but work to keep communities from developing more sustainable economies.[128]

Biogas advocates claim that anaerobic digesters reduce noxious odors and other forms of pollution, but the studies they cite rely on ideal conditions and rarely take into account downstream pollution. For example, digested manure is often used as a fertilizer, and the digestion process increases the water solubility of nutrients such as nitrogen and phosphorus, which leads to greater runoff and water pollution.[129] Manure spraying, overpowering smells, and chronic spills plague communities near digesters. A digester in Dane County, Wisconsin, had three separate spills over a five-month period, releasing 435,000 gallons of liquid manure into a dry ravine and creek, before an explosion left it inoperable in 2014.[130] Operators of an anaerobic digester in Weld County, Colorado, were forced to close their $115 million facility in 2017 after county commissioners suspended its permits due to hundreds of odor complaints by local residents.[131] The county is a major producer of livestock and dairy, but residents complained that the odors from the digester, which used manure as its main feedstock, were much worse than those from neighboring dairies and feedlots.[132] Residents have made similar complaints about noxious odors and pollution emanating from digesters located in Michigan,[133] Nebraska,[134] and Iowa,[135] among other places.

---

127. Roberts, *supra* note 122.
128. Most swine and poultry CAFOs are operated under a contract system in which the grower must use material such as feed provided by the integrator. Thus, their contribution to the local economy is limited.
129. NRCS, USDA, Conservation Practice Standard: Anaerobic Digester, Code 366 (2017) (366-CPS-1).
130. Jim Eichstadt, *Clear Horizons Manure Digester: Public Funds Wasted on Huge Fiasco*, Milkweed, Jan. 2015, at 9.
131. Jacy Marmaduke, *Waste-to-Energy Facility Brings Smelly Complications*, Coloradoan, Jan. 15, 2017, https://www.coloradoan.com/story/news/2017/01/16/waste--energy-facility-brings-smelly-complications/96538924/.
132. Grace Hood, *Fed Up With the Smell, Neighbors Want the Weld County Biogas Project Shut Down*, CPR News, Dec. 16, 2016, https://www.cpr.org/2016/12/16/fed-up-with-the-smell-neighbors-want-the-weld-county-biogas-project-shut-down/.
133. Phil Dawson, *Digester Told to Cover Smelly Lagoons in Holton*, 13 On Your Side, Aug. 16, 2019, https://www.wzzm13.com/article/news/digester-told-to-cover-smelly-lagoons-in-holton/69-d7715bd8-9041-4c6d-b2be-4dcb95fc77f0; Darren Cunningham, *Lagoons Leave "Stench" in Surrounding Communities, Meeting Planned*, Fox 17, Aug. 13, 2019, https://www.fox17online.com/2019/08/13/lagoons-leave-stench-in-surrounding-communities-meeting-planned/.
134. Jetske Wauran, *New Odor and Sewage Complaints About Big Ox Energy*, Siouxland News, June 14, 2018, https://siouxlandnews.com/news/local/new-odor-and-sewage-complaints-about-big-ox-energy.
135. Ian Richardson, *Odors, Citations Association With Other Big Ox Facilities, Including Riceville*, Globe Gazette, Oct. 10, 2018, https://globegazette.com/news/iowa/odors-citations-associated-with-other-big-ox-facilities-including-riceville/article_35d8647a-41f8-5594-b851-2571ffea4e87.html.

As discussed in Chapter IV.A.3, anaerobic digesters may reduce emissions when compared to conventional liquid manure management methods, but they are more expensive and less climate-friendly than dry manure and pasture-based management systems, which generate far less methane and are inherently less polluting. A report by *The Counter* found dozens of national, state, and local subsidies aimed at anaerobic digesters. These incentives benefit large-scale factory farms, not the public or the marginalized communities where they are often placed.[136] Rather than subsidizing highly polluting livestock systems, policymakers should help farms transition to dry manure management and managed rotational pasture-based systems. Unlike subsidies for anaerobic digesters, grazing-based dairy systems can benefit smaller-scale operations by reducing capital costs and giving operators a higher net income per cow. They also provide rural communities with a variety of co-benefits, including improved soil, air, and water quality; increased wildlife and livestock health; and enhanced property values and tourism economies.[137]

E. Greenhouse Gas Pricing

Carbon pricing for all greenhouse gases from agriculture could be a highly effective policy lever. While economic uncertainties make it difficult to predict precise impacts, a carbon price creates a broad signal affecting the decisions of most or all actors and can spur innovation toward lower greenhouse gas technologies and practices. Its broad reach and relative ease of administration make it an attractive policy tool. Governments can impose a greenhouse gas price through a carbon tax or fee, or through a cap-and-trade program. California created a cap-and-trade program that applies to CO_2 emitted by most economic sectors, and the Northeast states developed a cap-and-trade program for the power sector in the Regional Greenhouse Gas Initiative.[138] Policy discussions concerning carbon pricing should focus on the magnitude and growth rate of the price (or, equivalently, the size and speed of decrease of the cap), options for what to do with the income generated, timing for when the fee (or cap) should be applied, and whether and what exceptions should exist.

Various carbon pricing mechanisms can generate revenue that can be refunded to taxpayers, used as general revenues, used as an offset for other,

---

136. Jessica McKenzie, *The Misbegotten Promise of Anaerobic Digesters*, COUNTER, Dec. 3, 2019, https://thecounter.org/misbegotten-promise-anaerobic-digesters-cafo/.
137. NRCS, USDA, TECHNICAL NOTE NO. 1, PROFITABLE GRAZING-BASED DAIRY SYSTEMS 8-10 (2017).
138. See Guri Bang et al., *California's Cap-and-Trade System: Diffusion and Lessons*, 17 GLOBAL ENV'T POL. 12, 18-21 (2017), for a comparison of California's cap-and-trade system and the Regional Greenhouse Gas Initiative.

less popular taxes, or used to support particular projects. Given the long history of using public support to encourage change in the agricultural sector, policymakers should consider using carbon fee revenues to support a reduction of agricultural greenhouse gas emissions and to support practices that increase soil carbon storage. Allowing agricultural producers to earn revenue by storing soil carbon or by increasing biomass, especially if such payments were in lieu of current federal farm subsidies, could be an effective way to significantly cut emissions quickly while increasing the carbon sink. We discuss the option of voluntary markets for agricultural carbon credits, in which there is growing interest, below in Chapter VII.

Most discussions of carbon pricing focus on fossil fuel emissions and thus assume a fee would be placed only on CO_2, and not on other greenhouse gases, thus the term "carbon pricing." If this were the case, the impact on agriculture would be largely ineffective, as its primary climate change contribution is through nitrous oxide and methane. Thus, despite its name, all greenhouse gases should be included in the carbon pricing mechanisms to ensure that farmers do not shift practices to those with perhaps greater climate impact. For example, practices that use a bit more energy in order to rotate grazing animals or apply nitrogen fertilizer more precisely could be inappropriately discouraged if the price were not applied to all greenhouse gases. However, there are much greater difficulties in measuring precisely nitrous oxide and methane emissions, as well as greater implementation challenges (and opportunities). To date, there has been little study of this option, and we urge its close examination.

Given the difficulty of precisely measuring emissions of nitrous oxide and methane from agricultural operations, however, it would be difficult to have a precise fee applied to such emissions. Whether as an offset or within a cap or tax regime, it would be necessary to create methodologies that can model emissions based on practices, at least until precise measurement tools become available. Thus, the baseline for any greenhouse gas pricing system should be carefully examined.

Chapter VI. Public Policy Pathways Beyond USDA

> **Key Recommendations**
>
> - EPA should promulgate regulatory programs focused on the largest industrial agriculture facilities, which produce the majority of pollution but represent only about 10% or less of total operations. These regulations should require or support the reduction of methane and nitrous oxide emissions, and could include increased support for practices known to reduce emissions, prohibitions on certain activities or practices known to emit significant amounts of greenhouse gases and pollution, or direct emission limitations.
>
> - EPA should reform the renewable fuel standard to prohibit the conversion of nonagricultural land to grow renewable biomass and to ensure that the program only supports biofuels with substantial climate benefits. In particular, EPA must eliminate or reform the "aggregate compliance" approach to determining whether crops qualify as renewable biomass under the program and should instead ensure that biomass for renewable fuel is not grown on land cultivated after December 2007 and that the progrm does not directly or indirectly support land conversion.
>
> - The Corporate Average Fuel Economy program provides a model for reducing nitrous oxide emissions in fertilizers by encouraging greater sale of enhanced-efficiency fertilizers, which can increase the portion of nitrogen taken up by plants, leaving less excess nitrogen available to convert to pollution.
>
> - States and local governments should pass their own legislation designed to reduce emissions from agricultural operations. At a minimum, they should better target existing agricultural support and water pollution control programs to incentivize proven climate-friendly practices.
>
> - Federal and state legislators should consider adopting a fertilizer fee that could both encourage more judicious use of fertilizer and help fund training on climate-friendly agricultural practices.
>
> - States and local governments can also discourage carbon-intensive practices and encourage carbon-friendly practices through tax policy.

- Since many CAFOs are not truly independent of the large-scale integrators with which they contract, the Small Business Administration should make those CAFOs ineligible for SBA loans or loan guarantees, or at a minimum ensure that the CAFOs adhere to SBA requirements for eligibility, including size and affiliation standards.
- Policymakers should help animal feeding operations transition to dry manure management and well-managed pasture-based systems.
- Policymakers should not subsidize anaerobic digesters, which tend to benefit large-scale massive factory farms rather than the public or the marginalized communities where they are often placed. Digesters do nothing to reduce pollution or other health harms emanating from CAFOs, may in fact increase (rather than reduce) emissions, and are less cost-effective than other greenhouse gas reduction methods.
- Policymakers should consider using carbon fee revenues from state or federal greenhouse gas fee programs to support a reduction of agricultural greenhouse gas emissions and to support practices that increase soil carbon storage.
- All greenhouse gases should be included in carbon pricing mechanisms to ensure that farmers do not shift practices to those with greater climate impact.

Chapter VII.
Private- and Nonprofit-Sector Opportunities for Advancing Climate-Neutral Agriculture

There are a number of ways that the private and nonprofit sectors can boost carbon farming and help reduce net agricultural emissions. Farmers who adopt new and innovative practices designed to decrease net emissions will need access to capital—already a significant problem for many reformers—as well as incentives, training, practical and legal tools, and access to markets for their goods. Sustained funding and support for agricultural research will be critical, especially during periods when the executive branch is indifferent or hostile to publicly funded scientific research.

Private actors, especially those in the food sector, can adopt and support carbon-neutral goals for themselves and their supply chains, thereby incentivizing a shift to climate-neutral practices. They can also help fund research related to the benefits of such practices. Nonprofit organizations can also play a leading role, especially when the federal government is not engaged in incentivizing adoption of these practices. They can help develop and disperse tools for carbon farmers, including practical tools such as inexpensive methods for measuring soil carbon content, and legal tools such as conservation easements. Finally, some industry and environmental groups have significant enthusiasm for agricultural carbon markets as a method to encourage the adoption of climate-neutral practices. While voluntary offset markets in agriculture have the potential to reward farmers for improving their practices, they have several important limitations, and, as discussed below, enthusiasm for such markets must be measured.

A. Research

Research supporting organic agriculture in the United States was led by the private sector for many years due to the U.S. Department of Agriculture's (USDA's) general disregard for, and occasional hostility to, organic agriculture prior to the 1990s.[1] In the 1970s and 1980s, a number of private research organizations such as the Rodale Institute, the Aprovecho Institute, and the Michael Fields Agricultural Institute were created to conduct and support research into organic and ecological farming.[2] Their work to develop and proliferate new practices was instrumental in the growth of sustainable agriculture. Foundations and private donors should support the work such research organizations are conducting on climate-friendly practices, in addition to helping fund new organizations devoted to carbon farming. Private companies, alone or in conjunction with philanthropic or government organizations, could accelerate research into potentially marketable climate-friendly products such as methane-reducing feed additives, enhanced-efficiency fertilizers, improved perennial crops, and precision application and monitoring technologies. Once developed, these products could be marketed or potentially provided through incentive and subsidy programs.

B.   Financing Options

The seasonal nature of farming makes loans particularly important for farmers. In 2017, almost 1.3 million non-real estate loans were made to farmers.[3] To put that in perspective, fewer than 900,000 farm operations grossed $10,000 or more in agricultural sales that year.[4] The vast majority of agricultural loans are to pay for operating expenses, and while many of these loans are relatively small,[5] they are nonetheless critical for farmers to stay in business. Farmers pay for labor, equipment, seeds, and other expenses prior to harvest, which means farmers may have to wait months to receive any revenue at all. As a result, commercial farms often have hundreds of thousands of dollars of debt. The average family-owned commercial farm with outstanding loans paid about $54,000 in interest alone in 2017.[6]

Many of these loans are granted by small banks, some of which rely on agricultural loans for a substantial part of their business. These banks, called agricultural banks, have enjoyed much higher average rates of return on assets than other smaller banks in recent years, even as farm incomes have

---

1. URS NIGGLI ET AL., RESEARCH INSTITUTE OF ORGANIC AGRICULTURE, A GLOBAL VISION AND STRATEGY FOR ORGANIC FARMING RESEARCH 55-56 (2016).
2. *Id.*
3. FEDERAL RESERVE BANK OF KANSAS CITY, AGRICULTURAL FINANCE DATABOOK tbl.A-1 (2019).
4. Calculated by the authors using NATIONAL AGRICULTURAL STATISTICS SERVICE, USDA, 2017 CENSUS OF AGRICULTURE, U.S. NATIONAL LEVEL DATA 9 tbl.2 (2019).
5. The majority of loans were less than $25,000. FEDERAL RESERVE BANK OF KANSAS CITY, *supra* note 3.
6. Calculated by the authors using NATIONAL AGRICULTURAL STATISTICS SERVICE, USDA, 2017 CENSUS OF AGRICULTURE, FARM TYPOLOGY 4 tbl.1 (2020).

fluctuated.[7] However, agricultural lenders often hesitate to make loans to farmers using new or experimental practices, which can make it difficult for farmers to adopt innovative carbon-farming techniques, regardless of their actual exposure to risk.

The energy efficiency financing experience provides a possible model for encouraging financing for innovative practices. There, philanthropic support has often been critical for private and public financing of energy efficiency projects, which were new to lenders and thus underserved.[8] Similarly, the private philanthropic sector (either directly or through advocacy organizations) or USDA should support agricultural banks in lending to farms that use practices that are less well known and widely accepted. At a minimum, USDA and environmental organizations should ensure that agricultural banks are familiar with the benefits of carbon farming, which makes farms more resilient to weather disturbances and therefore exposes the lending institution to less risk. As is also done with energy efficiency and clean energy loans, nonprofits or foundations can guarantee or support private lending to extend its reach. Finally, as discussed further in Chapter V, Congress or state legislatures should create lending institutions, or existing ones could create specialty divisions, aimed at financing farms using climate-friendly practices. These could be public-private entities, with public support or loan guarantees, as exist to foster energy efficiency, or backed in part by philanthropic support. This would allow farmers throughout the country to receive loans regardless of whether their local banks are willing to finance carbon farming.

Private financing also has a role to play. While there has been a significant increase in venture capital funding for "ag-tech,"[9] most of the funding has focused on precision agriculture and a narrow range of practices. Philanthropists, impact investors, and foundations should instead focus investment on a broader range of carbon-farming practices.

C. Easements and Other Conservation Tools

In addition to federal conservation easement programs for farm owners, there are a number of other local, state, and national programs that com-

---

7. Nathan Kauffman & Matt Clark, *Farm Lending Activity Remains Robust*, Fed. Res. Bank Kan. City: Ag Fin. Databook, Apr. 25, 2016, https://www.kansascityfed.org/en/research/indicatorsdata/agfinancedatabook/articles/2016/04-21-2016/ag-finance-dbk-04-25-2016.
8. See, for example, the creation of the New York City Energy Efficiency Corporation (NYCEEC). NYCEEC, *Home Page*, https://nyceec.com/ (last visited Nov. 30, 2020). *See also, e.g.*, Carbon Trust, *Home Page*, https://www.carbontrust.com/home/ (last visited Nov. 30, 2020); Solar and Energy Loan Fund (SELF), *Home Page*, https://solarenergyloanfund.org/ (last visited Nov. 30, 2020).
9. *See* AgFunder, 2021 AgFunder AgriFoodTech Investment Report 3 (2021) (noting that startups raised over 30% more in 2020 than in 2019, and almost five times as much as in 2014).

pensate farm owners for implementing agricultural easements on their land. While these programs generally rely on public funds, either directly or through tax expenditures, they are often designed and administered by nonprofits. As such, the private sector can play an important role in adapting and expanding agricultural easement programs to support climate-friendly practices. Many organizations offering agricultural easements already recognize the environmental benefits of well-managed agricultural land, which can be significant (as discussed in Chapter V.B.5). Nonetheless, few land conservation organizations include climate change mitigation as one of their stated goals, even though it would allow them to more effectively manage land for sequestration.

Agricultural easements can be drafted to give both farmers and land conservation agencies greater flexibility to monitor and reduce net emissions. Conservation easements generally articulate their purposes, giving courts, conservation organizations, and landowners guidance on how to administer the easement under evolving conditions. Land conservation agencies and agricultural land trusts should incorporate climate change mitigation into easement purposes, ensuring that easement conditions encourage climate-friendly practices and that farmers' efforts to mitigate climate change do not conflict with their easements.[10] Indeed, ideally, easements would require implementation of basic climate-friendly practices, such as riparian buffers or cover crops, or at least discount the value of any agricultural easement that does not ensure the implementation of such practices.

Additionally, easements should be written to allow for ecological monitoring, scientific research, and publicly accessible data sources,[11] all of which are critical for improving land management.[12] Conservation organizations and agricultural land trusts should also use other legal tools outside of easements. By leasing land instead of offering permanent easements, for example, these organizations can carefully select farmers to manage their land, allowing them to develop long-term cooperative relationships with farmers dedicated to climate-friendly practices.[13]

Finally, as noted in Chapter IV, farmland offers great potential for the siting of renewable energy such as wind turbines and solar panels. Since on-farm energy and electricity contribute about 1% of total U.S. greenhouse gas

---

10. For example, conservation easements often prohibit new structures, including wind turbines and processing facilities for new agricultural products. Jessica Owley, *Conservation Easements at the Climate Change Crossroads*, 74 LAW & CONTEMP. PROBS. 199, 207-08 (2011).
11. Clauses that ensure open and easily available data should be included wherever possible.
12. Adena Rissman et al., *Adapting Conservation Easements to Climate Change*, 8 CONSERVATION LETTERS 68, 73 (2015).
13. FRED CHEEVER ET AL., PRIVATE LAND CONSERVATION IN THE FACE OF CLIMATE CHANGE (2013).

emissions, accelerating on-farm renewable energy and energy efficiency could be important. If larger-scale renewable energy development is directed to marginal farmland (where there is often the least opportunity for enhanced soil carbon sequestration)—and away from the most productive farmland—there can be a significant climate benefit. Since this is a new field, academics and not-for-profit organizations can help develop model leases, educate both farmers and energy developers about best practices, and advocate for supportive policies.[14]

D. Carbon Measurement Tools

One of the most important needs for advancing carbon farming is to improve our ability to assess the impacts of changed practices on carbon, both above ground and in the ground. Fortunately, there already exist several well-developed and widely accepted protocols for measuring total and changes in above-ground biomass (trees, shrubs, and annual plants, as well as plant debris). In contrast to below-ground stocks, above-ground biomass tends to be more easily quantified and verified through direct measurements—either non-destructively or in association with harvesting activities. In forests, the USDA Forest Service since 1930 has estimated total above-ground stocks in plots across the United States through its Forest Inventory and Analysis program, largely based on scaled estimates from direct measurements of standing trees.[15] Similarly, on croplands, above-ground biomass can be estimated based on harvest yields,[16] or through more novel approaches including remote sensing.[17]

By contrast, measuring soil carbon is a time-intensive, expensive, and complicated exercise. There are also few established protocols for measuring the precise greenhouse gas benefits of climate-friendly practices at a scale

---

14. See AMERICAN FARMLAND TRUST, SOLAR SITING GUIDELINES FOR FARMLAND (Sept. 2020), https://s30428.pcdn.co/wp-content/uploads/2020/01/AFT-solar-siting-guidelines-Jan-2020.pdf; AMERICAN FARMLAND TRUST, WHAT IS DUAL-USE SOLAR? (Mar. 2020), https://s30428.pcdn.co/wp-content/uploads/sites/2/2020/08/Dual-use-one-pager-web.pdf; American Farmland Trust & Blue Wave Solar webinar, Sustainable Development (May 27, 2020), https://bluewavesolar.com/bw-resources/webinar-sustainable-solar-development.
15. CHRISTOPHER WOODALL ET AL., METHODS AND EQUATIONS FOR ESTIMATING ABOVEGROUND VOLUME, BIOMASS, AND CARBON FOR TREES IN THE U.S. FOREST INVENTORY, USDA FOREST SERVICE NORTHERN RESEARCH STATION, Gen. Tech. Rep. NRS-88 (2010), https://doi.org/10.2737/NRS-GTR-88.
16. Stephen Prince et al., *Net Primary Production of U.S. Midwest Croplands From Agricultural Harvest Yield Data*, 11 ECOLOGICAL APPLICATIONS 1194–1205 (2001), https://doi.org/10.2307/3061021.
17. Salam Issa et al., *A Review of Terrestrial Carbon Assessment Methods Using Geo-Spatial Technologies With Emphasis on Arid Lands*, 12 REMOTE SENSING 2008 (2020), https://doi.org/10.3390/rs12122008; Robert Zomer et al., *Global Tree Cover and Biomass Carbon on Agricultural Land: The Contribution of Agroforestry to Global and National Carbon Budgets*, 6 SCI. REPORTS 29987 (2016), https://doi.org/10.1038/srep29987.

suitable for markets, making it difficult to pay farmers in carbon markets for implementing such practices. These challenges have slowed the development of agricultural carbon markets. Scientists are working to develop new, more efficient methods for measuring soil carbon content,[18] but the resources devoted to this problem are insufficient given the urgent need for a reliable and inexpensive way to test soil.

Some private companies, such as Indigo Agriculture and Nori, as well as private and NGO consortia such as the Ecosystem Service Market Consortium, are seeking to develop feasible and reliable protocols for determining the climate change impact of various practices.[19] These all rely on some mixture of sampling and modeling as well as regional approaches, and may be undermined by concerns of a conflict of interest. However, they offer a sense of the measurement (and policy) challenges carbon markets face and thus help identify current research needs and practical concerns that any measurement protocol must address.

Nonprofit organizations and universities should prioritize funding to develop and distribute cost-effective monitoring, measurement, and verification tools, while the private for-profit, not-for-profit, and philanthropic sectors should work with the research community to standardize measuring techniques. In turn, extension services and technical assistance providers should educate farmers about new developments in these tools to accelerate their adoption and acceptance.

E.   Carbon Markets

Carbon offset markets allow greenhouse gas polluters to pay another party to reduce emissions or sequester carbon instead of reducing their own emissions. Greenhouse gas emitters can purchase offsets for entirely voluntary purposes, such as to offset past emissions or otherwise reduce its carbon footprint (e.g., to sell "carbon-neutral" products) by making reductions in its supply chain. Relatedly, a greenhouse gas emitter that does not have a legal obligation to reduce its emissions, and may not currently be technologically

---

18. *See, e.g.*, Robert Pallasser et al., *A Novel Method for Measurement of Carbon on Whole Soil Cores*, in SOIL CARBON (Alfred Hartemink & Kevin McSweeney eds., Springer 2014); S. Billings et al., *Soil Organic Carbon Is Not Just for Soil Scientists: Measurement Recommendations for Diverse Practitioners*, 31 ECOLOGICAL APPLICATIONS 3 (2021), https://doi.org/10.1002/eap.2290.
19. *See* INDIGO AGRICULTURE, *About Indigo Ag*, https://www.indigoag.com/about (last visited Jan. 23, 2021); NORI, *About Us*, https://nori.com/company/about (last visited Jan. 23, 2021); ECOSYSTEM SERVICES MARKET CONSORTIUM, *About Us*, https://ecosystemservicesmarket.org/about-us/ (last visited Jan. 23, 2021).

Chapter VII. Private- and Nonprofit-Sector Opportunities 211

able to do so (e.g., airline flight emissions), may want to buy credits to offset its emissions levels (again, perhaps for marketing or recruitment purposes).

Alternatively, if allowed by a legal regime (such as certain cap-and-trade programs), a company that does have a legal obligation to reduce its emissions may wish to meet its compliance obligations by purchasing credits from agricultural producers who reduce their net emissions. These compliance offset programs raise significant concerns (in addition to measurement and verification issues discussed below)—including the creation of toxic pollutant hotspots—and are particularly problematic when fossil fuel emissions would be offset by agricultural soil carbon sequestration, which, among other challenges, is inherently impermanent.

Either voluntary or compliance offset markets stand in contrast to government or other programs that pay farmers to reduce net greenhouse gas emissions with no corresponding continuation of other emissions or statement about total emissions. These programs, such as the proposed carbon bank discussed above in Chapter V, while still facing measurement and other challenges, are entirely additional to other greenhouse gas reduction efforts. In all cases, the purchased reductions can help finance the transition to carbon farming, compensating farmers for sequestering carbon or reducing emissions.

Carbon offset markets were given a boost by the 2020 introduction[20] and 2021 re-introduction[21] of the Growing Climate Solutions Act, which would allow USDA to certify third-party carbon offset certifiers. Although some environmental organizations have argued that carbon markets should play a major role in decarbonizing agriculture,[22] it is unclear that carbon markets, especially at the current prices offered,[23] will be able to motivate widespread behavioral change among farmers without robust government support and regulation.[24] On the other hand, there appears to be significant interest among companies that have pledged to reduce their net emis-

---

20. S. 3894, 116th Cong. (2020).
21. S. 1251, 117th Cong. (2021).
22. Robert Parkhurst, *Carbon Markets in Agriculture Are the Next Big Thing*, Env't Def. Fund, Jan. 24, 2016, http://blogs.edf.org/growingreturns/2016/01/24/carbon-markets-in-agriculture-are-the-next-big-thing/.
23. Gosia Wozniacka, *Are Carbon Markets for Farmers Worth the Hype?*, Civil Eats, Sept. 24, 2020, https://civileats.com/2020/09/24/are-carbon-markets-for-farmers-worth-the-hype/; Institute for Agriculture and Trade Policy, Why Carbon Markets Don't Work for Agriculture 2 (2020).
24. Peter Alexander et al., *The Economics of Soil C Sequestration and Agricultural Emissions Abatement*, 1 Soil 331, 335 (2015) (noting weak demand for agricultural offsets absent government pressure).

sions to zero by 2050 to buy carbon credits from agriculture and elsewhere, with the market expected to expand 100 times by 2050.[25]

The market for agricultural compliance offsets in the United States as of 2021 is small and largely confined to rice production in California,[26] although agencies and others in California are also closely studying the potential for offset markets for manure management at confined animal production facilities. In addition, two private companies, Nori and Indigo Agriculture, have initiated carbon offset sales, paying farmers $15/ton (as of 2021) for carbon sequestered with differing conditions and program design.[27] As with public schemes to pay farmers for climate-friendly results (discussed above in Chapter V.B.3), any voluntary or compliance offset program will need to address questions of measurement, administrative feasibility, additionality, contract length, and compatibility with other programs, among other issues. As can be seen in Table 1, the few existing programs address these issues in different ways; their experience over the coming years may provide insights as to the most successful or credible approaches.

A key shortcoming of many offset programs that allow fossil emissions to be offset by increases in soil organic carbon is the false equivalency they make between permanent fossil carbon losses and shorter-term increases in soil organic carbon. In contrast to fossil carbon stocks, which would remain inert for long periods of time in the absence of anthropogenic activity, soil organic carbon naturally cycles on much shorter timescales. Thus, any gains in soil organic carbon incentivized through offset programs are perpetually vulnerable to decomposition, through which carbon re-enters the atmosphere. While rebuilding soil organic carbon can help mitigate climate change and help repay the debt of soil organic carbon losses from a history of agricultural activity, it cannot substitute for ongoing losses of fossil carbon. In addition to these natural processes, increases in soil or land carbon can be reversed by

---

25. Virginia Gewin, *As Carbon Markets Reward New Efforts, Will Regenerative Farming Pioneers Be Left in the Dirt?*, CIVIL EATS, July 27, 2021, at https://civileats.com/2021/07/27/as-carbon-markets-reward-new-efforts-will-regenerative-farming-pioneers-be-left-in-the-dirt/ (noting, in addition, the challenge of ensuring that credits are for "additional" efforts while also rewarding early adopters of climate-friendly practices). *See also* Lori Ioannou, *This Is a $15 Trillion Opportunity for Farmers to Fight Climate Change*, CNBC, June 12, 2019, at https://www.cnbc.com/2019/06/11/this-is-a-15-trillion-opportunity-for-farmers-to-fight-climate-change.htm.
26. Niina H. Farah, *Rice Growers on the Front Lines of U.S. Carbon Markets*, E&E NEWS, Jan. 20, 2016, https://www.eenews.net/stories/1060030839; Brian C. Murray, *Why Have Carbon Markets Not Delivered Agricultural Emission Reductions in the United States?*, CHOICES, 2d Quarter 2015, at 1.
27. Wozniacka, *supra* note 23. *See also* Gewin, *supra* note 25 (noting additional private sector agricultural carbon market platforms including TruCarbon, Bayer Carbon Initiative, B Carbon, Nutrien, and Ecosystem Service Market Consortium).

Table 1. Ecosystem Market Comparison

	Nori	Indigo Ag	Soil & Water Outcomes	ESMC
Acreage Min/Max	None	One-field min, no max	None	None
Contract Length	10 yrs	5 yrs	Annual with yearly renewal	Pilot – Annual Market Launch – Scope 1: 10 yrs; Scope 3: TBD
New Practice Requirement	Yes, with a look-back of up to 5 years during pilot phase	Yes, with a look-back of 2 growing seasons	Yes	Yes, but investigating potential of payments to producers already implementing conservation practices for Scope 3
Payment Schedule	End of month when offset credit is sold	50% yr 1, 20% yr 2, 10% yrs 3, 4, 5	Annually, split 50/50 – 1 shortly after signing, 1 after verification	Pilot – Annual Market- Launch - Annual to every 5 yrs depending on Scope for carbon 1 vs 3, respectively; annual for water quality.
Ability to Enroll Same Fields in Gov't Programs/ Other Markets	Designed to stack with both	Designed to stack with both, but other incentives cannot include payments for carbon credits or related assets (financing is okay)	No Note – payment for water quality and carbon outcomes	Designed to stack with gov't programs; individual fields cannot be in two market programs. Note – ESMC internally stacks carbon with GHG reductions, water quality, and water quantity.
Outcome Estimation	Soil sample reference network-based modeling (Soil Metrics) – cost incurred by Nori. Farmer has option to true-up via soil sampling – farmer incurs sampling cost.	Modeling (biogeochemical and statistical) + soil sampling. Indigo assumes cost (Indigo does not charge growers for anything)	Modeling, with 10% of fields subject to in-field soil and water sampling at no cost to farmers	Modeling (peer reviewed biogeochemical model) + soil sampling. ESMC assumes costs and includes in asset price to buyers.
Third Party Practice Verification	Minimum once every 3 years; standard audit procedure (review representative sample of receipts and invoices)	Random site visits and evidence checks, registry-approved methodology.	Yearly field visits, remote sensing	Scope 1– small subset of producers randomly selected for site visit + remoting sensing. Scope 3 –smaller subset of producers randomly selected for site visit +remote sensing.
Data Collected on Enrollment	Farm operational data – previous 10 years OR proprietary "Smart Defaults" option	Basic farmer info, field boundaries, and commitment to new practice(s)	Farm operational data – 2-3 years historical baseline plus 2-3 years of proposed practice change(s)	Scope 1 – detailed farm operational data Scope 3 – some operational data; Soil sampling and remote sensed data for both.
Penalty for Temporary Break in Practice Implementation	Farmer commits to make best effort to retain C stocks; not bound to any practice plan; not liable for *force majeure* C losses.	Payment pauses until soil carbon returns to previous level. Methodology prevents credits from being overestimated.	Breach of contract, farmer would not receive payment	Stall in soil carbon gains requires initial gains to be realized before additional credit issuance/payment; no consequences for dropping out of pre-market launch pilots
Enrollment Assistance	Supply Account Managers on-call; regular training; direct assistance with enrollment process	Customer success hotline or webchat options	Provided via staff and affiliates	Producer's preferred advisor (e.g. conservation district staff, CCAs) can be trained to assist; option to import data from 3rd party platform
Technical/ Agronomic Assistance	NA (but supply account managers include trained agronomists)	Free in-house agronomic guidance, supplemented with on-the-ground help	Free conservation agronomists on staff	Provided by ESMC's member organizations and partners (e.g. conservation district, CCAs, NGOs).

Source: Emily Bruner & Jean Brokish, Illinois Sustainable Ag Partnership, *Ecosystem Market Information: Background and Comparison Table* (2021) (reprinted with permission).

changes in land use practices.[28] The existing private markets use longer term contract arrangements to try to address this time frame issue.

Moreover, it is exceedingly difficult to precisely, accurately, and cost-effectively measure greenhouse gas reductions from altered farm practices. While more relaxed standards might be acceptable in a program where public or philanthropic funds are being used to obtain climate benefits, a much greater degree of certainty should be demanded for offset purchasers using agricultural net greenhouse gas reductions to claim credit for balancing off other fossil fuel emissions or to lower their own greenhouse-gas-reduction efforts. Soil carbon content on a single operation can vary substantially across depths, locations and seasons, even in seemingly uniform fields,[29] and soil carbon measurement tools are not yet standardized sufficiently to reliably ensure emissions reductions.[30] Verifying changes in soil carbon storage also requires time- and labor-intensive sampling to ensure that changes are significant and real relative to sources of spatial variation outside of management.[31] Even verified increases in carbon stocks can be misleading since they are often concomitant with increases in soil respiration (which is not included in greenhouse gas inventories) or other greenhouse gas emissions.[32] Nor does measuring total carbon stocks provide adequate information on persistence, which is dependent on soil chemistry, moisture, temperature, soil microbial community, and several other factors rarely included in carbon market measurement approaches.[33] Finally, any carbon offset market must include comprehensive measurements of a full greenhouse gas budget rather than only total carbon stocks since some practices can increase soil carbon or decrease methane emissions but increase nitrous oxide emissions. And since method-

---

28. Letter from Food & Water Watch et al. to Members of Congress (Oct. 15, 2020), https://foodandwaterwatch.org/sites/default/files/oppose_the_growing_climate_solutions_act_final_101520.pdf; INSTITUTE FOR AGRICULTURE AND TRADE POLICY, *supra* note 23.
29. Keith Paustian et al., *Quantifying Carbon for Agricultural Soil Management: From the Current Status Toward a Global Soil Information System*, 10 CARBON MGMT. 568, 571 (2019), https://doi.org/10.1080/17583004.2019.1633231 (accessed Jan. 22, 2020); INSTITUTE FOR AGRICULTURE AND TRADE POLICY, *supra* note 23.
30. Pete Smith et al., *How to Measure, Report and Verify Soil Carbon Change to Realize the Potential of Soil Carbon Sequestration for Atmospheric Greenhouse Gas Removal*, 26 GLOBAL CHANGE BIOLOGY 219-41 (2020); Cole D. Gross & Robert B. Harrison, *Quantifying and Comparing Soil Carbon Stocks: Underestimation With the Core Sampling Method*, 82 SOIL SCI. SOC'Y AM. J. 949-59 (2018); INSTITUTE FOR AGRICULTURE AND TRADE POLICY, *supra* note 23.
31. *See* Gross & Harrison, *supra* note 30.
32. Indeed, carbon dioxide emissions from soil respiration are "an order of magnitude greater than those from human activities, such as fossil fuel burning." Jennifer A.J. Dungait, *Organic Matter Turnover Is Governed by Accessibility Not Recalcitrance*, 18 GLOBAL CHANGE BIOLOGY 1781, 1786 (2012).
33. *See* Michael W.I. Schmidt, *Persistence of Soil Organic Matter as An Ecosystem Property*, 478 NATURE 49-56 (2011).

Chapter VII. Private- and Nonprofit-Sector Opportunities 215

ology is so important, any scheme must include full transparency related to sampling, quantification, uncertainties, and validation.

A 2021 report published by Environmental Defense Fund and the Woodwell Climate Research Center examined 12 different measurement, reporting, and verification (MRV) protocols (eight from the United States) applied to soil organic carbon credit valuation, and compared and contrasted these different approaches.[34] As discussed in the report, these various approaches and lack of standardization create a system where not all carbon credits are valued equally. Moreover, each approach faces challenges accounting for issues such as additionality, leakage, reversals, and permanence. Collectively, these challenges make it difficult to determine the overall climate benefits of these markets. Given these uncertainties, the report recommends continued research to better understand the true climate impact of carbon markets while limiting any carbon credit effort to voluntary value chain offsets, thus excluding offsets for emission compliance obligations. They argue that the most effective climate benefits continue to come from direct emissions reductions. "Consistent accounting and verification of direct emission reductions during agricultural production—reduced nitrous oxide emissions via improved nutrient management, reduced carbon dioxide emissions via reduced tractor use and reduced methane emissions from improved manure management—and from avoided land conversion is a less risky and permanent climate solution for supply chain and other public investment."[35]

Given these issues, organizers of offset systems should also explore alternative payment schemes that do not rely on having accurate or precise sequestration rates. For example, instead of paying for offsets per ton (as is generally the case), payments could be based on practices implemented per acre, with a price set by calculations of average benefits, or based on measurements of surrogate indicators. An added benefit of this approach is that it may allow schemes to account for co-benefits and more easily consider the full suite of impacts of particular practices rather than exclusively focusing on increasing soil organic carbon.

---

34. EMILY OLDFIELD ET AL., ENVIRONMENTAL DEFENSE FUND, AGRICULTURAL SOIL CARBON CREDITS: MAKING SENSE OF PROTOCOLS FOR CARBON SEQUESTRATION AND NET GREENHOUSE GAS REMOVALS (2021).
35. *Id.* at 4. *See also id.* at 2:
> We remain concerned that any end-use of carbon credits as an offset, without robust local pollution regulations, will perpetuate the historic and ongoing negative impacts of carbon trading on disadvantaged communities and Black, Indigenous and other communities of color. Carbon markets have enormous potential to incentivize and reward climate progress, but markets must be paired with a strong regulatory backing.

Many researchers and advocates have significant concerns about the use of offsets generally.[36] The existence of offsets can take pressure off for actual emission reductions;[37] they may not be truly additional or verifiable[38]; and they may make it easier for companies to claim carbon neutrality when in fact they are not.[39] Moreover, offsets can change the location of the pollution, increasing or slowing reductions in the area near the purchaser.[40] And even a climate change offset program can have local impacts because greenhouse gas pollution is often accompanied by air emissions more immediately harmful to human health.[41] Small-scale farmer organizations also criticize offset markets in agriculture as unreliable, detrimental to small- and medium-scale farmers, and likely to increase volatility in food prices.[42] Private market programs should seek to expressly address all these issues, as should legislatures and agencies considering regulations to help shape or oversee these private markets. Legislatures and agencies should carefully consider all of these factors when considering any carbon market program.

---

36. See, e.g., Lisa Song & James Temple, *The Climate Solution Actually Adding Millions of Tons of CO_2 Into the Atmosphere*, MIT TECH. REV. & PROPUBLICA (Apr. 2021) https://www.propublica.org/article/the-climate-solution-actually-adding-millions-of-tons-of-co2-into-the-atmosphere; TAMRA GILBERTSON & OSCAR REYES, CARBON TRADING: HOW IT WORKS AND WHY IT FAILS 11-12 (Dag Hammarskjöld Foundation, Critical Currents No. 7, 2009); Kevin Anderson, *The Inconvenient Truth of Carbon Offsets*, 484 NATURE 7 (2012); Lisa Song, *An Even More Inconvenient Truth*, PROPUBLICA, May 22, 2019, https://features.propublica.org/brazil-carbon-offsets/inconvenient-truth-carbon-credits-dont-work-deforestation-redd-acre-cambodia/; Lisa Song & James Temple, *A Nonprofit Promised to Preserve Wildlife. Then It Made Millions Claiming It Could Cut Down Trees*, MIT TECH. REV. & PROPUBLICA, May 2021, https://www.propublica.org/article/a-nonprofit-promised-to-preserve-wildlife-then-it-made-millions-claiming-it-could-cut-down-trees.
37. Letter from Food & Water Watch et al., *supra* note 28.
38. *Id.*; INSTITUTE FOR AGRICULTURE AND TRADE POLICY, *supra* note 23.
39. INSTITUTE FOR AGRICULTURE AND TRADE POLICY, *supra* note 23.
40. *Id.*
41. Letter from Food & Water Watch et al., *supra* note 28.
42. INSTITUTE FOR AGRICULTURE AND TRADE POLICY, FIVE REASONS CARBON MARKETS WON'T WORK FOR AGRICULTURE (2011); INSTITUTE FOR AGRICULTURE AND TRADE POLICY, *supra* note 23.

Key Recommendations

- Foundations and private donors should support the research private organizations are conducting on climate-friendly practices and help fund new organizations devoted to carbon farming. There is great need for research into publicly shared information about opportunities to reduce greenhouse gas emissions and increase carbon sequestration in soil, since now most research focuses elsewhere.

- USDA and nonprofit organizations should ensure that agricultural banks are familiar with the benefits of carbon farming, as increased resiliency exposes lending institutions to less risk.

- The private sector can play an important role in adapting and expanding agricultural easement programs to support climate-friendly practices. Agricultural easements should also require farmers to follow basic conservation practices.

- Funding is needed to develop, standardize, and distribute cost-effective monitoring, measurement, and verification techniques.

- Voluntary carbon markets can help finance the transition to carbon farming, compensating farmers for sequestering carbon or reducing emissions, but legislatures and agencies should proceed carefully given the myriad concerns about carbon markets and the inability to precisely, accurately, and cost-effectively measure greenhouse gas reductions from altered farm practices. In particular, policymakers should be very wary of offset markets that allow entities to buy agricultural carbon credits instead of implementing feasible reductions of their own greenhouse gas emissions.

Chapter VIII.
Off-Farm Food System Emission Reduction Opportunities

The food system encompasses the full life cycle of food. In addition to agriculture, this includes activities that take place off the farm—from the preplanting conversion of native grasslands and production of agricultural chemicals, for example, to the post-harvest distribution, consumption, and disposal of food.[1] The food system is responsible for approximately one third of both national and global greenhouse gas emissions.[2] We must approach the food system as a whole in order to craft laws and policies that address the system's full complement of social, nutritional, and environmental impacts.

A. Upstream: Greenhouse Gas Emissions From Farm Inputs

Conventional agriculture in the United States relies heavily on fossil fuels. Most commercial farms rely on energy-intensive equipment to perform a wide range of farm tasks, including weeding, planting, and harvesting, in order to reduce their labor needs. In addition, the manufacturing process for farm inputs—such as pesticides and, in particular, fertilizer—requires a substantial amount of energy. A detailed literature review in 2003 broke down the total energy requirements for agriculture in advanced economies using the following categories:[3]

- 36% for nitrogen fertilizer production

---

1. See Monica Crippa et al., *Food Systems Are Responsible for a Third of Global Anthropogenic GHG Emissions*, 2 NATURE FOOD 198-209 (2021); Sonja Vermeulen et al., *Climate Change and Food Systems*, 37 ANN. REV. ENV'T & RESOURCES 195, 198-202 (2012).
2. See Crippa et al., *supra* note 1; *see also* Vermeulen et al., *supra* note 1, at 195 (finding that food systems contribute 19-29% of global anthropogenic greenhouse gas emissions). GRAIN, an international research and advocacy organization, estimates that emissions from the food system are as high as 44%-57% of global emissions. GRAIN, *Commentary IV: Food, Climate Change, and Healthy Soils: The Forgotten Link, in* TRADE & ENV'T REV. 2013, at 19, 19-20 (United Nations Conference on Trade and Development 2013).
3. Mario Giampietro, *Energy Use in Agriculture, in* ENCYCLOPEDIA OF LIFE SCIENCES 4 (Nature Publishing Group 2003).

- 27% for on-farm fuel usage
- 15% for the manufacture of agricultural machinery
- 6% for irrigation
- 6% for pesticide production
- 5% for phosphorus and potassium fertilizer production
- 4% miscellaneous

We could recoup significant benefits from reducing the largest two upstream emitters of greenhouse gases—the production of nitrogen fertilizer and on-farm fuel usage—which together account for almost two-thirds of upstream emissions.

1. Emissions From Fertilizer Production

Nitrogen-based fertilizers accounted for 59% of total U.S. fertilizer consumption in 2010,[4] but were responsible for approximately 90% of emissions from fertilizer production.[5] A 2012 study found that fertilizer manufacturing contributed 282-575 million metric tons of carbon dioxide equivalent (MMT CO_2 eq.) globally, as compared to 5,120–6,116 MMT CO_2 eq. of direct emissions from agriculture.[6] That study further notes that nitrous oxide emissions from fertilizer application are about 40% of total agricultural emissions, suggesting that emissions from fertilizer manufacturing are equivalent to about 12-30% of the emissions from fertilizer application. Emissions data indicate that the average emission rate for fertilizer manufacturing scaled to production capacity in the United States is about 2.57 metric tons CO_2 eq. per metric ton of nitrogen (as ammonia) produced,[7] but the global range is very large, from 0.45 to 9.5 metric tons CO_2 eq. per

---

4. Jayson Beckman et al., U.S. Department of Agriculture, Agriculture's Supply and Demand for Energy and Energy Products 10 (2013) (EIB-112).
5. In 2011, ammonia production plants in the United States accounted for roughly 14% of the chemical manufacturing sector's total carbon footprint, or about 0.1% of total emissions. Their share is expected to rise, however. Globally, ammonia production is a major contributor to greenhouse gas emissions, representing as much as 5% of greenhouse gas emissions. While the United States accounts for only 6% of global ammonia production right now, the majority of new plants are being built in the United States or Canada. *See* Celeste LeCompte, *Fertilizer Plants Spring Up to Take Advantage of U.S.'s Cheap Natural Gas*, Sci. Am., Apr. 25, 2013, http://www.scientificamerican.com/article/fertilizer-plants-grow-thanks-to-cheap-natural-gas/.
6. Sonja J. Vermeulen et al., *Climate Change and Food Systems*, 37 Ann. Rev. Env't Resources 195–222 (2012).
7. EPA, 2017 National National Emissioins Inventory Data (2021); Nutrien, Fact Book 2018 (2018).

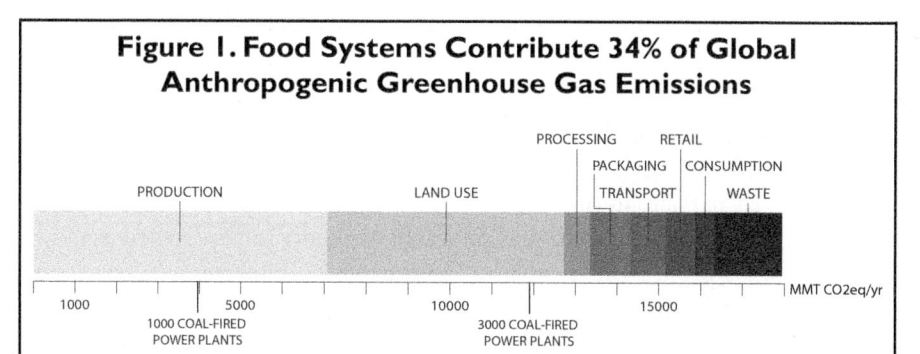

> **Figure 1. Food Systems Contribute 34% of Global Anthropogenic Greenhouse Gas Emissions**
>
> PRODUCTION | LAND USE | PROCESSING, PACKAGING, TRANSPORT | RETAIL, CONSUMPTION, WASTE
>
> 1000 | 5000 | 10000 | 15000 MMT CO2eq/yr
> 1000 COAL-FIRED POWER PLANTS | 3000 COAL-FIRED POWER PLANTS
>
> *Source:* Monica Crippa et al., *Food Systems Are Responsible for a Third of Global Anthropogenic GHG Emissions*, 2, NATURE FOOD 198-209 (2021).
>
> *Notes:* Production includes emissions from animal waste as fertilizer, animals in pasture, cultivation of food crops, drainage of organic soils for crop cultivation, CO_2 from urea fertilization, limestone and dolomite use, nitrogen-fixing crops, agricultural waste burning, manure management, and enteric fermentation. Land use includes emissions from deforestation and the degradation of organic soils (drainage and fires), including peatlands. Consumption includes emissions associated with household electricity and fuel use for food-related activities. Waste includes emissions from solid waste disposal and waste water treatment.

ton nitrogen produced, depending on the type of nitrogen fertilizer being manufactured.[8] These emissions are additional to emissions resulting from the application of fertilizer on croplands, meaning that the climate benefits of reducing fertilizer use, if accompanied by a commensurate reduction in fertilizer production, are significantly greater than indicated by direct emissions alone and that significant climate benefits may be achieved by tailoring the types of fertilizer manufactured.

New ammonia production facilities are approximately 30% more energy efficient than older ones, further indicating that this sector's emissions could be significantly reduced by modernizing production processes.[9] However, due to thermodynamic constraints to efficiency, emissions cannot drop much further through efficiency measures alone.[10]

In addition to greenhouse gas emissions with direct climate impacts, fertilizer manufacture and application also contribute to air and water pollution through emissions of ammonia. As discussed above in Chapter VI.A, the U.S.

---

8. A. KOOL ET AL., BLONK CONSULTANTS, LCI DATA FOR THE CALCULATION TOOL FEEDPRINT FOR GREENHOUSE GAS EMISSION OF FEED PRODUCTION AND UTILIZATION (2012); ROLF FRISCHKNECHT ET AL., SWISS CENTRE FOR LIFE CYCLE INVENTORIES, OVERVIEW AND METHODOLOGY (Ecoinvent Rep. 1) (2007).
9. INTERNATIONAL FERTILIZER INDUSTRY ASSOCIATION, FEEDING THE EARTH: ENERGY EFFICIENCY AND CO_2 EMISSIONS IN AMMONIA PRODUCTION 2 (2009).
10. *Id.*

Environmental Protection Agency (EPA) has the authority to impose emission limits on ammonia, which can—depending on the control technologies used—also reduce greenhouse gas emissions.[11] In 2017, nitrogen fertilizer plants in the United States accounted for 23% of total ammonia emissions attributed to large stationary sources.[12] Despite the fact that ammonia is a precursor to particulate matter pollution, which results in human health hazards and haze causing thousand of deaths annually in the United States,[13] EPA has not yet promulgated emission limitations on nitrogen fertilizer manufacturing.[14] Indeed, because ammonia is also often deposited onto aquatic ecosystems, where it leads to both water and air pollution,[15] including nitrous oxide emissions, EPA could also potentially limit these emissions under the Clean Water Act. Any of these actions would likely reduce the direct and indirect greenhouse gas emissions from fertilizer manufacturing.

In addition to the many opportunities to use fertilizer more efficiently outlined in Chapter IV above, there is also some promise in facilities that can produce nitrogen fertilizer from biomass instead of natural gas,[16] although any use of biomass as feedstock must address the land use impacts or carbon opportunity cost discussed in Chapter III above. Alternatively, proposed facilities could produce both electricity and fertilizer accompanied by carbon capture and storage (or reuse),[17] which could produce fertilizer with very low greenhouse gas emissions. The government should support research on such projects.

2. Fuel Economy Standards for Agricultural Equipment and Reduction of On-Farm Energy Use

Agriculture is estimated to consume at least 6% of the world's fossil fuel energy.[18] Nonetheless, EPA has yet to promulgate any standard for off-road diesel

---

11. *See* Gregg P. Macey, *Industrial Sector, in* Legal Pathways to Deep Decarbonization in the United States 301 (Michael B. Gerrard & John C. Dernbach eds., ELI Press 2019); Jessica Wentz & David Kanter, *Nitrous Oxide, in* Legal Pathways to Deep Decarbonization in the United States, *supra* at 916.
12. U.S. EPA, *National Emissions Inventory* (2017), https://edap.epa.gov/public/extensions/nei_report_2017/dashboard.html#table-db.
13. Nina Domingo et al., *Air Quality-Related Health Damages of Food*, 118 Proc. Nat'l Acad. Sci. 20 (2021), https://doi.org/10.1073/pnas.2013637118.
14. *See* 40 C.F.R. §60.16.
15. James Galloway et al., *The Nitrogen Cascade*, 53 Bioscience 341–56 (2003), https://doi.org/10.1641/0006-3568(2003)053[0341:TNC]2.0.CO;2.
16. *See, e.g.*, Yosuke Mikami et al., *Ammonia Production From Amino Acid-Based Biomass-Like Sources by Engineered Escherichia Coli*, 7 AMB Express 83 (2017).
17. *See, e.g.*, SCS Engineers, *Homepage*, http://www.scsengineers.com/ (last visited Nov. 30, 2020).
18. Ugo Bardi et al., *Turning Electricity Into Food: The Role of Renewable Energy in the Future of Agriculture*, 53 J. Cleaner Production 224, 226 (2013).

vehicles, despite progressively tightening its fuel economy standards for light-duty vehicles. Fuel efficiency for on-farm vehicles has consequently lagged. EPA should promulgate fuel economy standards for off-road diesel vehicles such as tractors to reduce their CO_2 emissions, which remain a significant source of on-farm emissions. Since turnover among off-road vehicles is slower than turnover among light-duty vehicles, however, significant improvements in emissions reduction will be slow.

Farm programs implemented by the U.S. Department of Agriculture (USDA) should also be designed to encourage farmers, preferably through incentives, to adopt less fuel-intensive practices. Tillage, for example, significantly increases CO_2 emissions from agricultural equipment, since plowing significantly increases a tractor's fuel requirements. A literature review found that tractors on no-till farms only emit one-sixth as much CO_2 eq. as tractors on farms practicing complete tillage.[19] Similarly, farmers and technology providers are developing systems to reduce the number of times tractors must go through fields by combining tasks, and are trying to increase the energy efficiency of motors, fans, pumps, and other farm equipment.

B. Downstream: Emissions From Food Processing, Packaging, Marketing, and Waste

Postproduction greenhouse gas emissions, while significant, have not been comprehensively catalogued in the United States.[20] The main contributors to emissions beyond the farm gate are energy expenditures associated with food processing, packaging, marketing, and distribution. Food waste contributes to emissions indirectly, through emissions resulting from the production, distribution, and marketing of the wasted food, and directly, through methane emissions from landfills. The food waste discussion below is limited to landfill emissions, since reductions in indirect contributions are susceptible to leakage and are difficult to track.[21] In contrast, efforts to divert food waste

---

19. Rattan Lal, *Carbon Emission From Farm Operations*, 30 ENV'T INT'L 981, 982 (2004).
20. *See* Rebecca Boehm et al., *A Comprehensive Life Cycle Assessment of Greenhouse Gas Emissions From U.S. Household Food Choices*, 79 FOOD POL'Y 67 (2018), *available at* https://www.sciencedirect.com/science/article/abs/pii/S0306919217310552?via%3Dihub, for a life-cycle assessment of greenhouse gas emissions from consumer food purchases that incorporates post-production emissions. Although it does not encompass household transportation, preparation, storage, or waste, the Boehm et al. analysis includes food production and transportation and wholesale, retail, and restaurant activity resulting from household purchases.
21. Food producers often shift their products to foreign or secondary markets in response to decreased consumer demand rather than decrease production. The dynamics of the U.S. cheese industry serve as an illustrative example. Cheese production has grown much more rapidly than domestic consumption since 2009, yet the industry has continued to expand by increasing exports, which tripled between 2007 and 2014, and through government purchasing programs that distribute surpluses to food banks

from landfills are relatively easy to monitor, and to implement, as several state and local governments have shown.

1. Processing, Packaging, Distribution, and Marketing Emissions

In 2006, the food processing sector emitted approximately 117 MMT CO_2 eq., making it one of only four industrial sectors in the United States responsible for more than 100 MMT CO_2 annually.[22] Mitigation within the food processing sector will largely depend on reducing energy intensity in addition to other cross-sector efforts, such as reducing reliance on fossil fuel energy sources. As a result, EPA and the U.S. Department of Energy should explore adopting more energy efficiency standards that would apply to appliances and processes within this sector.

2. Landfill Waste Emissions

Diverting food and agricultural waste from landfills provides an opportunity to significantly reduce greenhouse gas emissions.[23] Such a strategy could result in quick and powerful climate benefits. Organic matter, which includes food, wood, yard waste, and paper products, is the single largest component of landfills, constituting the majority of waste discarded in municipal waste

---

and nutrition assistance programs. *See* Mark O'Keefe, *Emerging Economies Will Drive Future Cheese Demand*, U.S. DAIRY EXPORTER BLOG, Feb. 25, 2016, http://blog.usdec.org/usdairyexporter/emerging-economies-will-drive-future-cheese-demand; Mark Fahey, *Americans Have an Insatiable Demand for Pizza Cheese*, CNBC, Oct. 10, 2016, http://www.cnbc.com/2016/10/04/best-cheeses-americans-have-an-insatiable-demand-for-pizza-cheese.html; Press Release, USDA, USDA Announces Plans to Purchase Surplus Cheese, Releases New Report Showing Trans-Pacific Partnership Would Create Growth for Dairy Industry (Oct. 11, 2016), https://www.usda.gov/media/press-releases/2016/10/11/usda-announces-plans-purchase-surplus-cheese-releases-new-report. Thus, it is unlikely that any decrease in demand as the result of reductions in food waste would result in an equivalent decrease in production.

22. SABINE BRUESKE ET AL., OAK RIDGE NATIONAL LABORATORY, U.S. MANUFACTURING ENERGY USE AND GREENHOUSE GAS EMISSIONS ANALYSIS 37 tbl.2.1-16 (2012).

23. It is sometimes argued that reducing food loss will result in reduced food production and distribution. *E.g.*, Craig Hanson et al., *What's Food Loss and Waste Got to Do With Climate Change? A Lot, Actually.*, WORLD RESOURCES INST., Dec. 11, 2015, http://www.wri.org/blog/2015/12/whats-food-loss-and-waste-got-do-climate-change-lot-actually. While intuitively this makes sense, there are a number of variables that make it impossible to predict what impact reduced domestic demand would have on land use, including funding for farm programs, support for biofuels, and fluctuations in global consumer demand and international commodity markets. Additionally, the amount of cropland and grazing land in the United States has stayed more or less constant since 1945, despite a radically higher supply of agricultural commodities gained through higher yields. If agriculture's land footprint remains unchanged in the face of such significant increases in supply, it appears unlikely to be affected by relatively low fluctuations in demand stemming from lower rates of food loss. *See also supra* note 21 (discussing how food producers often develop new markets in response to decreased demand rather than decrease production).

systems.[24] Food waste alone makes up more than 20% of the materials discarded.[25] Once in a landfill, organic matter decomposes without the presence of oxygen, releasing large amounts of methane as a result.[26] EPA estimates that organic matter in landfills was responsible for 17% of U.S. methane emissions in 2018.[27] A 2016 study, however, found that EPA underestimates the amount of municipal waste disposed of by a factor of two, indicating that the methane emissions from organic matter might actually be much higher.[28]

Food waste in landfills typically has a high moisture and organic matter content, making it an especially large contributor to methane emissions soon after disposal. As a result, food waste is responsible for as much as 90% of methane emissions from landfills during the initial years when they are less likely to be capped.[29] While reliable data on the sources of food waste are lacking, one industry-funded report estimates that residential food waste is responsible for 44% of post-farm food waste.[30] The commercial sector, which includes restaurants and grocery stores, is estimated to dispose of 44% of post-farm food waste, while waste from institutions and industry operations make up the remaining 12%.[31]

European countries have demonstrated that organics can be diverted from landfills in a cost-effective and environmentally beneficial way. The European Landfill Directive, passed in 1999, required members of the European Union to reduce biodegradable waste to 35% of 1995 levels by 2016.[32] Many Member States have gone beyond this requirement. A 2010 survey found that the majority of German households have access to an organic waste bin and many of them are required to use them.[33] Germany revised its national waste management law, the Circular Economy Act, in 2012 to require residents to sort organic waste for collection by 2015.[34] In 2016, a new law went

---

24. *See* EPA, Inventory of U.S. Greenhouse Gas Emissions and Sinks: 1990-2018, at 7-17 tbl.7-6 (2020) (EPA 430-R-20-002).
25. *Id.*
26. *Id.* at 7-4.
27. EPA, *Overview of Greenhouse Gases: Methane Emissions*, https://www.epa.gov/ghgemissions/overview-greenhouse-gases (last updated Sept. 8, 2020).
28. Jon Powell et al., Letter, *Estimates of Solid Waste Disposal Rates and Reduction Targets for Landfill Gas Emissions*, 6 Nature Climate Change 162, 162 (2016) (finding that the total amount of municipal waste disposed of in the United States was 115% higher than EPA's estimate in 2012).
29. Dana Gunders, Natural Resources Defense Council, Wasted: How America Is Losing Up to 40 Percent of Its Food From Farm to Fork to Landfill 14 (2012).
30. Business for Social Responsibility, Food Waste: Tier I Assessment 12 (2012).
31. *Id.*
32. *See generally* Council Directive 1999/31/EC, 1999 O.J. (L. 182).
33. Peter Krause et al., Umwelt Bundesamt, Compulsory Implementation of Separate Collection of Biowaste 3-4 (2015).
34. Kreislaufwirtschaftsgesetz [Circular Economy Act], Feb. 24, 2012, BGBl. I S.212, art. 11.

into effect in France banning supermarkets larger than 4,305 square feet in size from throwing away or destroying food.[35]

In 2016, EPA issued new rules requiring installation of systems to capture landfill gas (usually comprising half methane and half CO_2) at larger municipal waste landfills constructed after July 2014, and updated landfill gas capture systems for larger existing landfills constructed after 1987.[36] Yet even with this additional landfill capture, some significant gaps still remain: older and smaller landfills are not covered; there is a long time lag before full compliance will be required; and the landfill gas capture is not complete.

States and municipalities have also taken action to divert organic waste from landfills. Shifting waste to composting facilities converts the waste into useful material and results in negative net emissions.[37] In 2012, Vermont passed the Universal Recycling Law, which, among other things, enacted a complete ban on food waste in landfills.[38] The ban went into effect in 2020 and applies to all households and businesses. Similarly, in 2019, New York State enacted a food waste law, which takes effect on January 1, 2022, that requires large generators of food scraps to donate excess edible food and recycle (in composting, animal feed, digester facilities, or the like) all remaining food scraps if they are located within 25 miles of an organics recycler.[39] Currently in New York, 97% of food scraps are dumped in landfills, generating

---

35. Angelique Chrisafis, *French Law Forbids Waste by Supermarkets*, GUARDIAN, Feb. 4, 2016, https://www.theguardian.com/world/2016/feb/04/french-law-forbids-food-waste-by-supermarkets. *See* Proposition de Loi 632 du 9 décembre 2015 relative à la lutte contre le gaspillage alimentaire [Proposal of Law 632 of December 9, 2015, on the fight against food waste], Assemblée Nationale [French National Assembly], Dec. 9, 2015.
36. News Release, EPA, EPA Issues Final Actions to Cut Methane Emissions From Municipal Solid Waste Landfills (July 15, 2016), https://19january2017snapshot.epa.gov/newsreleases/epa-issues-final-actions-cut-methane-emissions-municipal-solid-waste-landfills-0_.html; Standards of Performance for Municipal Solid Waste Landfills, 81 Fed. Reg. 59332 (Aug. 29, 2016) (regulating new and modified landfills under the New Source Performance Standards program of the Clean Air Act); Emission Guidelines and Compliance Times for Municipal Solid Waste Landfills, 81 Fed. Reg. 59276 (Aug. 29, 2016) (regulating existing landfills under Clean Air Act §111(d)).
37. EPA, DOCUMENTATION FOR GREENHOUSE GAS EMISSION AND ENERGY FACTORS USED IN THE WASTE REDUCTION MODEL (WARM)—ORGANIC MATERIALS chs. 1-29 to 1-30 (2016).
38. VT. STAT. ANN. tit. 10, §6602(29) (West 2017). California, Connecticut, and Massachusetts have also passed legislation or promulgated regulations requiring commercial businesses to divert food waste from landfills under certain circumstances. CAL. PUB. RES. CODE §42649.81 (West 2017) (applies to businesses generating eight cubic yards of organic waste or more per week); CONN. GEN. STAT. ANN. §22a-226e (West 2017) (limits entities to no more than 52 tons of organic waste by 2020); MASS. REGS. CODE tit. 310, §§19.006, 19.017(3) (2017) (bans entities from disposing of more than one ton of food waste per week).
39. N.Y. ENVTL. CONSERV. LAW, §27-2203. *See also* N.Y. State Dep't of Environmental Conservation, *Food Donation and Food Scraps Recycling Law Overview* (2020), https://www.dec.ny.gov/chemical/114499.html.

Chapter VIII. Off-Farm Food System Emission Reduction Opportunities

over half of the state's methane emissions,[40] so shifting this waste to facilities that do not generate methane could be significant.

Cities can also take meaningful action. San Francisco passed an ordinance in 2009 requiring all businesses and households to sort organics for collection and composting.[41] San Francisco now collects more than 220,000 tons of organic waste each year, and it is considered the country's most successful composting program.[42] In 2014, the Seattle City Council also passed a mandatory composting ordinance.[43] Even though the ordinance limits fines for noncompliance to $1 for residents and $50 for commercial businesses,[44] composting collection rates went up significantly after the law went into effect in 2015.[45]

Congress should design legislation banning food waste in landfills; Vermont's Universal Recycling Law provides an ideal model. Failing national action, states and municipal governments should adopt similar laws. Given that waste from retail establishments is estimated to make up almost half of the total waste, laws that address only this portion of the waste could still have a significant impact.

*****

While what happens on cropland, pasture or grazing land, or animal production facilties is the centerpiece of the food system, activities off the farm contribute as much as the on-farm greenhouse gas emissions. Fortunately, just as there are many proven opportunities to make U.S. agriculture climate-neutral, there are also many demonstrated ways to significantly reduce off-farm emissions. And, as with climate-neutral farming, those opportunities will multiply once all those in the food chain set their mind to achieving this goal and develop further possibilities. Advocates should urge, and policy should mandate, that all sectors in the food system quickly move toward climate neutrality.

---

40. N.Y. League of Conservation Voters, *Working to Solve New York's Food Waste Problem* (2018), https://nylcv.org/news/working-solve-new-yorks-food-waste-problem/; N.Y. State Methane Reduction Plan (2017), https://www.dec.ny.gov/docs/administration_pdf/mrpfinal.pdf.
41. S.F., Cal., Ordinance 100-09 (June 9, 2009).
42. This includes yard waste in addition to food waste.
43. Sean Kennedy, *In Seattle, Compost Your Food Scraps—Or Else*, CNN, Oct. 3, 2014, http://www.cnn.com/2014/09/24/politics/seattle-composting-law/.
44. SEATTLE, WASH., MUN. CODE §§21.36.082(C), 21.36.083(B) (2016).
45. Sara Bernard, *Why Seattle Still Has a Huge Garbage Problem*, GRIST, June 15, 2015, http://grist.org/cities/why-seattle-still-has-a-huge-garbage-problem/.

Key Recommendations

- While what happens on cropland, pasture or grazing land, or animal feeding operations is the centerpiece of the food system, activities off the farm contribute as much as on-farm greenhouse gas emissions.

- EPA should update air pollution limits for fertilizer plants under its Clean Air Act authority to help reduce emissions from nitrogen fertilizer production.

- The government should support research aimed at reducing greenhouse gas emissions during fertilizer production.

- EPA should promulgate fuel economy standards for off-road diesel vehicles such as tractors to reduce their carbon dioxide emissions, which remain a significant source of on-farm emissions.

- USDA farm programs should encourage farmers, preferably through incentives, to adopt less fuel-intensive practices and more energy-efficient equipment.

- EPA and the U.S. Department of Energy should explore adopting more energy efficiency standards that would apply to appliances and processes within the agriculture and food processing sectors, such as irrigation pumps, ventilation fans, or cleaning and drying equipment.

- Americans waste about one-third of food that is produced. The vast majority of the waste is dumped in landfills, where it rots and emits methane, making landfills one of the larger sources of methane emissions in the United States. To prevent this, Congress should ban food waste in landfills and support food donations, food waste recycling, and development of composting and other recycling facilities. Failing federal action, states and municipalities should adopt such laws.

Chapter IX.
Changing Consumption Patterns

Just as the federal government uses farm programs to influence what farmers grow, it also uses dietary recommendations, labeling systems, and procurement policies to influence what people consume. The private sector also affects consumption patterns through advertising, labels, and menu options. Changes to consumption patterns provide synergistic support for changes to agricultural systems and could thus play an important role in reducing global emissions.[1] Below we provide a brief overview of some of the ways that the government and private sector can encourage positive change in the food system by influencing consumption choices.

A. Dietary Guidelines

Federal dietary guidelines, updated by the U.S. Department of Agriculture (USDA) and the U.S. Department of Health and Human Services (HHS) every five years, are much more than the federal government's recommendations regarding nutrition and diet. They also dictate how government agencies teach nutrition; determine what students, seniors, and other recipients of government-funded meals are fed; and guide government-funded research and nutrition projects.[2] Due to the guidelines' tremendous impact, environmentalists and sustainable food advocates have sought to incorporate sustainability goals into them, with mixed success. In the 2010 dietary guidelines, the "guiding principles" encouraged the development and expansion of "sustainable agriculture and aquaculture practices" for the first time.[3] In 2015,

---

1. Michael A. Clark et al., *Global Food System Emissions Could Preclude Achieving the 1.5° and 2°C Climate Change Targets*, 370 SCIENCE 705-08 (2020) ("[R]educing [greenhouse gas] emissions from the global food system will likely be essential to meeting the 1.5° or 2°C target."); ROBERT LEMPERT ET AL., CENTER FOR CLIMATE AND ENERGY SOLUTIONS, PATHWAYS TO 2050: ALTERNATIVE SCENARIOS FOR DECARBONIZING THE U.S. ECONOMY 6 tbl.1, 21, & 30 tbl.AB-1 (2019) (Dietary changes such as lower meat and dairy consumption are capable of achieving a dramatic drop in net agricultural greenhouse gas emissions.).
2. *See* 7 U.S.C. §5341(a)(1) ("Each such report . . . shall be promoted by each Federal agency in carrying out any Federal food, nutrition, or health program.").
3. USDA & HHS, DIETARY GUIDELINES FOR AMERICANS: 2010, at 57 (2010).

the Dietary Guidelines Advisory Committee[4] tried to build on this brief nod to sustainability by recommending that the guidelines incorporate sustainability in their dietary recommendations.[5] Secretary of Agriculture Tom Vilsack and Secretary of HHS Sylvia Burwell, who shared joint responsibility over the guidelines, responded with a letter stating that the inclusion of sustainability as a goal in the guidelines was beyond their statutory authority,[6] and thus sustainability was not included as a guiding principle or goal. The 2020 dietary guidelines again did not incorporate any sustainability goal.[7]

Advocates have persuasively argued that this is an incorrect interpretation of the enabling legislation, which requires the guidelines to be "based on the preponderance of the [current] scientific and medical knowledge."[8] The statute also establishes a nutrition monitoring and related research program, which, according to the law, should include information about "food supply and demand determinations."[9] Because sustainability is crucial to the long-term viability of the country's food supply, advocates argue, the statute gives USDA and HHS the authority to consider it as a factor in the guidelines. Though the statute may not require consideration of sustainability, it is likely a court would find it within the agencies' authority should they decide to incorporate it as a guiding principle or goal. Such a move would not be without precedent. Not only did the 2010 U.S. dietary guidelines include sustainability in its guiding principles, but those of several other countries, including Brazil, Denmark, the Netherlands, and Sweden, explicitly acknowledge the interdependence of healthy diets and environmental sustainability.[10]

Though the 2015 dietary guidelines did not incorporate sustainability in its recommendations, the Dietary Guidelines Advisory Committee's 2015 report did emphasize the connection between environmental sustainability and healthy diets, defining "sustainable diets" as a "pattern of eating that

---

4. The Dietary Guidelines Advisory Committee is a group of medical and scientific experts that issues a detailed report to HHS and USDA on the latest scientific evidence regarding health and nutrition prior to each iteration of the dietary guidelines.
5. USDA & HHS, Scientific Report of the 2015 Dietary Guidelines Advisory Committee pt. D ch. 5 (2015).
6. Tom Vilsack & Sylvia Burwell, *2015 Dietary Guidelines: Giving You the Tools You Need to Make Healthy Choices*, USDA Blog, Feb. 11, 2017, https://www.usda.gov/media/blog/2015/10/6/2015-dietary-guidelines-giving-you-tools-you-need-make-healthy-choices.
7. USDA & HHS, Dietary Guidelines for Americans: 2020 (2020).
8. 7 U.S.C. §5341(a)(2). *E.g.*, Michele Simon, My Plate, My Planet: Food for a Sustainable Nation—Statutory Authority for Sustainability in the Dietary Guidelines for Americans: Legal Analysis (2015).
9. 7 U.S.C. §5302.
10. Ministry of Health of Brazil, Dietary Guidelines for the Brazilian Population 18-19, 31-32 (2d ed. 2014); Megha Cherian, *Sustainability: A Growing Factor in Dietary Guidelines?*, Global Citizen, May 11, 2016, https://www.globalcitizen.org/en/content/sustainability-growingfactor-in-dietary-guidelines/.

promotes health and well-being and provides food security for the present population while sustaining human and natural resources for future generations."[11] Among the issues the Committee recommended integrating into the guidelines were land and water use, soil fertility, biodiversity loss, and greenhouse gas emissions.[12] The Committee's review of the literature on population-level dietary patterns and long-term food sustainability found "a moderate to strong evidence base" that increasing the consumption of healthy plant-based foods would reduce the environmental impact of the average U.S. diet. It is likely that the 2015 dietary guidelines would have recommended the reduced consumption of carbon-intensive meat as a result of these findings had the agencies found it to be within their authority to include sustainability as a factor. Brazil's dietary guidelines, for example, encourage the use of minimally processed plant-based foods over animal products to reduce greenhouse gas emissions and deforestation.[13] (Reduced consumption of red meat was recommended in 2015, but for health reasons, not for its impact on sustainability.)

Both USDA and HHS must acknowledge their legal authority to include sustainability as a factor in the guidelines. The next dietary guidelines should follow the scientific consensus on nutrition and sustainable food systems and encourage a healthy diet focused on minimally processed foods and reduced consumption of carbon-intensive meat.[14] By incorporating sustainability into the guidelines, USDA and HHS could quickly and effectively decrease the carbon intensity of the American diet. An industry-funded food and health

---

11. USDA & HHS, *supra* note 5.
12. *Id.* at 1-2.
13. Ministry of Health of Brazil, *supra* note 10, at 31-32.
14. Walter Willett et al., *Food in the Anthropocene: the EAT–Lancet Commission on Healthy Diets From Sustainable Food Systems*, 393 The Lancet 447-92 (2019) ("Improved production practices are less effective than a shift to healthy diets in abating food-related greenhouse-gas emissions because most emissions are associated with production of animal source foods whose characteristics, such as enteric fermentation in ruminants, have little potential for change. Increasing shift toward more plant-based diets will enable food production to stay within the climate change boundary."); Intergovernmental Panel on Climate Change, Global Warming of 1.5°C. An IPCC Special Report on The Impacts of Global Warming of 1.5°C Above Pre-Industrial Levels and Related Global Greenhouse Gas Emission Pathways, in the Context of Strengthening the Global Response to the Threat of Climate Change, Sustainable Development, and Efforts to Eradicate Poverty 327 (Valérie Masson-Delmotte et al. eds., 2018) ("dietary shifts could contribute one-fifth of the mitigation needed to hold warming below 2°C, with one-quarter of low-cost options"); Tim Searchinger et al., *Creating a Sustainable Food Future: A Menu of Solutions to Feed Nearly 10 Billion People by 2050*, in Creating a Sustainable Food Future 2 (World Resources Institute 2018) ("Closing the land and GHG mitigation gaps requires that, by 2050, the 20 percent of the world's population who would otherwise be high ruminant-meat consumers reduce their average consumption by 40 percent relative to their consumption in 2010."); Laura Wellesley et al., Chatham House, The Royal Institute of International Affairs, Changing Climate, Changing Diets: Pathways to Lower Meat Consumption viii (2015) ("we cannot avoid dangerous climate change unless consumption trends change").

survey conducted in 2020 found that more than half of American consumers claim that food sustainability is important and a third say that sustainability has a "real impact" on their purchases but over "6 in 10 find it hard to know whether their food choices are environmentally sustainable."[15] The same survey shows that about 40% of consumers (and a higher portion of younger consumers) are familiar with the "MyPlate" graphic that embodies the U.S. dietary guidelines, suggesting that if the guidelines included and were clear about sustainability, they could have a significant impact on food consumption patterns. These sustainability goals could also be advanced by addressing health concerns, as the same survey shows that most American consumers are seeking to eat more fruits and vegetables for health reasons, which generally have a much lower climate impact than conventionally produced animal products. Thus, further revision to the dietary guidelines to incorporate a focus on sustainability would help Americans learn about the environmental consequences of their food choices while immediately affecting what millions of Americans eat each day, encouraging healthy and climate-friendly options.

B. Federal Procurement and Food Assistance

Federal procurement policies and food assistance programs can also help increase the availability and accessibility of more healthy and sustainable food. Each year, the federal government provides food to 30 million students through the National School Lunch Program, 703,000 seniors through the Commodity Supplemental Food Program, and 84,000 Indian households through the Food Distribution Program on Indian Reservations Program, in addition to the millions of federally funded meals provided at military facilities, hospitals, prisons, and elsewhere.[16] Few of these programs require meals to adhere to dietary guidelines or to take into consideration the climate impact of food choices, and funding for these programs is insufficient to provide high-quality healthy meals.[17] Changes to these practices and pro-

---

15. INTERNATIONAL FOOD INFORMATION COUNCIL FOUNDATION, 2017 FOOD AND HEALTH SURVEY (2020).
16. FOOD AND NUTRITION SERVICE, USDA, SUMMARY OF ANNUAL DATA, FY 2015-2019, https://www.fns.usda.gov/pd/overview (last updated July 10, 2020).
17. For example, the federal government provides schools only $0.33 for paid lunches and $3.41 for free lunches through the National School Lunch Program. School Nutrition Association, *School Meal Trends & Stats*, https://schoolnutrition.org/AboutSchoolMeals/SchoolMealTrendsStats/ (last visited Nov. 30, 2020). Schools are eligible for additional funding of up to $0.08 per meal under certain conditions. *Id. See also* Emily Welker et al., *The School Food Environment and Obesity Prevention: Progress Over the Last Decade*, 5 CURRENT OBESITY REP. 145 (2016) (noting that schools in the National School Lunch Program generally have insufficient funding for the kitchen equipment, training, technical assistance, and labor needed to meet nutritional standards).

grams can have a significant impact on climate change, the environment, and public health.

For example, expanding federal funding for food assistance programs can lead to changes in consumption patterns, which can, in turn, lead to changes in production practices. The diets of many Americans are largely composed of inexpensive processed and ultra-processed foods,[18] which are associated with a high incidence of preventable, diet-related diseases including heart disease, depression, diabetes, and cancer.[19] However, people tend to choose healthier options when they have more time and money to purchase and prepare food.[20] This link between healthy choices and increased resources is further illustrated by a 2016 analysis conducted by the National Health and Nutrition Examination Survey which found that raising monthly Supplemental Nutrition Assistance Program (SNAP) benefits by only $30 per person would increase the consumption of healthy foods, while decreasing the consumption of fast food.[21] And a 2019 study found that the federal government could eliminate the substantial disparity in nutritional quality between foods consumed by low- and high-income households by increasing SNAP benefits by 15%.[22] Ensuring that people have the resources to acquire healthy and sustainable food—which, as discussed above, the dietary guidelines should encourage them to consume—could contribute to substantial reductions in emissions and an improvement in public health.

Requiring federally provided meals to adhere to dietary guidelines and to consider the carbon footprint of the food provided—while increasing funding for the procurement and preparation of healthy and sustainable food by the government—could also have significant climate and health benefits. The federal government has recognized as much, publishing a detailed set of guidelines on health and sustainability for federal food service.[23] The *Food Service Guidelines for Federal Facilities* "provides specific food, nutrition, facility efficiency, environmental support, community development, food

---

18. Jennifer Poti et al., *Is the Degree of Food Processing and Convenience Linked With the Nutritional Quality of Foods Purchased by US Households*, 101 AM. J. CLINICAL NUTRITION 1251 (2015); David Ludwig, Commentary, *Technology, Diet, and the Burden of Chronic Disease*, 305 JAMA 1352 (2011).
19. Anaïs Rico-Campà et al., *Association Between Consumption of Ultra-Processed Foods and All Cause Mortality: SUN Prospective Cohort Study*, 365 BRIT. MED. J. 1, 6 tbl.1 (2019).
20. *See* Nathan A. Rosenberg & Nevin Cohen, *Let Them Eat Kale: The Misplaced Narrative of Food Access*, 45 FORDHAM URB. L.J. 1091, 1113-20 (2018) (discussing the upstream causes of unhealthy food consumption).
21. PATRICIA M. ANDERSON & KRISTIN F. BUTCHER, CENTER ON BUDGET AND POLICY PRIORITIES, THE RELATIONSHIPS AMONG SNAP BENEFITS, GROCERY SPENDING, DIET QUALITY, AND THE ADEQUACY OF LOW-INCOME FAMILIES' RESOURCES 1, 3, 5-14 (2016).
22. Hunt Allcott et al., *Food Deserts and the Causes of Nutritional Inequality*, 134 Q.J. ECON. 1793 (2019).
23. DEPARTMENT OF HEALTH AND HUMAN SERVICES, AN ANCILLARY REPORT ON THE FOOD SERVICE GUIDELINES FOR FEDERAL FACILITIES 4 (2017).

safety, and behavioral design standards" for food services at federal facilities. The intent was to incorporate these guidelines in the federal procurement process, including them in requests for proposals so that agencies would consider these standards when evaluating bids for contracts for food service at federal facilities.[24] The guidelines—which were updated in 2017—include food and nutrition standards that align with the 2015 dietary guidelines, as well as with Executive Order 13693, *Planning for Federal Sustainability in the Next Decade*, which directs federal agencies to improve their environmental performance.[25] Accordingly, the guidelines aim to "increase the availability of healthier food and beverages in federal food service facilities so that consumers can more readily choose healthier options," and to "ensure that environmentally responsible practices are conducted in federal food service facilities," among other goals.[26] Ensuring that federal contracts for the procurement of food fully adhere to these guidelines could greatly contribute to improved health and environmental outcomes.

Notably, there is precedent for affording preferential treatment to sustainable products. The Federal Acquisition Regulations (FAR)—a regulatory system that governs all acquisitions by executive agencies—contains a part on "Environment, Energy" and other matters.[27] Included within this part are subparts on "Sustainable Acquisition Policy" and "Contracting for Environmentally Preferable Products and Services," among others.[28] Pursuant to these provisions, the EPA has a program through which it "leverages the significant federal purchasing power to prevent pollution, realize lifecycle costs savings, and increase US industry competitiveness."[29] Agencies should follow suit and leverage their purchasing power to provide healthier and more sustainable food choices throughout federal programs and facilities. Another example is a preference for procuring unprocessed, local products in child nutrition programs, a preference that originated from the 2008 Farm Bill's provisions directing USDA to pass regulations encouraging institutions participating in child nutrition programs to purchase local agricultural products.[30] Three years later, USDA issued a rule allowing these institutions to apply a geographic preference in the procurement of unprocessed, local

---

24. *Id.* at 4-5.
25. *Id.* at 8.
26. *Id.* at 9, 21.
27. Federal Acquisition Regulations Part 23.
28. *Id.*
29. EPA, *About the Environmentally Preferable Purchasing Program*, https://www.epa.gov/greenerproducts/about-environmentally-preferable-purchasing-program (last visited Jan. 23, 2021).
30. Food, Conservation, and Energy Act of 2008, Pub. L. No. 110-234, §1102, 122 Stat. 923, 1125-26.

agricultural products.[31] Congress should pass legislation expanding on this concept, explicitly allowing schools participating in child nutrition programs to give a preference to climate-friendly agricultural products. Modeled on Massachusetts' local preference law, which requires state agencies to give preference to food products grown or produced in Massachusetts, such a law could provide carbon farmers with an enormous new market.[32]

Alternatively, the federal government could establish a greenhouse gas reduction target for food purchases, similar to greenhouse gas reduction targets in the procurement of energy supplies and in transportation systems. For example, a bill proposed in Maryland in 2021 requires the state to develop a methodology to estimate the greenhouse gas emissions of food and beverages purchased, establish a 2022 baseline of annual emissions associated with these purchases, and reduce that amount by 25% by 2030.[33]

State and local governments can also take action to influence consumers' food choices. For example, they can participate in the Good Food Purchasing Program (GFPP), a program that works with large institutions such as school districts to encourage them to leverage their purchasing power to advance "five core values: local economies, environmental sustainability, valued workforce, animal welfare and nutrition."[34] Adherence to the GFPP could lead to more climate-friendly and healthier consumption patterns.

C. Private-Sector Strategies

1. Certification Systems and Supply Chain Commitments

Certification is another method that may help influence consumer choice and thus could encourage the growth of carbon farming. Organic certification has helped create a price premium for organic products—which as noted earlier in Chapter IV often are more climate-friendly—leading to increased investment and innovation in the field.[35] As a result, organic food has grown from 1% of the market in 1997 to almost 5% of the market in 2014.[36] It is not

---

31. Geographic Preference Option for the Procurement of Unprocessed Agricultural Products in Child Nutrition Programs, 76 Fed. Reg. 22603 (Apr. 22, 2011) (codified at 7 C.F.R. pts. 210, 215, 220, 225-226).
32. Mass. Gen. Laws Ann. Ch. 7, §23B (West 2012).
33. Maryland HB 317, Sec. 14-410.1 (2021). Similar bills have been introduced in New York and Washington, D.C. *See* Sen. Bill S740 (2021 New York), Bill 24-0018 (2021 Washington, D.C.).
34. Center for Good Purchasing, *About the Program*, https://goodfoodpurchasing.org/program-overview/ (last visited Jan. 23, 2021).
35. *See* Eric Toensmeier, The Carbon Farming Solution 369 (Brianne Goodspeed & Laura Jorstad eds., 2016).
36. USDA Economic Research Service (ERS), *Organic Market Overview*, https://www.ers.usda.gov/topics/natural-resources-environment/organic-agriculture/organic-market-overview/ (last updated Sept. 10,

clear whether the organic certification program has driven more sustainable practices in conventional agriculture, although some argue that it has helped spur increased interest in "natural" and other indications of sustainability. More importantly, perhaps, this certification system and the growth of this approach demonstrate the feasibility of such farming, laying the groundwork for government programs that incentivize sustainable practices.

By way of example, several companies and private organizations,[37] such as the Rainforest Alliance and Nespresso, already have, or are in the process of developing, certifications for carbon-neutral coffee.[38] Environmental groups and other nonprofit organizations have also developed certifications for sustainably produced food. For example, the Carbon Underground has developed a regenerative agriculture standard known as the Soil Carbon Initiative that can be adopted by farmers and that will measure certain outcomes, rather than practices.[39] Another nonprofit started the Regenerative Organic Certification that focuses on "the health of the planet, animal welfare, and social fairness."[40] Standards and certifications such as these could help boost interest and investment in climate-friendly practices.[41]

Similarly, other food processing companies should commit to providing carbon-neutral products, which will require them to work with their supply chain and help producers through subsidies, price premiums, technical assistance, and other means to change practices to achieve this goal. A recent report examining the adoption of sustainable practices found that "[l]arge food companies are looking at everything from warehousing and transportation to actual production practices as a way to reduce emissions and promote carbon drawdown through farming."[42] However, "regenerative production and values based supply chains look like an emerging market: highly frag-

---

2020).

37. A publicly administered national certification system would have several advantages; however, the federal government is unlikely to develop one without prior successful private initiatives. The first organic certification agency in the United States, California Certified Organic Farmers, was created in 1973, 17 years before Congress established a national organic certification system with the passage of the Organic Foods Production Act of 1990.

38. *Project Profile: Sustainable Climate-friendly Coffee (CO_2 Coffee)*, RAINFOREST ALLIANCE, July 31, 2016, http://www.rainforest-alliance.org/work/climate/projects/oaxaca-carbon-coffee; *Every Cup of Nespresso Coffee Will Be Carbon Neutral by 2022*, NESPRESSO, Sept. 17, 2020, https://nestle-nespresso.com/news/every-cup-of-nespresso-coffee-will-be-carbon-neutral-by-2022.

39. JENNIFER O'CONNOR, GUIDELIGHT STRATEGIES, BARRIERS FOR FARMERS & RANCHERS TO ADOPT REGENERATIVE AG PRACTICES IN THE US 122 (2020).

40. *Id.*

41. It remains to be seen whether environmental concerns will motivate consumers to purchase certified products. Research indicates that organic food consumers are largely motivated by health and taste. Renée Hughner et al., *Who Are Organic Consumers? A Compilation and Review of Why People Purchase Organic Food*, 6 J. CONSUMER BEHAV. 94, 101-03 (2007).

42. *Id.*

mented, lacking consistent data and information, and dependent on personal relationships. Ultimately, this lack of adequate access to appropriate supply chains is inhibiting the growth and development of a robust regenerative food economy."[43] There is thus an opportunity for the private sector to help drive this change and incentivize the adoption of climate-friendly practices in their supply chains.

2. Plant-Forward Alternatives

More than one-third of all calories consumed in the United States (before the COVID-19 pandemic) were from foods prepared away from home.[44] Studies show that people tend to consume more calories and meat when eating out.[45] Given this environment, consumers are unlikely to choose climate-friendly meals unless they are easy to find, attractive, and inexpensive. The average restaurant menu, whether fast-food or sit-down, principally offers carbon-intensive meat options for entrées.[46] Restaurants should offer an expanded range of climate-friendly options, helping to make climate-friendly diets more convenient and affordable.[47] In response to the detriments of "commodity-based diets" high in corn, wheat, soy, and animal products,[48] local governments, through their own purchasing, advertising, or public support, should encourage a wider range of whole or minimally processed plant-based options at restaurants. Doing so would support restaurants that market more vegetarian options, as well as other climate-friendly options such as meat and dairy products from integrated crop-livestock systems with demonstrated climate benefits.

A shift to a minimally processed plant-focused diet not only benefits public health, but also has enormous environmental and climate benefits. There is a growing interest in plant-based meat products, which mimic the taste and feel of conventional meat, and "cultured meat" products, which are produced from cellular agriculture and are molecularly identical to conventional meat.

---

43. *Id.* at 86.
44. Michelle J. Saksena et al., USDA Economic Research Service America's Eating Habits: Food Away From Home 5 (2018) (EIB-196).
45. Jessica E. Todd et al., USDA, The Impact of Food Away From Home on Adult Diet Quality 7-8 (Economic Research Report No. 90, 2010), https://www.ers.usda.gov/webdocs/publications/46352/8170_err90_1_.pdf.
46. In fact, "entrée" was generally used to refer to a "substantial meat course" in the United States until the Second World War. Dan Jurafsky, The Language of Food: A Linguist Reads the Menu 30 (2014).
47. Such a development would likely require significant consumer demand and pressure. *See* Karen Ganz et al., *How Major Restaurant Chains Plan Their Menus: The Role of Profit, Demand, and Health*, 32 Am. J. Preventative Med. 383 (2007).
48. *See* Ludwig, *supra* note 18.

Already numerous companies are marketing plant-based meat products, the most well-known in 2020 being the Impossible Burger and Beyond Burger, and many cell-based alternative meats are claimed to be close to market. Although more research is needed, initial studies have found lower life-cycle greenhouse gas emissions, land use requirements, and water use for these alternatives, and so there could be significant beneficial impacts from a shift from livestock meat to these alternatives.[49]

Such a shift to more plant-forward consumption would reduce methane emissions from cattle and manure, nitrous oxide emissions from animal feed production, and deforestation and grassland conversion for cattle pasture and animal feed cropland. A 2018 large meta-analysis looking at a wide range of indicators found that "meat, aquaculture, eggs, and dairy use ~83% of the world's farmland and contribute 56 to 58% of food's different emissions, despite providing only 37% of our protein and 18% of our calories."[50] It further found that the "impacts of the lowest-impact animal products typically exceed those of vegetable substitutes" and provides "new evidence for the importance of dietary change."[51] In addition to urging a wide range of approaches to reduce production impacts, such as those discussed earlier in this book, the study found that "[c]ommunicating average product impacts to consumers enables dietary change and should be pursued."[52]

Notably, with the market for these meat alternatives at almost one billion dollars in 2020 and growing fast, the conventional meat industry is seeking to impose barriers to such products. For example, in 2018, the U.S. Cattlemen's Association petitioned USDA to redefine "beef" and "meat" to exclude any product that is not from "animals born, raised, and harvested in the traditional manner."[53] Also in 2018, Missouri became the first state to enact a general prohibition on the use of "meat" to describe products not derived from slaughtered animals,[54] and by 2019, fourteen other states followed suit.[55] In addition, federal legislators introduced in 2019 a bill that would require plant-based and cell-cultured meat products to carry the word

49. Peter Newton & Daniel Blaustein-Rejto, *Social and Economic Opportunities and Challenges of Plant-Based and Cultured Meat for Rural Producers in the US*, 5 Frontiers Sustainable Food Sys. (2021).
50. Joseph Poore & Thomas Nemecek, *Reducing Food's Environmental Impacts Through Producers and Consumers*, 360 Science 987, 990 (2018).
51. *Id.* at 990-91.
52. *Id.* at 991.
53. Petition for the Imposition of Beef and Meat Labeling Requirements: To Exclude Products Not Directly Derived From Animals Raised and Slaughtered From the Definition of "Beef" and "Meat,' U.S. Cattleman's Association (USCA), Feb. 9, 2018.
54. Mo. Rev. Stat. §265.494(7).
55. Environmental Health State Bill Tracking Database, National Conference of State Legislatures (Dec. 8, 2020), https://www.ncsl.org/research/environment-and-natural-resources/environmental-health-legislation-database.aspx#map.

"imitation" immediately before or after the name of the food, alongside a disclaimer that indicates the food is not derived from, and does not contain meat.[56] While many of these laws are likely vulnerable to legal challenges based on the First Amendment,[57] they still impede the growth of this more climate-friendly opportunity and policymakers should oppose the imposition of such market barriers.

*****

What we eat directly affects what we grow, which in turn has enormous environmental and climate consequences. Thus, any effort to reduce agriculture's carbon footprint should consider these strategies to influence consumer choice and to incentivize a shift to healthier, more sustainable food purchases. Our ability to meet our climate change goals depends on it.

> **Key Recommendations**
>
> - Food choice and diet offer important opportunities to reduce the climate change impact of agriculture and the food sector. Federal and state policies and procurement actions shape people's dietary choices far more than realized, now often promoting high-climate-change-impact foods.
>
> - The dietary guidelines' enabling statute gives USDA and HHS the authority to consider sustainability as a factor in the guidelines.
>
> - U.S. dietary guidelines should follow the scientific consensus on nutrition and sustainable food systems and encourage a healthy diet focused on minimally processed foods and reduced consumption of greenhouse gas-intensive meat.
>
> - Agencies should leverage their purchasing power to procure healthier and more sustainable food choices throughout federal programs and facilities.

---

56. H.R.4881-The Real Meat Act of 2019.
57. *See, e.g.*, Turtle Island Foods, SPC v. Richardson, 425 F. Supp. 3d 1131 (W.D. Mo. 2019).

- The private sector should create certification systems for sustainable agricultural practices and adopt climate-friendly practices in their supply chains.
- Restaurants should offer an expanded range of climate-friendly options, helping to make climate-friendly diets more convenient and affordable.
- Policymakers should oppose the imposition of market barriers to meat or dairy alternatives, such as labeling restrictions or mandates.

Chapter X.
Conclusion

Agriculture is responsible for more than 10% of U.S. greenhouse gas emissions, or, fully aggregating all agricultural impacts together, over 20%. At the same time, the full food system contributes about one third of total emissions. These emissions are entirely avoidable. The climate-friendly agricultural practices described in this book are proven to significantly reduce greenhouse gas emissions from farming, ranching, and livestock production. In addition, agriculture is unique among major sectors of the economy in possessing the potential not only to reduce emissions, but also to remove carbon from the atmosphere and sequester it. By both reducing emissions and increasing carbon sequestration, U.S. agriculture can become climate neutral.

Curbing climate change is not the only reason that policymakers and producers should support agricultural practices that reduce emissions or increase soil carbon. Virtually all of these practices—including, for example, the use of cover crops, managed rotational grazing, agroforestry, silvopasture, and improved manure management—also provide other environmental benefits such as cleaner water or better wildlife habitat. In addition, these practices make agricultural operations more resilient to changes in weather patterns that will come with climate change. Many of these practices are also cost effective and profitable, especially once established.

Agriculture is a highly subsidized industry that continues to be shaped by public programs and support. Policymakers should use these pathways to encourage the widespread adoption of climate-friendly agricultural practices, which will lead to a more climate-resilient food supply, healthier rural communities, less erratic and extreme weather, and a more stable and vibrant farm sector. There are tremendous opportunities for the U.S. Department of Agriculture (USDA) to fund additional agricultural research, development, and extension to advance climate goals. Congress and USDA can also take advantage of the department's support programs, including crop insurance, commodity payments, conservation, credit, and trade programs to spur the adoption of climate-friendly practices. Outside of USDA, a number of federal agencies should be enlisted to reduce net agricultural emissions through reforms in regulatory strategies, tax policy, lending and subsidy programs, and greenhouse gas pricing. The private and nonprofit sectors also have a significant role to play in research, education, and market development as well as in encouraging and leveraging governmental action.

Climate change presents the most significant threat to agriculture and human well-being in the world today. While change has often been slow in the agricultural sector, policymakers now have a real opportunity to realize climate neutrality in agriculture, while improving other environmental attributes, rural communities, and the lives of agricultural workers. Policymakers and others should rise up to this challenge. Our future depends on it.

Index

adaptation, 36, 65, 116, 130–31, 154, 184
Agricultural Conservation Easement Program (ACEP), 155–56
agricultural practices & systems, 1–9, 12–13, 15–16, 18, 35–61, 63–109, 111–82, 187–89, 192–96, 201–17, 219–20, 222–23, 231–34, 236–37
    agroecology, 28, 72, 98
    agroforestry, 58, 63, 64, 67–69, 71, 107, 169, 171–73, 209
    carbon farming, 72, 121, 144, 151–52, 177, 184, 211, 217
    climate-friendly, 115
    conventional, 73–74, 170, 174, 192, 219
    diversified vs. monoculture, 2, 31, 66, 69, 71–73, 107, 118, 161, 171
    grazing, 64, 91–93, 106, 109, 179, 241
    organic, 73–74, 118, 151, 236
    perennial, 169, 171–76, 179
    regenerative, 212, 236
    sustainable, 28, 114, 119, 122–23, 140, 148–50, 154, 206, 229, 236
Agricultural Research Service (ARS), 114–15
Agricultural Resource Management Survey (ARMS), 21, 124, 194
Agricultural Risk Coverage (ARC), 141–42
Agriculture and Food Research Institute (AFRI), 114, 173
Agriculture Resilience Act, 165–66
alley cropping, 64, 68, 70, 72, 105–06, 124, 138, 144, 152, 170–72
ammonia, 164, 185–86, 189, 221–22
anaerobic, 98–99, 109, 150, 184, 198–201
animal feed, 5, 8, 36–38, 42–46, 60, 71, 75, 90, 93, 100, 190, 195, 221, 231
animal feeding operations (AFOs), 27, 29, 63, 94–97, 151, 178, 183, 185–86, 195, 199, 204, 228
antibiotics, 95, 101
antimicrobial, 95
aquaculture, 229
Biden, Joe, 150–51, 176
biochar, 64, 83
biofuels, 8, 26, 38, 45, 66, 70, 74–76, 107, 190–91, 203, 224
    aggregate compliance, 190–91, 203
biogas, 99, 198, 200
biomass, 1, 4, 58–59, 66–67, 69, 72, 75, 81, 88–89, 99, 107, 191, 202–03, 209, 222
biosolids, 82–83

breeds, 70, 172
Building Blocks program, 78, 134–35
Bureau of Land Management (BLM), 113, 167–68
byproducts, 74–75, 89–90, 108
carbon bank, 144, 211
carbon farming, 4, 31, 70, 111, 124, 165, 184, 235
carbon markets, 7, 210–12, 215, 217
carbon offset markets, 210–11
carbon opportunity cost, 49–51, 53, 61, 75, 222
carbon pricing, 201–02, 204
carbon sequestration, 5, 50, 59, 61, 66, 68, 79, 89, 93, 98, 107, 113, 135, 144–45, 152, 178, 196, 215, 241
Census of Agriculture, 15, 18
certification, 235–36
Clean Air Act (CAA), 150, 182–88, 226, 228
Clean Water Act (CWA), 188
climate change, 60, 77, 141, 242
   climate-friendly, 32, 111, 116, 236
   climate hubs, 131, 134
   climate-neutral, 3, 103–04, 106, 116, 118, 134–35, 163–64, 177, 227, 242
Commodity Credit Corporation, 12, 143–44, 174
commodity crops, 81, 84–85, 163
compost, 94, 227
Comprehensive Environmental Response, Compensation, and Liability Act (CERCLA), 189–90
concentrated animal feeding operations (CAFOs), 4, 27, 29–30, 42, 55–56, 94–96, 102, 108, 112, 121, 164, 178, 185–89, 197, 199–200, 204
conservation, 63–64, 72, 76–77, 79–81, 102, 105–06, 119, 128–29, 132–33, 145–49, 151–59, 172–73, 196, 207–08, 226–27
conservation compliance, 123, 141, 157–58
conservation easements, 154, 196, 205, 208
Conservation Innovation Grants, 119
conservation practices, 128, 132–34, 139–41, 149, 152, 154, 171
Conservation Reserve Enhancement Program (CREP), 147
Conservation Reserve Program (CRP), 145–46, 152–53
   CLEAR30, 146–47
   Continuous CRP (CCRP), 146
Conservation Stewardship Program (CSP), 132
conservation technical assistance, 132–33
conversion, 90, 190
Cooperative Extension Service (CES), 129–31
Coronavirus Food Assistance Program (CFAP), 143

Index

Corporate Average Fuel Economy (CAFE), 192
credits, agricultural, 136, 160–61
crop insurance, 137–38, 166
data, 15–16, 25, 27, 42, 44–45, 47–48, 50, 64, 66, 78, 80, 123–24, 128, 159–60, 220–21
    current population survey, 15, 18
    data collection and analysis, 123
digesters, 99, 150, 199, 201, 204
discrimination, 27, 160–61
diversified farming, 66, 72, 81, 107, 118, 174
drought, 3, 36, 92
easements, 155, 196, 207–08, 217
ecological leftovers (*see also* livestock), 75, 90
Economic Research Service (ERS), 1, 12, 18, 20, 24, 38, 86, 117–18, 123–26, 194, 235
Ecosystem Service Market Consortium, 210, 212
electricity, 102, 199, 222
Emergency Planning and Community Right-to-Know Act (EPCRA), 185, 189, 190
emissions, 38–48, 51, 53–55, 57, 79, 83, 85–87, 91–94, 96, 99–102, 154–56, 183–85, 187–88, 198–99, 219–26
    agricultural emissions, 4, 6–7, 35, 38–39, 41–43, 46–48, 54–56, 60–61, 63, 87, 90, 104, 121, 123, 126–27
    methane and nitrous oxide emissions, 181, 183, 189, 203
energy, 102–03, 151, 154, 157, 194, 198–200, 207, 219–22, 224, 226, 228–29, 234
energy efficiency, 224, 228, 207
enteric fermentation, 41–42
Environmental Quality Incentives Program (EQIP), 148–49, 151–53
Environmental Working Group (EWG), 96, 100, 145–46, 154
erosion, 71, 159, 93, 89
exports, 1, 143, 162, 223–24
externalities 4, 72, 111
Farm Bill, 119–20, 127–29, 133, 135, 137, 139, 141–42, 145–47, 149, 151–57, 159, 162, 164, 166, 171–72
    2008 Farm Bill, 123, 127, 151, 155, 157, 177, 234
        Section 1619, 127–29
    2014 Farm Bill, 120, 135, 137, 141, 151, 155, 157, 162, 171
    2018 Farm Bill, 119, 133, 135, 139, 141–42, 145–46, 152–56, 159, 162
Farm Bill Law Enterprise, 127
Farm Credit System (FCS), 136, 160–62, 170
Farm Service Agency (FSA), 136, 145–47, 156–58, 160–62, 170, 197

farmers, 11, 13, 15, 25–27, 78, 80–82, 84, 86, 88, 137–38, 151–53, 160–61, 205–06, 211–12, 236
    black and non-white, 6, 13–15, 23, 26–27, 31–33, 125
    indigenous, 15
farms, 11–33, 36, 91, 93, 100, 102, 111–79, 206–07, 213, 219, 221–23, 225, 227
    dairy farms, 96, 150, 198
    lifestyle and paper farms, 13, 15–17, 19, 32, 164
farmworkers, 22, 25–26, 32, 112
Federal Crop Insurance Corporation (FCIC), 140–41, 171
fertility, 91
fertilizers, 40, 84–86, 106, 122, 170, 194, 220–21
    4Rs, 84
    enhanced-efficiency fertilizer, 192, 203, 206
    slow-release fertilizer, 85
finance, 11, 18, 24, 174, 206
floods, 87, 138
food security, 128, 157
food system, 2, 15, 69, 116, 118, 147, 221
food waste, 223, 225
forage, 70
Good Farming Practices, 138
Government Accountability Office (GAO), 138
grains (*see also* Kernza and rice), 71, 75, 190
grazing (*see also* livestock), 59, 64, 70, 88, 90–93, 98, 109, 166–68, 201
    management-intensive grazing, 91
    multi-paddock grazing, 91–92
    prescribed grazing, 64, 91, 106
    rotational grazing, 91–93, 109, 179, 241
Great Depression, 14, 163
Growing Climate Solutions Act, 211
hazardous air pollutant (HAP), 186
Healthy Forests Reserve Program, 156
heat stress, 35
herbicide, 68, 78
herd health, 55, 93, 109
Homestead Act, 22
Indigo Agriculture, 210, 212
inequality, 2–3, 24, 233
intensification, 72, 82
Intergovernmental Panel on Climate Change (IPCC), 51–53, 67, 88, 93, 96, 231
Internal Revenue Service, 15, 17, 196
irrigation, 85–86, 121

Index

Irrigation Innovation Consortium, 121
Kernza, 71, 121, 172
labor, 12–13, 23
landfills, 224–25
legumes, 37
Leopold Center, 122
life-cycle assessment, 50, 93, 223
livestock (*see also* grazing), 3, 51, 59, 70, 90–91, 93, 95–98, 100, 108, 195, 199
maize, 188
Market Facilitation Program (MFP), 135, 143, 171
markets, 1, 70, 86, 135, 144, 162–63, 171, 184, 210–13, 216, 235
measurement, 70, 209–10
meat, 46, 231, 238–39
methane, 42–43, 52, 54–55, 83, 99–101, 181, 184, 194, 225–27
methane inhibitors, 100–01
microorganisms, 41, 87
mitigation, 64–65, 69, 78–79, 81, 86, 91, 93–94, 96, 100, 116, 184
models, 39, 55–56, 73, 91, 146, 166, 186, 192–93, 203, 209
National Agricultural Statistics Service (NASS), 123
National Agroforestry Center, 172-73
national ambient air quality standards (NAAQS), 183, 185
National Cover Crop Initiative, 120
National Institute of Food and Agriculture, 114–15, 119, 131
National School Lunch Program, 232
National Sustainable Agriculture Coalition, 119
Natural Resources Conservation Service (NRCS), 63–64, 102, 119, 129, 132–33, 139, 148–49, 151–59, 161, 167, 172, 200–01
New Deal, 13–14, 22–23, 175
nitrification, 87
nitrogen, 59, 82, 84–86, 91, 188, 192, 194, 220, 222
nitrous oxide, 3, 40, 53, 68, 85–86, 98, 100, 104, 181, 185, 187
Nori, 210, 212–13
nutrients, 6, 68–69, 72–73, 82, 85, 95, 98, 100, 105–06, 148, 158
Office of Inspector General (OIG), 157, 197
Organic Farming Research Foundation, 118, 206
pasture, 88, 90, 93
payments for ecosystem services, 165, 175
perennial agriculture, 64, 69–72, 76, 81, 107, 121, 168–70, 172–74
    alley cropping, 64, 68, 70, 72, 105–06, 124, 138, 144, 152, 170–72
    perennial crops, 37, 66, 69–71, 81, 89, 107, 119, 121, 161, 169–70, 172–76, 206
    perennial practices, 169, 171–76, 179
pesticides, 30, 74, 112

pests, 131, 148
pollinators, 147
precision agriculture, 85
price loss coverage, 141–42, 171
private sector, 7, 117, 120, 130, 169, 235
procurement, 232, 235
productivity, 59, 69–70, 90–91, 94, 116, 127
profitability, 13, 19, 82, 112, 115, 160, 172, 201, 237
public health, 77, 112, 177, 230, 233
ranchers, 89, 236
rangeland, 59, 91, 166, 168
regenerative agriculture, 4, 63, 140, 212, 236
Regional Greenhouse Gas Initiative, 201
renewable fuel standard, 75, 182, 190
research
   budget, 50, 114–15, 118, 124, 132–33, 135, 155–56, 163, 233
   public funding, 115, 117–18
   research agencies, 125
resilience, 71, 165–66, 184
Resource Conservation and Recovery Act (RCRA), 190
restoration, 49, 168, 190
rice, 57, 82, 87–88, 108, 212
Risk Management Agency (RMA), 137–41
Rodale Institute, 4, 73, 80, 206
rotation, 37, 81, 97, 138
runoff, 68–69, 87, 94, 188, 193
rural, 11–12, 16–17, 24–25, 27–30, 74, 101, 115, 151, 189, 199, 238
silvopasture, 37, 64, 68, 89, 105–06, 108, 116, 144, 152, 171, 241
Small Business Administration, 160, 181
Sodbuster, Sodsaver, and Swampbuster, 156–57
soil amendments, 83, 87, 89, 93, 109
soil health, 94, 122, 148, 219
soy, 70
Specialty Crop Block Grant Program, 174
stakeholders, 11–33, 130
states, 44, 46–48, 63–64, 71, 75, 81, 83, 121–22, 129, 163, 166, 184–85, 193–97, 226–27, 238
subsidies, 7, 22, 27, 112, 126, 135–37, 143, 149, 164–66, 170–71, 182, 198, 206
Supplemental Nutrition Assistance Program, 224, 233
Sustainable Agriculture Research and Education (SARE) Program, 119
Sustainable Agriculture Systems Program, 115
tariff bailout, 27

Index

taxes, 17, 20, 193, 195–96, 202
    assessment, 17
    benefits, 17, 195–96
    expenditures, 17, 193, 196, 208
    reform policy, 181
tillage, 78–80, 82, 170, 223
    no-tillage, 71, 78–80, 98, 129
trade, 11, 27, 45, 162, 172–73, 201, 211, 214, 216, 219
uncertainty, 129
underestimates, 47, 49, 52, 55, 214
U.S. Department of Agriculture (USDA), 12–13, 20, 45, 55, 63, 73, 111, 181, 223, 229, 241
U.S. Environmental Protection Agency (EPA), 40–43, 45–49, 51–52, 54–57, 68, 76, 82–83, 94–98, 150, 181–91, 198–99, 203, 222–26, 228, 234
U.S. Forest Service, 113, 167–68, 172
U.S. Geological Survey, 159
vaccine, 101
value added product producer grant, 174
Vilsack, Tom, 139, 230
water quality, 2–3, 29, 89, 93, 146, 148, 184, 196
wealth, 19
weather, 11, 35, 86, 152
weeds, 36, 77, 80, 98, 131, 142
wildfire, 26

Notes

Notes

Notes

Notes